T0138723

Generalized Barycentric Coordinates in Computer Graphics and Computational Mechanics

Generalized Barycentric Coordinates in Computer Graphics and Computational Mechanics

edited by

Kai Hormann
N. Sukumar

CRC Press is an imprint of the
Taylor & Francis Group, an **Informa** business
AN A K PETERS BOOK

CRC Press
Taylor & Francis Group
6000 Broken Sound Parkway NW, Suite 300
Boca Raton, FL 33487-2742

Printed on acid-free paper
Version Date: 20170918

International Standard Book Number-13: 978-1-4987-6359-2 (Hardback)

Library of Congress Cataloging-in-Publication Data

Names: Hormann, Kai, editor. | Sukumar, N., editor.
Title: Generalized barycentric coordinates in computer graphics and computational mechanics / edited by Kai Hormann and N. Sukumar.
Description: Boca Raton : Taylor & Francis, CRC Press, 2017.
Identifiers: LCCN 2016056577 | ISBN 9781498763592 (hardback : alk. paper).
Subjects: LCSH: Barycentric coordinates. | Mechanics. | Computer graphics--Mathematics. | Center of mass.
Classification: LCC QA556 .G284 2017 | DDC 516/.16--dc23
LC record available at https://lccn.loc.gov/2016056577

Visit the Taylor & Francis Web site at
http://www.taylorandfrancis.com

and the CRC Press Web site at
http://www.crcpress.com

Printed and bound in the United States of America by
Edwards Brothers Malloy on sustainably sourced paper

Contents

Preface

Interpolating given discrete data with continuous functions in one or more variables is a fundamental problem in diverse fields of sciences and engineering. Barycentric coordinates, which were introduced by Möbius [282] in 1827, still provide perhaps the most convenient way to linearly interpolate data prescribed at the vertices of a d-dimensional simplex. Barycentric interpolation is widely used in computer graphics, whereas such interpolating (basis) functions are also adopted as trial and test approximations in finite element and boundary element methods. Starting with the seminal work published by Wachspress [407] in 1975,* the ideas of barycentric coordinates and barycentric interpolation have been extended in recent years to arbitrary polygons in the plane and general polytopes in higher dimensions, which in turn has led to novel solutions in applications like mesh parametrization, image warping, mesh deformation, and finite element and boundary element methods. This book summarizes the latest developments and applications of *generalized barycentric coordinates* in computer graphics and computational mechanics.

The advent of *mean value coordinates* [144] in 2003 was a turning point in the sustained interest and further development of generalized barycentric coordinates. This construction generated renewed attention to Wachspress coordinates, and led to many new pathways in geometry (polygonal mesh) processing and polygonal finite element computations. Realizing this trend, we co-organized (together with Gautam Dasgupta and Eitan Grinspun) a workshop in 2012 that was supported by the U.S. National Science Foundation, on *Barycentric Coordinates in Geometry Processing and Finite/Boundary Element Methods*, which was held at Columbia University in New York. There was broad participation at the workshop from both communities to foster synergy between the two fields. This book is envisioned as the second step in the partnership of researchers from computer graphics and computational mechanics. We are hopeful that the contents of this book will be beneficial to both the uninitiated undergraduate or graduate student as well as the experienced researcher who is well-versed in generalized barycentric coordinates.

This book is divided into three sections: Section I (Chapters 1–5) is on the theoretical foundations of generalized barycentric coordinates; and Sections II (Chapters 6–10) and III (Chapters 11–15) are on its applications in computer graphics and computational mechanics, respectively. There exist many distinct constructions for generalized barycentric coordinates; an overview with comparisons and contrasts of known generalized barycentric coordinates is presented in Chapter 1.

*A revised and extended version of this book [406] was published in 2016.

The mathematical theory of simplicial finite elements is well-established; however, for polygons in 2D and polyhedra in 3D, the relationship between the shape of the polytope and the interpolation properties on them are not yet fully understood. Theoretical interpolation estimates for these coordinates, with supportive numerical experiments, are presented in Chapter 2. Besides interpolation within polygonal domains, realizing continuous linearly precise interpolants over smooth domains—with so-called transfinite barycentric coordinates (Chapter 3)—is also of broad interest. In many computer graphics applications, bijective mappings between simple polytopes are needed, and one route to achieve this is through composite bijective mappings, which are discussed in Chapter 4. The smooth and discrete Laplacian on manifolds has a rich history in mathematical theory and numerical computations. The preservation of the properties of the continuous operator on discrete grids is desirable, which forms the foundation for many numerical discretizations that are based on discrete exterior calculus and mimetic schemes. A primer on the Laplacian is provided in Chapter 5.

Applications of generalized barycentric coordinates in computer graphics and geometry processing are covered in Section II. One of the fundamental concepts in computer graphics for enhancing the visual quality of a rendered triangle mesh is texture mapping. This requires us to first compute a suitable mesh parameterization of the 3D mesh over a 2D domain, which can be done efficiently using generalized barycentric coordinates (Chapter 6). Complex barycentric coordinates naturally arise from the identification of $\mathbb{R}^2$ with the complex plane $\mathbb{C}$ and provide a convenient framework for deforming planar shapes and images intuitively by enclosing the region of interest with a polygonal cage and moving the cage vertices (Chapter 7). An alternative approach to shape deformation is given by S-patches (Chapter 8), which generalize the idea of parametric Bézier surfaces from triangular and quadrilateral to arbitrary polygonal domains and can be extended to arbitrary dimensions. Generalized barycentric coordinates further play a key role in the representation of primal-dual triangulations, with applications in mesh optimization and finite element methods (Chapter 9), and in the computational design and analysis of self-supporting masonry structures (Chapter 10).

New and emerging polyhedral formulations and their applications to second- and fourth-order elliptic partial differential equations (PDEs) are emphasized in Section III. In solid mechanics, use of polyhedral finite element formulations is appealing for pervasive fracture and fragmentation simulations (Chapter 11) and to realize extremely large deformations in numerical simulations (Chapter 12). For such applications, simulations on polyhedral meshes can outperform existing tetrahedral finite element simulation capabilities. For scattered sets of points in $\mathbb{R}^d$, there exist meshfree generalized barycentric coordinates known as maximum-entropy coordinates. The essentials on their construction with applications in solid mechanics and biomechanical simulations are presented in Chapter 13. In recent years, new computational methods on polygonal and polyhedral meshes have been developed that do not require the explicit computation of the basis functions in the interior of the polytope: among these, a boundary element method that uses PDE-aware (Trefftz-based) basis functions (Chapter 14) and the *virtual element method* (Chapter 15), which provides a variational foundation for mimetic finite-difference schemes, are prominent.

A few words on the notation adopted in this book. To ensure uniformity and to facilitate understanding, we use common notation for the most frequently used concepts related to generalized barycentric coordinates that appear in computer graphics and computational mechanics. Throughout the book, Ω is a general bounded domain in $\mathbb{R}^d$, with $\partial\Omega$ its boundary and $\bar{\Omega}$ its closure. When Ω is specifically a bounded polytope in $\mathbb{R}^d$, we use P for the open set, and the vertices of P are $\boldsymbol{v}_1, \ldots, \boldsymbol{v}_n$. A generic point in $\bar{P}$ is denoted by $\boldsymbol{x}$, and the i-th generalized barycentric coordinate is $\phi_i \colon \bar{P} \to \mathbb{R}$. In general, we follow the convention of using boldface for vector quantities, to distinguish them from scalar quantities; for example, $\boldsymbol{x} = (x_1, \ldots, x_d) \in \mathbb{R}^d$.

Our thanks go to Rick Adams at CRC Press for initiating this book project and to Jessica Vega and Marcus Fontaine at CRC Press for their assistance during various stages of the publishing process. Most importantly, we are very grateful to all contributors, for their time and effort in preparing the chapters with diligence and in a timely manner. Their contributions have shaped this book.

Kai Hormann
Lugano, Switzerland

N. Sukumar
Davis, CA, USA

Contributors

Dmitry Anisimov
Università della Svizzera italiana
Lugano, Switzerland

Marino Arroyo
Universitat Politècnica de Catalunya
Barcelona, Spain

Alexander G. Belyaev
Heriot-Watt University
Edinburgh, UK

Joseph E. Bishop
Sandia National Laboratories
Albuquerque, USA

Andrea Cangiani
University of Leicester
Leicester, UK

Heng Chi
Georgia Institute of Technology
Atlanta, USA

Pierre-Alain Fayolle
University of Aizu
Aizuwakamatsu, Japan

Vitaliy Gyrya
Los Alamos National Laboratory
Los Alamos, USA

Andrew Gillette
University of Arizona
Tucson, USA

Bruno Lévy
Inria Nancy Grand Est
Villers-lès-Nancy, France

Oscar Lopez-Pamies
University of Illinois
Urbana-Champaign, USA

Gianmarco Manzini
Los Alamos National Laboratory
Los Alamos, USA

Pooran Memari
LIX, CNRS, École Polytechnique,
Université Paris Saclay
Palaiseau, France

Glaucio H. Paulino
Georgia Institute of Technology
Atlanta, USA

Alexander Rand
CD-adapco
Austin, USA

Scott Schaefer
Texas A&M University
College Station, USA

Teseo Schneider
Università della Svizzera italiana
Lugano, Switzerland

Oliver J. Sutton
University of Leicester
Leicester, UK

Cameron Talischi
McKinsey & Company
Chicago, USA

Etienne Vouga
University of Texas
Austin, USA

Max Wardetzky
Georg-August-University
Göttingen, Germany

Ofir Weber
Bar-Ilan University
Ramat Gan, Israel

Steffen Weißer
Saarland University
Saarbrücken, Germany

I

Theoretical Foundations

CHAPTER 1

Barycentric Coordinates and Their Properties

Dmitry Anisimov

Università della Svizzera italiana, Lugano, Switzerland

CONTENTS

BARYCENTRIC COORDINATES are commonly used in computer graphics and computational mechanics to represent a point inside a simplex as an affine combination of the simplex's vertices. We show how they can be generalized to arbitrary polytopes and present the most known constructions of these generalized barycentric coordinates in 2D and 3D.

1.1 INTRODUCTION

It was known since the days of the *Peripatetic School* and usually attributed to *Archimedes* (c. 287 BC–c. 212 BC) [116] that a lever $[v_1, v_2]$ with two weights w_1 and w_2 attached to its ends is balanced when a fulcrum is placed at the point $x \in [v_1, v_2]$ such that

$$w_1 l_1 = w_2 l_2, \tag{1.1}$$

where $l_1 = x - v_1$ and $l_2 = v_2 - x$ (see Figure 1.1). Equation (1.1) is called the *law of the lever* and the point of balance x is called the *center of mass* of this lever or its *barycenter* (from Ancient Greek βάρος = "weight" and κέντρον = "center"). The weights w_1 and w_2 are often called homogeneous, because multiplying them with a common non-zero scalar α does not change the equation. Rearranging terms, (1.1) can be written in the form

$$w_1(v_1 - x) + w_2(v_2 - x) = 0 \tag{1.2}$$

and further as

$$w_1 v_1 + w_2 v_2 = Wx, \qquad W = w_1 + w_2.$$

Choosing the scalar $\alpha = \frac{1}{W}$, we can define the normalized weights $\phi_1 = \alpha w_1$ and $\phi_2 = \alpha w_2$ and write the barycenter x as an affine combination of the ends of the lever with these weights,

$$\phi_1 + \phi_2 = 1, \tag{1.3}$$

$$\phi_1 v_1 + \phi_2 v_2 = x. \tag{1.4}$$

The normalized weights ϕ_1 and ϕ_2 are called the *barycentric coordinates* of the point x with respect to the segment $[v_1, v_2]$.

While the problem above is about finding the barycenter x for the given weights, it is also interesting to study the opposite problem. Given the end points v_1 and v_2

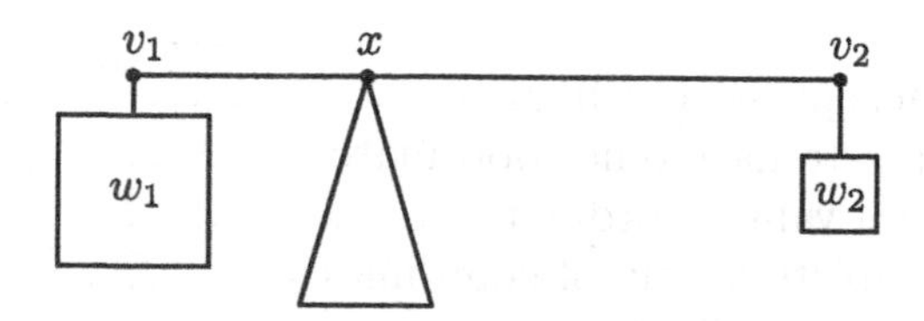

Figure 1.1 Law of the lever.

of an arbitrary segment and some point x along this segment, how do we find the barycentric coordinates ϕ_1 and ϕ_2 of x with respect to this segment? It turns out that they are uniquely determined by (1.3) and (1.4) as ratios of lengths,

$$\phi_1 = \frac{l_2}{l}, \qquad \phi_2 = \frac{l_1}{l},$$

where $l = l_1 + l_2 = v_2 - v_1$ is the length of the segment.

1.1.1 Barycentric coordinates for simplices

In 1827, the German mathematician *August Ferdinand Möbius* (1790–1868) [282] considered the problem of finding barycentric coordinates with respect to an arbitrary d-simplex in $d \in \mathbb{N}$ dimensions. For example, a 1-simplex is a line segment in 1D, a 2-simplex is a triangle in 2D, and a 3-simplex is a tetrahedron in 3D.

Given a non-degenerate d-simplex Δ with $d+1$ vertices $\boldsymbol{v}_1, \ldots, \boldsymbol{v}_{d+1} \in \mathbb{R}^d$, we search for $d+1$ functions $\boldsymbol{\phi} = [\phi_1, \ldots, \phi_{d+1}] \colon \Delta \to \mathbb{R}^{d+1}$, which satisfy the *partition of unity property*

$$\sum_{i=1}^{d+1} \phi_i(\boldsymbol{x}) = 1 \qquad \forall \boldsymbol{x} \in \Delta \tag{1.5}$$

and the *linear reproduction property*

$$\sum_{i=1}^{d+1} \phi_i(\boldsymbol{x}) \boldsymbol{v}_i = \boldsymbol{x} \qquad \forall \boldsymbol{x} \in \Delta. \tag{1.6}$$

The functions $\phi_1, \ldots, \phi_{d+1}$ are called *barycentric coordinates* with respect to Δ.

Analogously to the one-dimensional case in the previous section, it turns out that these barycentric coordinates are uniquely determined by (1.5) and (1.6), and Möbius shows that they are ratios of volumes,

$$\phi_i(\boldsymbol{x}) = \frac{V_i(\boldsymbol{x})}{V}, \qquad i = 1, \ldots, d+1, \tag{1.7}$$

where $V_i(\boldsymbol{x}) = \mathrm{Vol}[\boldsymbol{v}_1, \ldots, \boldsymbol{v}_{i-1}, \boldsymbol{x}, \boldsymbol{v}_{i+1}, \ldots, \boldsymbol{v}_{d+1}]$ are the volumes of the corresponding d-simplices and $V = V_1(\boldsymbol{x}) + \cdots + V_{d+1}(\boldsymbol{x}) = \mathrm{Vol}[\boldsymbol{v}_1, \ldots, \boldsymbol{v}_{d+1}]$ is the volume of Δ and does not depend on $\boldsymbol{x}$.

1.1.2 Generalized barycentric coordinates

To the best of our knowledge, Kalman [221] was the first to propose a generalization of barycentric coordinates to convex polyhedra, and in recent years there has been a growing interest in the problem of finding barycentric coordinates with respect to arbitrary polytopes. Throughout this book, we consider a non-degenerate polytope P with $n \geq d+1$ vertices $\boldsymbol{v}_1, \ldots, \boldsymbol{v}_n \in \mathbb{R}^d$, which is viewed as an open set. We denote the boundary of this polytope by ∂P and its closure by $\bar{P}$.

Definition 1.1. Given the polytope P, the n functions $\boldsymbol{\phi} = [\phi_1, \ldots, \phi_n]\colon \bar{P} \to \mathbb{R}^n$ are called *generalized barycentric coordinates* if they satisfy the *partition of unity property*

$$\sum_{i=1}^{n} \phi_i(\boldsymbol{x}) = 1 \qquad \forall \boldsymbol{x} \in \bar{P} \tag{1.8}$$

and the *linear reproduction property*

$$\sum_{i=1}^{n} \phi_i(\boldsymbol{x})\boldsymbol{v}_i = \boldsymbol{x} \qquad \forall \boldsymbol{x} \in \bar{P}. \tag{1.9}$$

For $n = d+1$, the only functions that satisfy the properties in Definition 1.1 are the ϕ_i in (1.7). For $n > d+1$, however, the ϕ_i are no longer uniquely determined, which is the reason for the existence of the different constructions of generalized barycentric coordinates that we review in the remainder of this chapter.

In addition to the defining properties (1.8) and (1.9), it is often desirable for the functions ϕ_i to have a few extra properties,

- *Non-negativity*: $\phi_i(\boldsymbol{x}) \geq 0$ for any $\boldsymbol{x} \in \bar{P}$; (1.10a)
- *Lagrange property*: $\phi_i(\boldsymbol{v}_j) = \delta_{ij}$, where δ_{ij} is the Kronecker delta; (1.10b)
- *Linearity on the boundary*: ϕ_i is linear on each facet of P; (1.10c)
- *Smoothness*: $\phi_i \in C^\infty$. (1.10d)

We remark that all these properties are satisfied in the case $n = d+1$ for the linear functions ϕ_i in (1.7).

In general, there are two types of generalized barycentric coordinates. On the one hand, there are coordinates with a closed form. Analogously to (1.2), the closed-form definition is often based on certain *weight functions* $\boldsymbol{w} = [w_1, \ldots, w_n]\colon P \to \mathbb{R}^n$, which satisfy

$$\sum_{i=1}^{n} w_i(\boldsymbol{x})(\boldsymbol{v}_i - \boldsymbol{x}) = \boldsymbol{0} \qquad \forall \boldsymbol{x} \in P. \tag{1.11}$$

These weight functions are also called *homogeneous coordinates*, because normalizing them gives the generalized barycentric coordinates

$$\phi_i(\boldsymbol{x}) = \frac{w_i(\boldsymbol{x})}{W(\boldsymbol{x})}, \qquad W(\boldsymbol{x}) = \sum_{j=1}^{n} w_j(\boldsymbol{x}), \qquad i = 1, \ldots, n. \tag{1.12}$$

On the other hand, there are computational coordinates that can only be obtained numerically, for example, by solving an optimization problem.

1.2 2D COORDINATES

We first focus on the 2D case and consider a simple polygon P. Without loss of generality, we assume the vertices $\boldsymbol{v}_i$ of this polygon to be given in a counterclockwise direction and we treat the vertex indices cyclically, that is, $\boldsymbol{v}_{i+kn} = \boldsymbol{v}_i$

for $i \in \{1, \ldots, n\}$ and $k \in \mathbb{Z}$. Note that all the coordinates in this section, except for Hermite and complex, are generalized barycentric coordinates in the sense of Definition 1.1.

The earliest generalizations of barycentric coordinates in 2D were closed-form constructions and restricted to convex polygons. In this setting, the key objective is finding smooth and positive weights w_i, which satisfy (1.11). It is then known [149] that the normalized weights ϕ_i in (1.12) are well-defined generalized barycentric coordinates with respect to P and satisfy all properties in (1.10). However, most of these constructions lead to negative weights w_i at certain points inside non-convex polygons. Even worse, the denominator W may vanish, so that the ϕ_i are not necessarily well-defined for all $\boldsymbol{x} \in P$. So far, no closed-form construction of positive weights for arbitrary non-convex polygons is known, and even if it exists, the resulting coordinates would not be more than C^0 at concave corners [10]. We discuss different closed-form coordinates in Sections 1.2.1–1.2.7.

It was later realized that ϕ_i can also be obtained numerically as the solution of an optimization problem subject to the constraints given by the required properties of the coordinates. This optimization problem can either be global or local, and we present different kinds of such computational coordinates in Sections 1.2.8–1.2.10.

If the points $\boldsymbol{v}_i$ are not given as vertices of a polygon, but rather as scattered points, Definition 1.1 and properties (1.10b) and (1.10d) still make sense. In this setting, we can define generalized barycentric coordinates ϕ_i with respect to the set $\Pi = \{\boldsymbol{v}_1, \ldots, \boldsymbol{v}_n\} \in \mathbb{R}^2$ of $n \geq 3$ scattered points, and we review three different constructions in Sections 1.2.11–1.2.13.

We also give a short overview of coordinates that generalize Definition 1.1 in some other way. In particular, we briefly discuss Hermite and complex coordinates in Sections 1.2.14 and 1.2.15.

1.2.1 Wachspress coordinates

Wachspress [407] was one of the first who suggested a generalization of barycentric coordinates to a polygon with $n > 3$ vertices. Later, Meyer et al. [275] propose a simple local formulation of these *Wachspress coordinates*, which is given by the normalization (1.12) of the weight functions

$$w_i = \frac{\cot \gamma_{i-1} + \cot \beta_i}{r_i^2}, \qquad i = 1, \ldots, n,$$

where γ_{i-1} and β_i are the angles shown in Figure 1.2 and $r_i = \|\boldsymbol{v}_i - \boldsymbol{x}\|$. These coordinates are rational functions with numerator and denominator of degrees at most $n-2$ and $n-3$, respectively, which are the minimal possible degrees [412].

For strictly convex polygons, it is clear that all $w_i(\boldsymbol{x}) > 0$ for any $\boldsymbol{x} \in P$, so that Wachspress coordinates are well-defined and satisfy all properties in (1.10). They are also affine invariant. For non-convex polygons, the coordinates are not well-defined at some points in the polygon's interior, because the denominator W vanishes, but they can be generalized to weakly convex polygons [264].

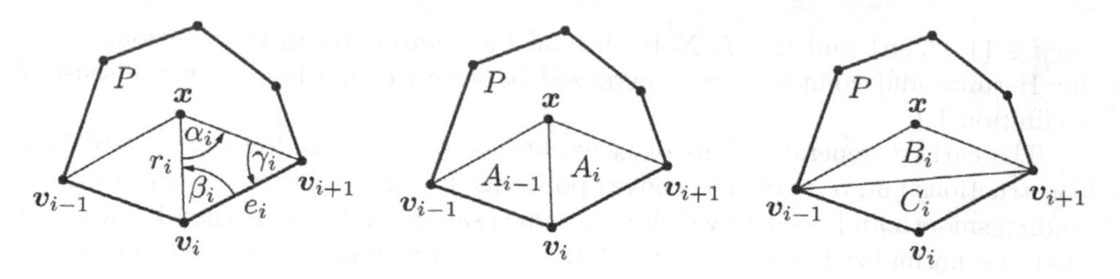

Figure 1.2 Notation used for signed angles, distances, and signed areas in a polygon P.

1.2.2 Discrete harmonic coordinates

Discrete harmonic coordinates [136, 309] arise from the standard piecewise linear finite element approximation to the Laplace equation and are given by the weight functions

$$w_i = \cot \beta_{i-1} + \cot \gamma_i, \qquad i = 1, \dots, n,$$

where β_{i-1} and γ_i are the angles shown in Figure 1.2. Interestingly, it turns out that for all polygons, whose vertices lie on a common circle, these coordinates are identical to Wachspress coordinates and therefore possess the same properties [149]. For other strictly convex polygons, they are still well-defined and satisfy all properties in (1.10), except for non-negativity. For non-convex polygons, the denominator W of these coordinates vanishes at some points in the polygon's interior, so that they are not well-defined.

1.2.3 Mean value coordinates

The derivation of discrete harmonic coordinates suggests that different properties of harmonic functions can be exploited to derive other generalized barycentric coordinates. Floater [144] considers the circumferential mean value property of a harmonic function $u\colon P \to \mathbb{R}$, which states that for any disc $B = B(\boldsymbol{x}, r) \subset P$ of radius $r > 0$, centered at $\boldsymbol{x}$, with boundary ∂B,

$$u(\boldsymbol{x}) = \frac{1}{2\pi r} \int_{\boldsymbol{y} \in \partial B} u(\boldsymbol{y}) d\boldsymbol{y}. \tag{1.13}$$

He then shows that applying (1.13) to a piecewise linear function leads to *mean value coordinates.* These coordinates are given by the weight functions

$$w_i = \frac{\tan(\alpha_{i-1}/2) + \tan(\alpha_i/2)}{r_i}, \qquad i = 1, \dots, n,$$

where α_{i-1} and α_i are the angles shown in Figure 1.2 and $r_i = \|\boldsymbol{v}_i - \boldsymbol{x}\|$.

Hormann and Floater [200] show that mean value coordinates are well-defined even for sets of nested simple polygons and everywhere in the plane. They are positive inside the kernel of a star-shaped polygon and satisfy properties (1.10b) and (1.10c). Moreover, mean value coordinates are positive inside any quadrilateral, similarity invariant, and smooth, except at the vertices of P, where they are only C^0.

To derive *positive mean value coordinates* for non-convex polygons, Lipman et al. [251] use the transfinite description of mean value coordinates [217] and restrict the integration to the part of ∂P that is visible from $\boldsymbol{x}$. These coordinates can be efficiently evaluated by exploiting the GPU, but they are only C^0 along certain lines inside the polygon.

1.2.4 Complete family of coordinates

Floater et al. [149] show that for arbitrary real functions $c_i \colon P \to \mathbb{R}$, the weights

$$w_i = \frac{c_{i-1}A_i - c_i B_i + c_{i+1}A_{i-1}}{A_{i-1}A_i}, \qquad i = 1, \ldots, n, \tag{1.14}$$

with the signed areas A_i and B_i shown in Figure 1.2, satisfy property (1.11). Therefore, the task of finding generalized barycentric coordinates simplifies to finding functions c_i such that the weights in (1.14) are positive, or, if not, at least sum up to a non-vanishing denominator W, without our having to worry about properties (1.8) and (1.9), which are satisfied by construction, and the resulting coordinates are as smooth as the functions c_i. Moreover, any set of generalized barycentric coordinates can be expressed in terms of the weights in (1.14) with the proper choice of c_i.

An interesting special case of this *complete family of coordinates* is given by $c_i = \|\boldsymbol{v}_i - \boldsymbol{x}\|^p$ for any $p \in \mathbb{R}$. In this case, the choices $p = 0$, $p = 1$, and $p = 2$ give Wachspress, mean value, and discrete harmonic coordinates, respectively. Surprisingly, the only coordinates for this choice of c_i, which are non-negative for any convex polygon, are Wachspress and mean value coordinates [149]. For any p, the coordinates are similarity invariant, and a simple geometric interpretation can be found in [232, Appendix A].

An alternative geometric construction for general weight functions w_i, which satisfy (1.8) and (1.9) and are non-negative by construction, is based on power diagrams [78]. However, the resulting *power coordinates* are not necessarily smooth.

1.2.5 Metric coordinates

Malsch et al. [265, 375] construct the so-called *metric coordinates*, which are given by the weight functions

$$w_i = \frac{A_{i-2}}{C_{i-1}q_{i-2}q_{i-1}} - \frac{B_i}{C_i q_{i-1} q_i} + \frac{A_{i+1}}{C_{i+1}q_i q_{i+1}}, \qquad i = 1, \ldots, n,$$

where A_i, B_i, and C_i are the areas shown in Figure 1.2, and

$$q_i = r_i + r_{i+1} - e_i$$

with $r_i = \|\boldsymbol{v}_i - \boldsymbol{x}\|$ and $e_i = \|\boldsymbol{v}_{i+1} - \boldsymbol{v}_i\|$. The functions q_i are non-negative everywhere in the plane, which guarantees a non-vanishing denominator W for all

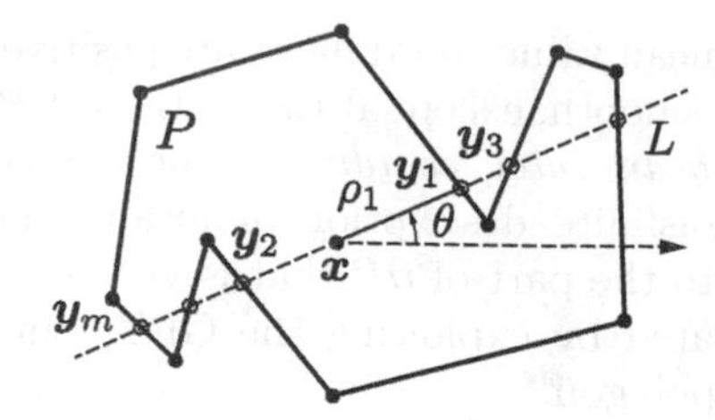

Figure 1.3 Notation used for the Gordon–Wixom interpolant.

simple polygons [200]. Metric coordinates are not necessarily positive inside convex polygons, satisfy properties (1.10b) and (1.10c), and they are smooth, except at the vertices of P, where they are only C^1. These coordinates are not well-defined for polygons with three consecutive collinear vertices.

1.2.6 Poisson coordinates

Taking inspiration from the derivation of mean value coordinates, Li et al. [247] derive *Poisson coordinates* from the Poisson integral formula for a harmonic function $u\colon P \to \mathbb{R}$, which states that for any disc $B = B(r) \subset P$ of radius $r > 0$, centered at the origin, with boundary ∂B and any point $\boldsymbol{x} \in B$,

$$u(\boldsymbol{x}) = \frac{1}{2\pi r} \int_{\boldsymbol{y} \in \partial B} \frac{r^2 - \|\boldsymbol{x}\|^2}{\|\boldsymbol{x} - \boldsymbol{y}\|^2} u(\boldsymbol{y}) d\boldsymbol{y}. \tag{1.15}$$

Poisson coordinates extend mean value coordinates, because the circumferential mean value theorem (1.13) is a special case of (1.15) when $\boldsymbol{x}$ is at the center of B. They are well-defined for any simple polygon P and possess the same properties as mean value coordinates. Moreover, following a specific construction, Poisson coordinates are proved to be pseudo-harmonic, that is, they reproduce harmonic functions on n-dimensional balls (see [94] for more details on pseudo-harmonic coordinates). The closed-form expressions of the Poisson weights w_i are rather lengthy and can be found in [247, Section 4.1].

1.2.7 Gordon–Wixom coordinates

Consider a line L through $\boldsymbol{x} \in P$ along some direction at a given angle θ with respect to a fixed axis, which intersects ∂P at the points $\boldsymbol{y}_1, \ldots, \boldsymbol{y}_m$ (see Figure 1.3). Summing over all m intersections, Belyaev et al. [37, 38] extend *Gordon–Wixom coordinates* [171] from convex to arbitrary simple polygons and show that they can be obtained as basis functions from the transfinite interpolant (see Chapter 3 for more details)

$$f(\boldsymbol{x}) = \frac{1}{2\pi} \int_0^{2\pi} \left(\sum_{i=1}^{m} \frac{\varepsilon_i f(\boldsymbol{y}_i)}{\rho_i} \Big/ \sum_{i=1}^{m} \frac{\varepsilon_i}{\rho_i} \right) d\theta, \tag{1.16}$$

where $f(\boldsymbol{y}_i)$ are data sampled from a piecewise-linear function f given along the polygon's boundary, $\rho_i = \|\boldsymbol{x} - \boldsymbol{y}_i\|$, and $\varepsilon_i \in \{-1, 0, 1\}$ with $\varepsilon_i = -1$ if the ray from $\boldsymbol{x}$ to $\boldsymbol{y}_i$ approaches $\boldsymbol{y}_i$ from the outside of P, $\varepsilon_i = 0$ if it is tangent to ∂P at $\boldsymbol{y}_i$, or $\varepsilon_i = 1$ if it approaches $\boldsymbol{y}_i$ from the inside of P. These coordinates satisfy properties (1.10b) and (1.10c), but they can be negative inside non-convex polygons. Manson et al. [266] extend (1.16) and present *positive Gordon–Wixom coordinates* inside any simple polygon, which are as smooth as the boundary. The analytic expressions of these coordinates are rather lengthy, and we refer the reader to [266, Section 3.2].

1.2.8 Harmonic coordinates

As observed in [149], one way of acquiring generalized barycentric coordinates with respcct to any simple polygon is by solving the Laplace equation

$$\Delta \boldsymbol{\phi} = \mathbf{0} \tag{1.17}$$

subject to suitable Dirichlet boundary conditions. These boundary conditions are given by a set of piecewise linear functions $\boldsymbol{g} = [g_1, \ldots, g_n] \colon \partial P \to \mathbb{R}^n$ such that every g_i has the Lagrange property (1.10b). These so-called *harmonic coordinates* satisfy all properties in (1.10).

The classical way [215] to approximate these coordinates is by discretizing (1.17) over the space of piecewise linear functions with respect to a triangulation of $\bar{P}$. Alternatively, harmonic coordinates can also be approximated using the boundary element method [342], the complex variable boundary method [417], and the method of fundamental solutions [269].

1.2.9 Maximum entropy coordinates

Arroyo and Ortiz [15] and Sukumar [372] show independently that generalized barycentric coordinates can be obtained as the solution of a constrained optimization problem based on the principle of maximum entropy [211]. These *maximum entropy coordinates* are non-negative, but the Lagrange property (1.10b) holds only for strictly convex polygons. It was later realized in [380] that, using prior distributions [212, 230, 357], the constrained optimization can be modified to be

$$\max_{\boldsymbol{\phi}(\boldsymbol{x}) \in \mathbb{R}^n_+} - \sum_{i=1}^{n} \phi_i(\boldsymbol{x}) \ln \frac{\phi_i(\boldsymbol{x})}{\omega_i(\boldsymbol{x})}$$

subject to (1.8) and (1.9), where $\omega_i \colon P \to \mathbb{R}_+$ is a prior estimate for ϕ_i, which can be viewed as a weight function (see Chapter 13 for further details).

Hormann and Sukumar [202] show how to construct ω_i such that maximum entropy coordinates satisfy all properties in (1.10), except smoothness, for any simple polygon. Regarding smoothness, these coordinates are proven to be C^0-continuous [379] for any set of C^k prior functions with $k \geq 0$ and assumed to be as smooth as the priors. Unlike harmonic coordinates, maximum entropy coordinates do not require solving a global optimization problem, but can rather be evaluated locally at any point $\boldsymbol{x} \in P$ using Newton's method that converges quadratically.

1.2.10 Local coordinates

Zhang et al. [433] propose to minimize the sum of total variations of the coordinate functions ϕ_i,

$$\min_{\phi} \sum_{i=1}^{n} \int_P \|\nabla \phi_i\|,$$

subject to the constraints given by (1.8), (1.9), (1.10a), (1.10b), and (1.10c), where the total variation of ϕ_i is the L_1-norm of $\|\nabla \phi_i\|$ in P. The solution of this convex minimization problem gives the so-called *local coordinates*, which are at least C^0. The name stems from the fact that the numerically computed function values are very close to zero in a large part of P. The exact support of these coordinates, however, is not known.

1.2.11 Affine coordinates

Among all generalized barycentric coordinates with respect to a scattered set of points Π, Waldron [408] suggests considering those with minimal L_2-norm, which are uniquely defined as the *affine* functions

$$\phi_i(\boldsymbol{x}) = (\boldsymbol{x} - \boldsymbol{c}) \cdot ((VV^*)^{-1}(\boldsymbol{v}_i - \boldsymbol{c}))^T + \frac{1}{n}, \qquad i = 1, \ldots, n,$$

where $\boldsymbol{c} = \frac{1}{n}\sum_{i=1}^{n} \boldsymbol{v}_i$ is the barycenter of Π, $V = (\boldsymbol{v}_1 - \boldsymbol{c}, \ldots, \boldsymbol{v}_n - \boldsymbol{c})$ is a $2 \times n$ matrix, and V^* is the adjoint of V. These coordinates are non-negative inside a convex region that contains $\boldsymbol{c}$, and they are equal to $\frac{1}{n}$ at $\boldsymbol{c}$. They are smooth and well-defined everywhere in the plane, but, in general, the Lagrange property (1.10b) holds only in the case $n = 3$ (see Section 1.1.1).

1.2.12 Sibson coordinates

Let V be the Voronoi diagram [404] of the set Π and C_i be the Voronoi cell in V that contains $\boldsymbol{v}_i$. Further let $V_{\boldsymbol{x}}$ be the Voronoi diagram of $\Pi \cup \{\boldsymbol{x}\}$ and $C_{\boldsymbol{x}}$ be the Voronoi cell in $V_{\boldsymbol{x}}$ that contains $\boldsymbol{x}$ (see Figure 1.4). Intersecting the cells C_i with $C_{\boldsymbol{x}}$, we can define the weight functions

$$w_i = \mathrm{Area}[C_i \cap C_{\boldsymbol{x}}], \qquad i = 1, \ldots, n. \tag{1.18}$$

These areas were originally proposed by Sibson [360, 361] with the emphasis on natural neighbor interpolation, where the natural neighbors of some $\boldsymbol{x}$ are all points from Π for which $w_i(\boldsymbol{x}) \neq 0$.

Normalizing the weights in (1.18) as in (1.12) defines the *Sibson coordinates.* These coordinates are well-defined over the convex hull of Π, have local support,

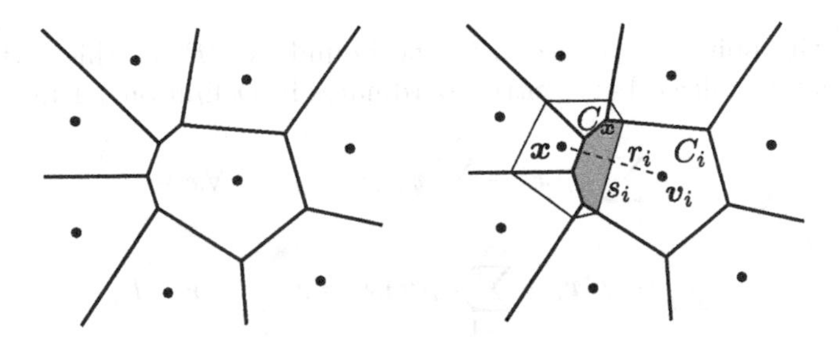

Figure 1.4 Voronoi diagram of a set of scattered points (left) and notation used for the construction of Sibson and Laplace coordinates (right).

and satisfy the Lagrange property (1.10b). They are C^1, except at $\boldsymbol{v}_i$, where they are only C^0. A simple computational algorithm for Sibson coordinates can be found in [371] and further details in [376].

1.2.13 Laplace coordinates

The concept of natural neighbors (see Section 1.2.12) is also used for the definition of so-called *Laplace coordinates* [34, 103, 370], which are given by the weight functions

$$w_i = \begin{cases} \frac{s_i}{r_i}, & \text{if } i \in I_{\boldsymbol{x}}, \\ 0, & \text{otherwise,} \end{cases} \qquad i = 1, \ldots, n,$$

where s_i is the length of the edge of $C_{\boldsymbol{x}}$ that is contained in C_i, $r_i = \|\boldsymbol{v}_i - \boldsymbol{x}\|$, and $I_{\boldsymbol{x}} \subset \{1, \ldots, n\}$ is the subset of indices of the natural neighbors of $\boldsymbol{x}$ (see Figure 1.4).

These coordinates have the same properties as Sibson coordinates, but they are only C^0 along the boundary of their support. Interestingly, for polygons, whose vertices lie on a circle, Sibson, Laplace, Wachspress, and discrete harmonic coordinates are all identical [375]. A simple computational algorithm for Laplace coordinates can be found in [371] and further details in [377].

1.2.14 Hermite coordinates

To interpolate not only boundary data, but also derivative information along the boundary, we can complement the functions ϕ_i with a second set of n functions ψ_i and substitute the Lagrange property (1.10b) with the Hermite property

$$\begin{aligned} \phi_i(\boldsymbol{v}_j) &= \delta_{ij}, & \psi_i(\boldsymbol{x}) &= 0 & \forall \boldsymbol{x} &\in \partial P, \\ \frac{\partial \phi_i}{\partial \boldsymbol{\nu}}(\boldsymbol{x}) &= 0 \quad \forall \boldsymbol{x} \in \partial P, & \frac{\partial \psi_i}{\partial \boldsymbol{\nu}}(\boldsymbol{x}) &= \delta_{ij} & \forall \boldsymbol{x} &\in [\boldsymbol{v}_j, \boldsymbol{v}_{j+1}], \end{aligned} \tag{1.19}$$

where $\boldsymbol{\nu}$ is the unit vector normal to the boundary ∂P. In this setting, the key properties of generalized barycentric coordinates in Definition 1.1 take on the form

$$\sum_{i=1}^{n} \phi_i(\boldsymbol{x}) + \sum_{i=1}^{n} \psi_i(\boldsymbol{x}) = 1 \qquad \forall \boldsymbol{x} \in \bar{P}, \tag{1.20}$$

$$\sum_{i=1}^{n} \phi_i(\boldsymbol{x})\boldsymbol{v}_i + \sum_{i=1}^{n} \psi_i(\boldsymbol{x})\boldsymbol{\nu}_i = \boldsymbol{x} \qquad \forall \boldsymbol{x} \in \bar{P}, \tag{1.21}$$

where $\boldsymbol{\nu}_i$ is the unit normal of the edge $[\boldsymbol{v}_i, \boldsymbol{v}_{i+1}]$. Note that (1.20) is essentially the same as (1.8), because $\psi_i(\boldsymbol{x}) = 0$.

To the best of our knowledge, Lipman et al. [252] were the first to derive coordinates ϕ_i and ψ_i that satisfy (1.20) and (1.21) from Green's third identity. These *Green coordinates* have a closed form, but do not satisfy (1.19). By fitting a bivariate polynomial in a weighted least squares sense to values and derivatives given along the polygon's boundary, Manson and Schaefer [267] construct *moving least squares coordinates*, which are defined for arbitrary, even disconnected polygons and have a closed form. Another closed-form solution for Hermite coordinates with property (1.19) for arbitrary simple polygons is presented by Li et al. [248], who derive *cubic mean value coordinates* from the mean value property of biharmonic functions. Weber et al. [418] extend harmonic coordinates to *biharmonic coordinates*, which satisfy (1.19), but do not posses a closed form.

1.2.15 Complex coordinates

If we interpret P as a subset of $\mathbb{C}$ and view its vertices $\boldsymbol{v}_i = (x_i, y_i)$ as complex numbers $z_i = x_i + \mathrm{i}\, y_i$, then we can reformulate generalized barycentric coordinates as complex functions [415, 416]. These *complex coordinates* are very useful in the context of planar shape deformation (see Chapter 7 for more details). The complex setting also reveals that Green coordinates (see Section 1.2.14) can alternatively be derived from Cauchy's integral formula [415].

1.2.16 Comparison

To compare the generalized barycentric coordinates from this section, except for Hermite and complex, we summarize some of their properties in Table 1.1 and show contour plots of some coordinate functions for a convex as well as a concave polygon in Figures 1.5–1.8.

In particular, Table 1.1 lists the domain for which the coordinates are well-defined and whether or not they have a closed-form definition, are non-negative (1.10a), and satisfy the Lagrange property (1.10b). In addition, the table shows how smooth the coordinates are over the interior of the respective domain. The partition of unity (1.8) and linear reproduction (1.9) properties are not included in the table, because all the coordinates comply with them. Moreover, we do not include the extra property (1.10c) since it is satisfied by all the coordinates defined for polygons and does not apply to coordinates for scattered points.

Coordinates (Section)	Valid domain	Closed form	Non-negativity	Lagrange property	Smooth-ness
WP (1.2.1)		✓	✓	✓	C^∞
DH (1.2.2)		✓		✓	C^∞
CF (1.2.4)		✓		✓	C^∞
MV (1.2.3)		✓		✓	C^∞
PM (1.2.3)		✓	✓	✓	C^0
MT (1.2.5)		✓		✓	C^∞
PS (1.2.6)		✓		✓	C^∞
GW (1.2.7)		✓		✓	C^0
PG (1.2.7)		✓	✓	✓	C^0
HM (1.2.8)			✓	✓	C^∞
ME (1.2.9)			✓	✓	C^k
LC (1.2.10)			✓	✓	C^0
AF (1.2.11)		✓			C^∞
SB (1.2.12)		✓	✓	✓	C^1
LP (1.2.13)	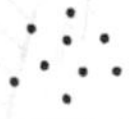	✓	✓	✓	C^0

Table 1.1 Properties of 2D generalized barycentric coordinates.

For a better comparison we group all coordinates by the first two properties. The first group includes Wachspress (WP), discrete harmonic (DH), and the complete family (CF) of coordinates. These coordinates are well-defined for convex polygons only and have a closed form. The second group consists of mean value (MV), positive mean value (PM), metric (MT), Poisson (PS), Gordon–Wixom (GW), and positive Gordon–Wixom (PG) coordinates that are well-defined for arbitrary simple polygons and have a closed form. In the third group we list harmonic (HM), maximum entropy (ME), and local (LC) coordinates that are also well-defined for arbitrary simple polygons, but can be obtained only numerically. The last group

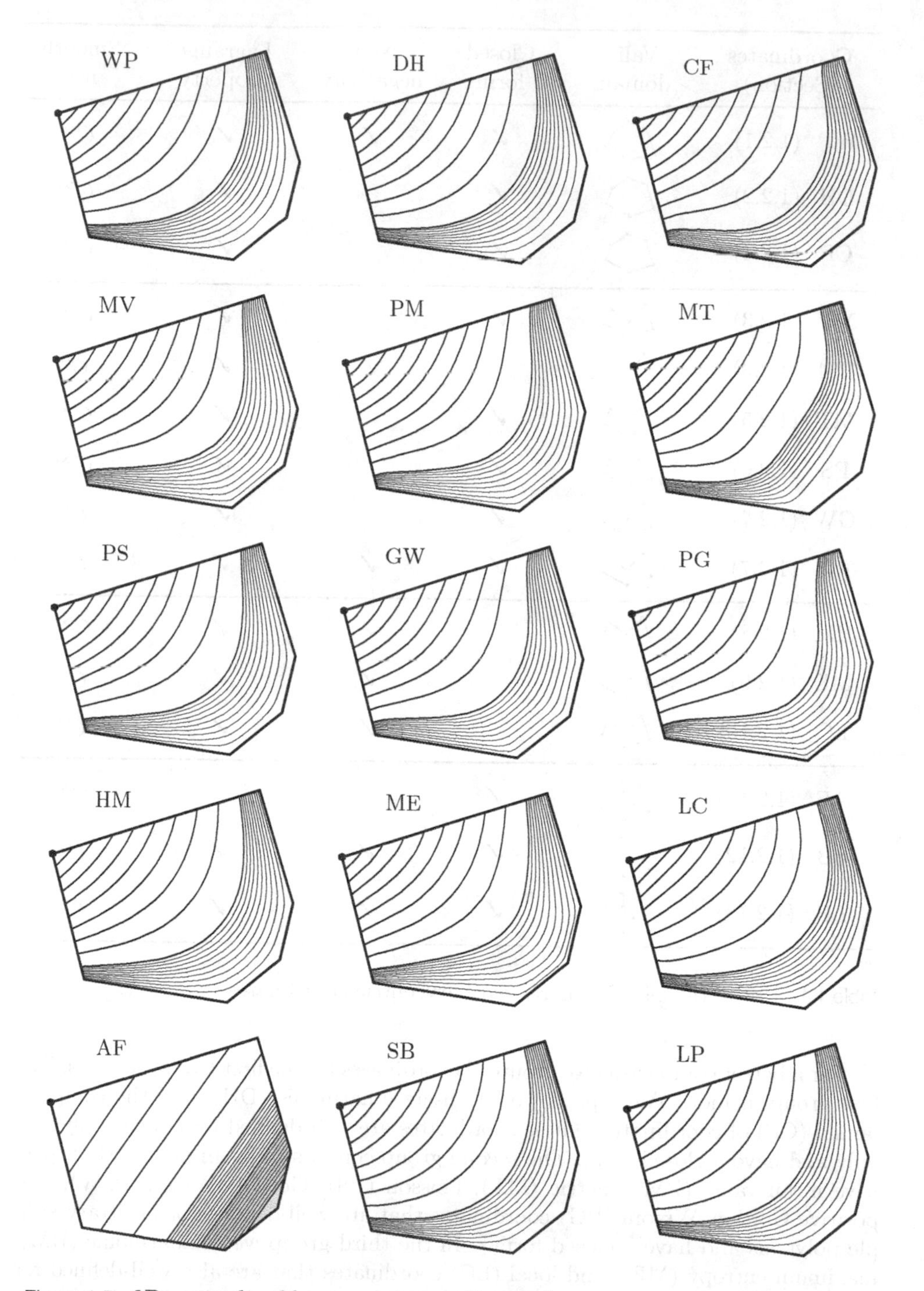

Figure 1.5 2D generalized barycentric coordinates for a convex polygon with respect to the marked vertex.

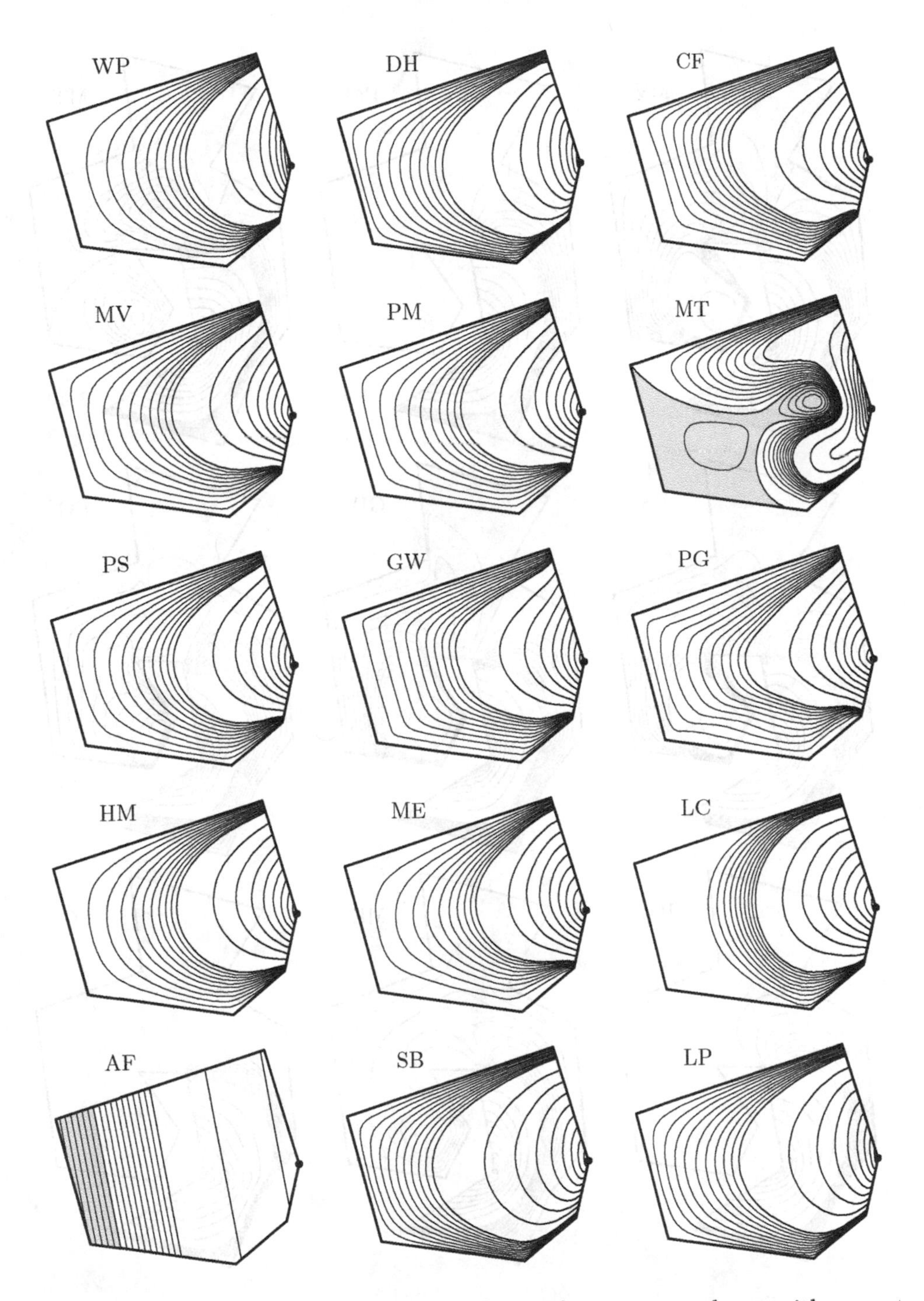

Figure 1.6 2D generalized barycentric coordinates for a convex polygon with respect to the marked vertex.

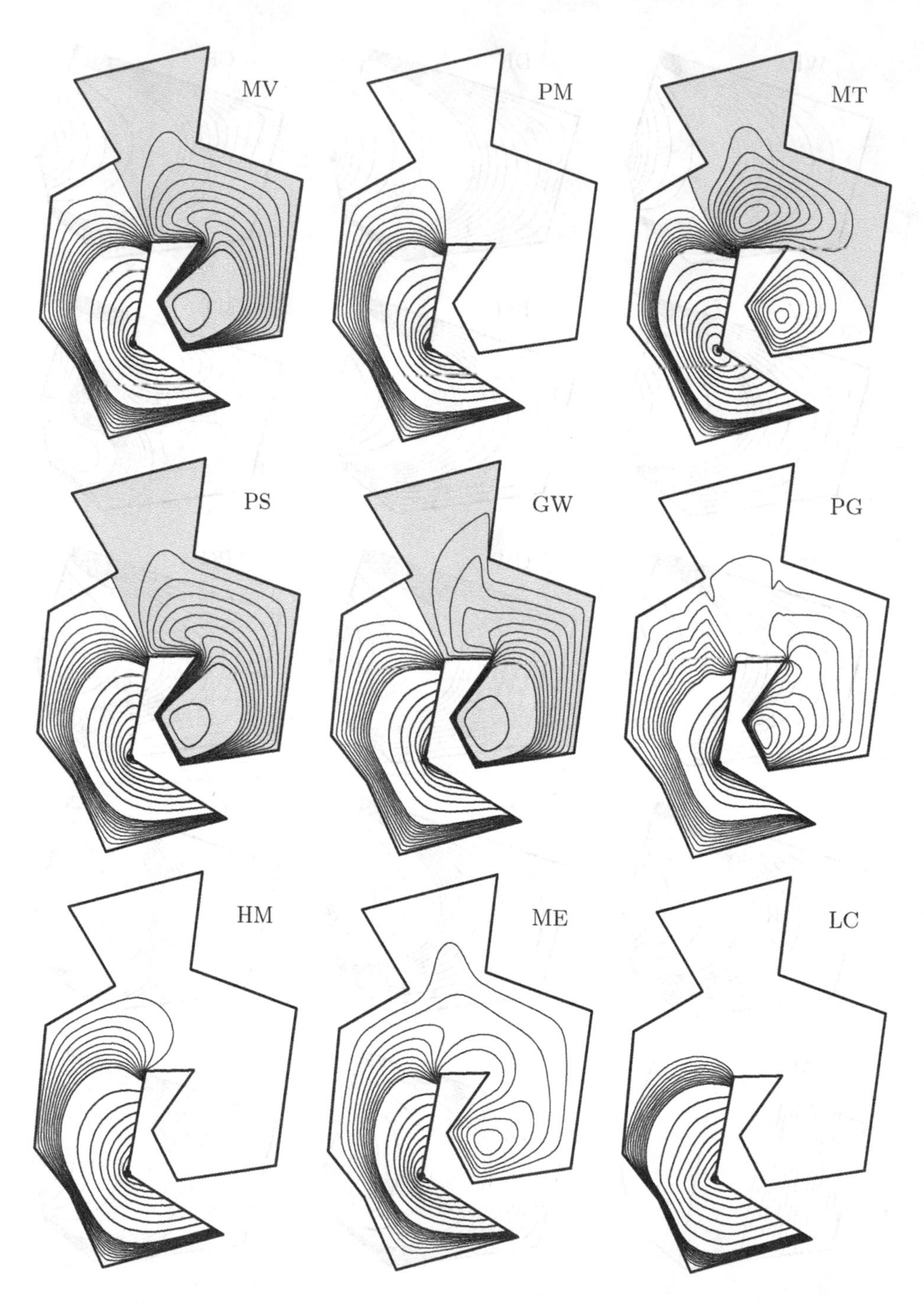

Figure 1.7 2D generalized barycentric coordinates for a concave polygon with respect to the marked vertex.

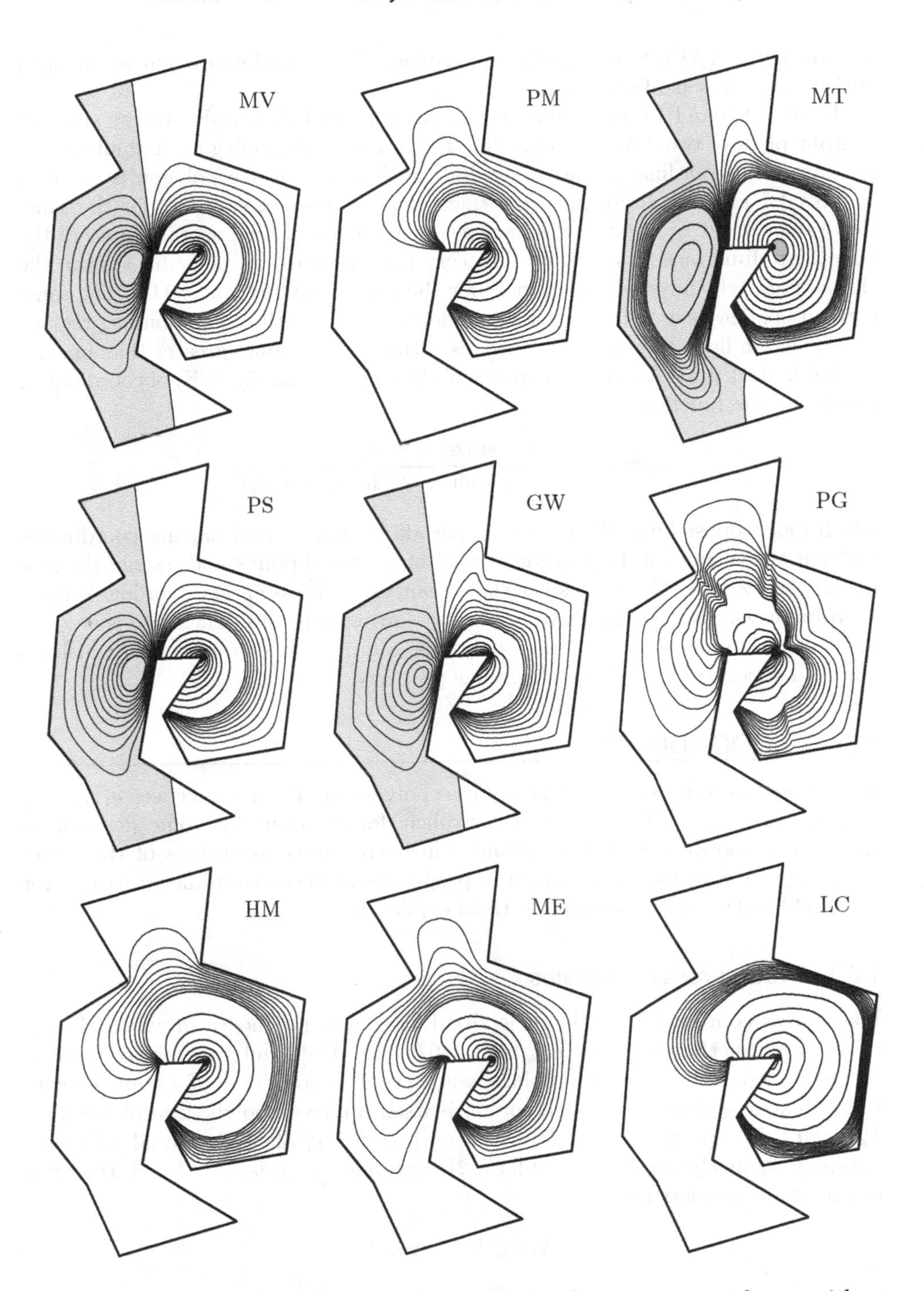

Figure 1.8 2D generalized barycentric coordinates for a concave polygon with respect to the marked vertex.

includes affine (AF), Sibson (SB), and Laplace (LP) coordinates that are defined with respect to sets of scattered points.

Figures 1.5 and 1.6 provide a visual comparison of all coordinates by showing contour plots of two different basis functions for a convex polygon. A similar comparison of all coordinates that are well-defined for a concave polygon is given in Figures 1.7 and 1.8. In all plots the thick curves represent contours for the function values $0.0, 0.1, \dots, 1.0$ and allow for a comparison of the general shapes of the coordinate functions. In addition, the decay towards zero can be deduced from the thin curves, which represent contours for the values $0.01, 0.02, \dots, 0.09$. The same contour spacing is used for negative function values, and the corresponding regions are shaded in light gray. Likewise, regions with function values greater than one are shaded in dark gray. As a representative of the complete family (CF) of coordinates, we choose the function

$$c_i = 2r_i \frac{\sin \alpha_{i-1} + \sin \alpha_i}{\sin \alpha_{i-1} + \sin \alpha_i + \sin(\alpha_{i-1} + \alpha_i)},$$

which was proposed in [149, Section 5]. For affine, Sibson, and Laplace coordinates we treat the vertices of the polygon as a set of scattered points and restrict the plot to the convex hull. The plots show that mean value, Poisson, and Gordon–Wixom coordinates can become negative inside non-convex polygons. The same is true for metric coordinates, also for convex polygons, and for affine coordinates, which do not even necessarily satisfy the Lagrange property.

1.3 3D COORDINATES

We now focus on the 3D case and consider a polyhedron P with n vertices $\boldsymbol{v}_1, \dots, \boldsymbol{v}_n$ and m faces $f_1, \dots, f_m$. We present explicit formulations for some generalized barycentric coordinates in this setting, which are direct extensions of the corresponding 2D constructions. Since the properties of these coordinates carry over from 2D to 3D, we do not mention them explicitly.

1.3.1 Wachspress coordinates

Wachspress coordinates can be generalized to convex polyhedra in which all vertices have exactly three incident faces [407, 411, 413], and to arbitrary convex polyhedra using the concept of polar duals [218]. The polar dual of a bounded convex polyhedron that contains the origin is itself a convex polyhedron of the form $P_d = \{\boldsymbol{y} \in \mathbb{R}^3 \mid \boldsymbol{y} \cdot \boldsymbol{z} \leq 1 \, \forall \boldsymbol{z} \in P\}$. Since each vertex $\boldsymbol{v}_i$ of P has a dual polygonal pyramid $\triangle_i$ in P_d (see Figure 1.9 for a 2D example), translating P such that $\boldsymbol{x}$ is at the origin and letting

$$w_i = \mathrm{Vol}[\triangle_i], \qquad i = 1, \dots, n$$

leads to *3D Wachspress coordinates* after the normalization in (1.12). Note that the constructions in [411, 413] also extend to higher dimensions.

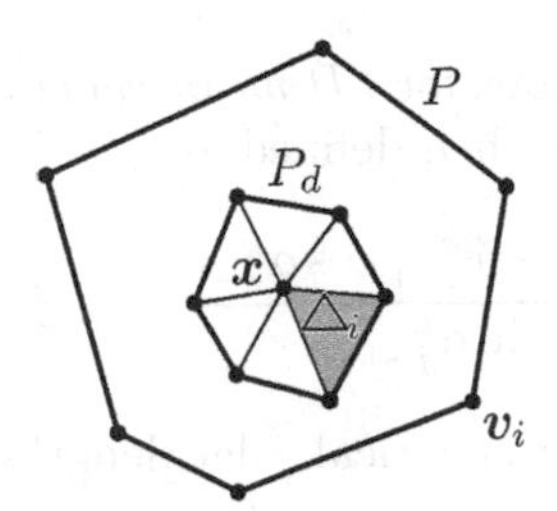

Figure 1.9 Polar dual P_d of a hexagon P centered at $\boldsymbol{x}$, where the triangle $\triangle_i$ is dual to the vertex $\boldsymbol{v}_i$.

1.3.2 Discrete harmonic coordinates

It was first mentioned in [276] that *3D discrete harmonic coordinates* exist, but no exact formula was given. Later, Ju et al. [216] show that for convex polyhedra with triangular faces these coordinates are given by the weight functions

$$w_i = \sum_{j \in N_i} h_j \cot \beta_j, \qquad i = 1, \ldots, n,$$

where $N_i \subset \{1, \ldots, n\}$ is the subset of the indices of the vertices in the one-ring neighborhood of $\boldsymbol{v}_i$, β_j is the dihedral angle between the faces $[\boldsymbol{v}_i, \boldsymbol{v}_j, \boldsymbol{v}_{j^+}]$ and $[\boldsymbol{x}, \boldsymbol{v}_j, \boldsymbol{v}_{j^+}]$, and $h_j = \|\boldsymbol{v}_j - \boldsymbol{v}_{j^+}\|$ (see Figure 1.10).

Interestingly, unlike their 2D counterparts (see Section 1.2.2), discrete harmonic and Wachspress coordinates are not identical for polyhedra, whose vertices lie on a common sphere. However, Ju et al. [216, Section 3.2.2] introduce *Voronoi coordinates*, which are identical to Wachspress coordinates on such polyhedra.

1.3.3 Mean value coordinates

Floater et al. [150] and Ju et al. [217] independently generalize mean value coordinates to arbitrary polyhedra with triangular faces. Consider a face $f_k = [\boldsymbol{v}_1^k, \boldsymbol{v}_2^k, \boldsymbol{v}_3^k]$ of P. By projecting f_k onto the unit sphere centered at $\boldsymbol{x}$ we obtain a triangular

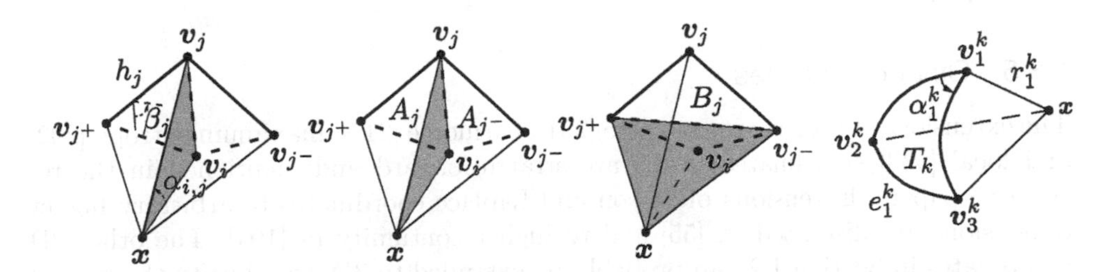

Figure 1.10 Notation used for signed angles, distances, and signed volumes in a polyhedron with triangular faces.

wedge T_k. The weight functions for *3D mean value coordinates* with respect to the vertices $\boldsymbol{v}_j^k$ of the face f_k are then defined as

$$w_j^k = \frac{e_j^k - e_{j-1}^k \cos\alpha_{j+1}^k - e_{j+1}^k \cos\alpha_{j-1}^k}{r_j^k \sin\alpha_{j+1}^k \sin\alpha_{j-1}^k}, \qquad j = 1,\dots,3, \quad k = 1,\dots,m,$$

with the angles α_j^k and the spherical edge lengths e_j^k as shown in Figure 1.10 and $r_j^k = \|\boldsymbol{v}_j^k - \boldsymbol{x}\|$. Details regarding the generalization of this construction to polyhedra with arbitrary faces can be found in [145].

1.3.4 Complete family of coordinates

Ju et al. [216] show that for convex polyhedra with triangular faces, Wachspress, discrete harmonic, and mean value coordinates can all be unified in a simple framework, which closely resembles the one in Section 1.2.4. This *complete family of 3D coordinates* is given by the weight functions

$$w_i = \sum_{j \in N_i} \frac{c_{j,j+}}{A_j} + \sum_{j \in N_i} \frac{c_{i,j} B_j}{A_j A_{j-}}, \qquad i = 1,\dots,n,$$

where $c_{i,j} \colon P \to \mathbb{R}$ are arbitrary real functions, A_j and B_j are the signed volumes shown in Figure 1.10, and $N_i \subset \{1,\dots,n\}$ is the subset of indices of the vertices in the one-ring neighborhood of $\boldsymbol{v}_i$. The name stems from the fact that this framework includes all possible coordinates for polyhedra with triangular faces. For example, given the signed angle $\alpha_{i,j}$ formed by $\boldsymbol{v}_i - \boldsymbol{x}$ and $\boldsymbol{v}_j - \boldsymbol{x}$, the choices

$$\begin{aligned}
c_{i,j} &= 2 - \left(\frac{\|\boldsymbol{v}_i - \boldsymbol{x}\|}{\|\boldsymbol{v}_j - \boldsymbol{x}\|} + \frac{\|\boldsymbol{v}_j - \boldsymbol{x}\|}{\|\boldsymbol{v}_i - \boldsymbol{x}\|} \right) \cdot \cos\alpha_{i,j}, \\
c_{i,j} &= \|(\boldsymbol{v}_i - \boldsymbol{x}) \times (\boldsymbol{v}_j - \boldsymbol{x})\|^2, \\
c_{i,j} &= \|(\boldsymbol{v}_i - \boldsymbol{x}) \times (\boldsymbol{v}_j - \boldsymbol{x})\| \cdot \alpha_{i,j}
\end{aligned}$$

lead to Wachspress, discrete harmonic, and mean value coordinates, respectively.

Alternatively, all non-negative coordinates for convex polyhedra can be represented as *3D power coordinates* [78], and both approaches can be extended to convex polytopes in higher dimensions.

1.3.5 Other coordinates

The extensions of positive mean value [251], harmonic [215], maximum entropy [202] and local [433] coordinates to 3D are straightforward and mentioned in the respective papers. Extensions of Sibson and Laplace coordinates to arbitrary higher dimensions are discussed in [55] and to higher continuity in [194]. The other 2D coordinates in Section 1.2 can probably be extended to 3D, too, but to the best of our knowledge, this has not been done so far. Apart from Wachspress coordinates, the only other coordinates that have been explicitly extended to arbitrary higher dimension are affine coordinates [408].

CHAPTER 2

Shape Quality for Generalized Barycentric Interpolation

Andrew Gillette*

University of Arizona, Tucson, USA

Alexander Rand

CD-adapco, Austin, USA

CONTENTS

AT A TIME WHEN generalized barycentric coordinates are gaining popularity in an ever-growing set of application contexts, there is a need for a clear understanding of the relationship between the interpolation properties of various coordinate types and the geometries on which they are defined. Consider, for

*Supported in part by the U.S. National Science Foundation under grant DMS-1522289.

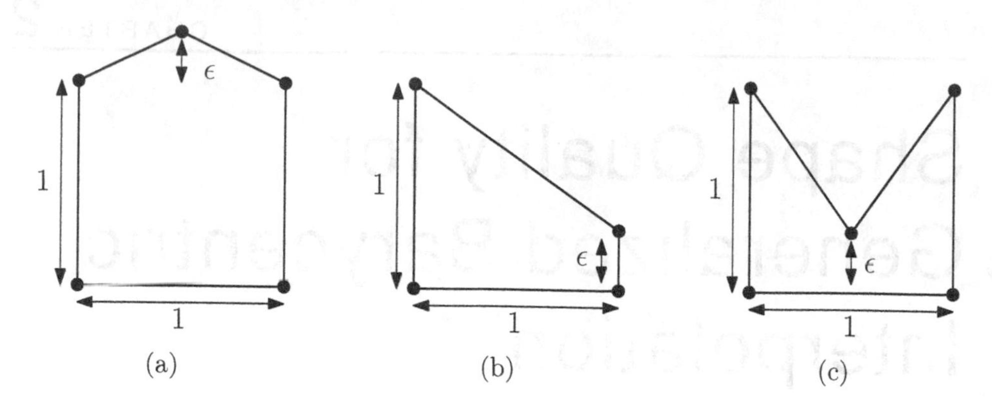

Figure 2.1 Three families of polygons with degenerate geometry as $\epsilon \to 0$ are shown: (a) an interior angle approaching 180°; (b) an edge approaching length 0; (c) a vertex approaching the interior of a non-adjacent edge.

instance, an academic or industrial code employing generalized barycentric coordinates that fails to give a numerical output when applied to some polygonal or polyhedral mesh. Could the premature termination of the simulation be averted by changing the coordinates that are used? Or by improving the "quality" of the mesh elements in some way? Or neither?

2.1 INTRODUCTION

In this work, we present both practical computational evidence and theoretical mathematical analyses for a variety of circumstances in which generalized barycentric interpolation will predictably succeed. Our primary proxy for the success of interpolation is a sharp upper bound on $||\nabla\phi_i||_\infty$, the maximum gradient of a generalized barycentric coordinate function, in terms of the relative positioning of the vertices $\{\boldsymbol{v}_1, \ldots, \boldsymbol{v}_n\}$ of the polygon or polyhedron P on which ϕ_i is defined. In computer graphics applications, control of gradients is essential to avoid visual artifacts. In computational mechanics applications, the gradient of the solution to a partial differential equation is often a quantity of interest in itself; in Poisson's equation, for instance, the gradient of the solution contributes to the energy norm (i.e., the H^1-norm) that is used to measure error. Further, we seek the broadest possible class of polygons or polyhedra over which our gradient estimates hold, as this puts the least restrictive requirements on the domain geometry.

To get a sense of the subtleties that can arise in this kind of analysis, consider three ways in which polygonal geometry can become "degenerate," as shown in Figure 2.1: collinearity of consecutive vertices, short edge length relative to diameter, and close proximity between a vertex and a non-adjacent edge. We carry out a numerical experiment to test five kinds of generalized barycentric coordinates on these representative "low-quality geometric domains." Boundary values for the generalized barycentric coordinates are derived from a simple quadratic function,

ϵ	HM	WP	MV	DH	LS-1	LS-2	LS-3	ME-T	ME-U
0.1600	1.3e0	2.6e-1	6.0e-2	2.3e-1	1.8e-1	2.2e-2	5.4e-2	7.9e-2	5.1e-1
0.0400	1.3e0	1.3e0	1.1e-1	1.5e0	3.6e-1	5.4e-2	1.2e-1	1.6e-1	2.2e0
0.0100	1.3e0	3.1e0	1.3e-1	3.9e0	4.4e-1	6.5e-2	1.4e-1	1.9e-1	5.1e0
0.0025	1.3e0	6.4e0	1.3e-1	8.3e0	4.5e-1	6.7e-2	1.4e-1	2.0e-1	9.3e0
0.0000	1.3e0	—	1.3e-1	—	4.5e-1	6.8e-2	1.5e-1	2.0e-1	$\star$

Table 2.1 Comparison among the harmonic (HM), Wachspress (WP), mean value (MV), discrete harmonic (DH), moving least squares (LS-$*$), and maximum entropy (ME-$*$) coordinates over the family of domains depicted in Figure 2.1a. The harmonic coordinates were approximated using a fine triangular mesh to provide a benchmark value; the H^1-seminorm of the interpolant is reported. For the other coordinates, the reported value is the H^1-seminorm of the difference between the interpolant and the harmonic interpolant.

$u(x, y) = 0.77223(x - 0.331)^2 + 1.1123(y + 0.177344)^2$, with some arbitrary coefficients selected to avoid degeneracy or symmetry during interpolation. First, we compute an approximation to the harmonic coordinates using a fine triangular mesh (see Section 1.2.8). Since harmonic coordinates provide an interpolant with minimal H^1-seminorm by Dirichlet's principle, this establishes a good benchmark interpolant.

We then compute interpolants using the following coordinates: Wachspress, mean value, discrete harmonic, moving least squares with linear, quadratic, and cubic weights, and maximum entropy with a "triangle inequality" prior (constructed from edge weight functions that vanish along each edge of the polygon and are strictly positive elsewhere) and the uniform prior. All of these coordinates are defined in Section 1.2, except for the moving least squares coordinates, which are described in the discussion below.

Table 2.1 shows the result of our experiment for the family of domains shown in Figure 2.1a. Here, the limiting case, when $\epsilon = 0$, looks like a square with a hanging node, but we formally treat it as a pentagon with two adjacent collinear sides. The H^1-seminorm of the approximation to the harmonic interpolant is reported in the first column, demonstrating that this value does not grow as $\epsilon \to 0$. For other coordinates, the H^1-seminorm of the difference between the interpolant and the harmonic interpolant is reported. Here, the well-known advantage of mean value coordinates over Wachspress coordinates is evident: the Wachspress gradients grow with large interior angles whereas mean value coordinates do not. Discrete harmonic coordinates exhibit a similar behavior as they, like Wachspress coordinates, are not well-defined when the domain is not strictly convex.

Maximum entropy coordinates [202] are defined both as $\epsilon \to 0$ and when $\epsilon = 0$; however, the choice of prior affects their behavior in the limit. In the case of the uniform prior (ME-U), the value at the peak vertex jumps from 1 to 1/3 when the geometry changes from a pentagon to a square, breaking the Lagrange property.

Note that if the ME-U coordinates are used throughout an entire mesh, the resulting interpolant will be conforming, that is globally C^0. However, the value of the ME-U interpolant along adjacent collinear edges may not be the same as the interpolant produced by other kinds of generalized barycentric coordinates. We flag this special case with $\star$ in Table 2.1. On the other hand, when using the "triangle inequality" prior (ME-T), the Lagrange property is satisfied when $\epsilon = 0$ and the success of the interpolant as $\epsilon \to 0$ is confirmed by the data.

Moving least squares coordinates [267] are defined in terms of a minimization problem, weighted by the distance from a point to the boundary. The weight function $W(\boldsymbol{x}, t) = \|\boldsymbol{p}'(t)\|/\|\boldsymbol{p}(t) - \boldsymbol{x}\|^\alpha$ ensures that the coordinates have the boundary values encoded by $\boldsymbol{p}$ on each edge; the larger the exponent α, the more weight is given to smoothness of the coordinates near the boundary relative to smoothness on the interior of the domain. As reported in the LS-α columns of Table 2.1, for $\alpha = 1, 2, 3$, the moving least squares interpolant exhibits small error compared to the harmonic interpolant as the geometry degenerates to a square, with $\alpha = 2$ producing the smallest error. From this perspective, the quadratic weight seems to do the best mimicking harmonic coordinates by balancing smooth enforcement of boundary requirements with interior smoothness; the other weight choices overly prefer one of these objectives to the detriment of the interpolation error. Note that moving least squares coordinates may take on negative values, which can present challenges in graphics applications. In regards to the error analysis presented here, however, non-negativity of the interpolant is not required.

The results of our experiments for the domains shown in Figures 2.1b and 2.1c are provided in Tables 2.2 (page 34) and 2.4 (page 37), respectively. All the results are discussed more thoroughly in Section 2.5.

2.2 DECONSTRUCTING THE *A PRIORI* ERROR ESTIMATE

The typical *a priori* error analysis of a finite element method begins with a characterization of the partial differential equation to be solved as an equation between a coercive, continuous bilinear form on $V \times V$ and a bounded linear operator L on V, where V is the function space from which u is sought [70, 369]. We fix the common setting of $V = H_0^1(\Omega)$, the space of H^1 functions on domain Ω, with zero boundary conditions on $\partial\Omega$. Cea's lemma then asserts the existence of a constant C_C such that

$$\|u - u_h\|_{H^1(\Omega)} \leq C_C \min_{v_h \in V_h} \|u - v_h\|_{H^1(\Omega)}. \tag{2.1}$$

Here, $u \in V$ is the unknown continuous solution to the problem, u_h is the approximation to u computed by the finite element method, and V_h is a finite-dimensional subspace of V from which u_h is sought. Hence, (2.1) establishes that the error between u_h and u is within a constant multiple of the best possible error when looking for a solution in V_h.

The next step in the analysis is to bound the error in the particular case of $v_h := Iu$, where $I\colon V \to V_h$ is an interpolation operator. A subtle but crucial point is that I takes as input the entire unknown function u to produce Iu and thus can employ quantities such as $u(\boldsymbol{v}_i)$ in its definition. The finite element solution u_h, on the other hand, is the output of a solution operator whose only inputs are the bilinear form, the linear operator V_h, and the boundary conditions. For linear interpolation derived directly from barycentric or generalized barycentric coordinates, the interpolation error estimate has the form

$$\|u - Iu\|_{H^1(\Omega)} \leq C_{\mathrm{IE}}\, h\, |u|_{H^2(\Omega)}, \tag{2.2}$$

where h is the maximum diameter of an element in a mesh of Ω. The constant C_{IE} in (2.2) depends both on the interpolation method (i.e., the definition of Iu) and on the shape of the elements used in the mesh. This dependence is quite subtle and often misunderstood.

Analysis leading to estimate (2.2) can be simplified via two observations. First, the error can be decomposed into individual mesh elements rather than the entire solution domain. Second, assuming that the interpolation method is affine invariant, all estimates can be rescaled to diameter 1 elements. Thus, it is sufficient to establish

$$\|u - Iu\|_{H^1(P)} \leq C_{\mathrm{IE}}\, |u|_{H^2(P)}, \tag{2.3}$$

for any polytope P of diameter 1 that is similar to an element in the mesh of Ω.

For interpolation via a set of generalized barycentric coordinates $\boldsymbol{\phi}$, the interpolation operator $I\colon H^2(P) \to \operatorname{span}\{\boldsymbol{\phi}\}$ is defined by

$$Iu := \sum_{i=1}^{n} u(\boldsymbol{v}_i)\phi_i. \tag{2.4}$$

In order to prove (2.3), we may only assume that $u \in H^2(P)$; however, the formula for Iu in (2.4) requires point values for u at vertices of P. We can appeal to the Sobolev embedding theorem [4, 70, 240] to ensure that u is in fact a continuous function and hence that $u(\boldsymbol{v}_i)$ is well-defined. We state a special case of the theorem that is also called Morrey's inequality, which we have tailored to the relevant setting.

Theorem 2.1 (Sobolev embedding/Morrey's inequality). *Let $k > d/2$. For any polygonal or polyhedral domain P, there exists a constant C_{SE} depending only on k and d such that for any $u \in H^k(P)$,*

$$\|u\|_{L^\infty(\overline{P})} \leq C_{\mathrm{SE}}\, \|u\|_{H^k(P)}.$$

Moreover, there is a continuous function in the $L^\infty(\overline{P})$ equivalence class of u.

The constant C_{SE} depends, in general, on the geometry of P and it is possible to find a sequence of polytopes $\{P_i\}_{i=1}^{\infty}$ for which the associated sequence of constants grows without bound. In these cases, the sequence of boundaries $\{\partial P_i\}_{i=1}^{\infty}$ is typically approaching some kind of cusp, fractal, or other kind of "degenerate" geometry. Accordingly, we require that P belong to some class of polytopes whose geometry is controlled via shape quality metrics and then seek a uniform bound on C_{SE} for this class.

We will also make use of the Bramble–Hilbert lemma [69], which estimates how well a function u can be approximated by polynomials in relevant Sobolev norms. It can be established uniformly over convex domains [117, 402] and can be extended to non-convex domains that are "star-shaped with respect to a ball" [70]. We state the convex version here.

Theorem 2.2 (Bramble–Hilbert). *There exists a constant C_{BH} depending only upon d such that for any convex polytope $P \subset \mathbb{R}^d$ of diameter 1 and any $u \subset H^2(P)$, there exists an affine function p_u such that,*

$$\|u - p_u\|_{H^1(P)} \leq C_{\mathrm{BH}}\, |u|_{H^2(P)}. \tag{2.5}$$

We now state five additional estimates that are either helpful or essential in proving (2.3). All of these are stated in terms of the spatial dimension d for unit diameter domains.

Uniform Sobolev Embedding: $$\sum_{i=1}^{n} |u(\boldsymbol{v}_i)| \leq C_{\mathrm{USE}}\, \|u\|_{H^2(P)}. \tag{2.6}$$

Bounded Gradients in L^∞: $$|\nabla \phi_i|_{L^\infty(P)} \leq C_{\nabla\phi}. \tag{2.7}$$

Bounded Gradients in H^1: $$\|\phi_i\|_{H^1(P)} \leq C_{H^1}. \tag{2.8}$$

Bounded Interpolation: $$\|Iu\|_{H^1(P)} \leq C_{\mathrm{BI}}\, \|u\|_{H^2(P)}. \tag{2.9}$$

Interpolation Error: $$\|u - Iu\|_{H^1(P)} \leq C_{\mathrm{IE}}\, |u|_{H^2(P)}. \tag{2.10}$$

Each estimate is of interest when the associated constant C can be bounded above *uniformly* over some class $\mathfrak{p}$ of polygons or polyhedra. In such cases, we say the estimate "holds over $\mathfrak{p}$" and we have the following implications.

Proposition 2.3. *Let $\mathfrak{p}$ be a class of polygons or polyhedra with unit diameter.*

(i) If (2.7) holds over $\mathfrak{p}$, then (2.8) holds over $\mathfrak{p}$.

(ii) If (2.6) and (2.8) hold over $\mathfrak{p}$, then (2.9) holds over $\mathfrak{p}$.

(iii) If (2.9) holds over $\mathfrak{p}$, then (2.10) holds over $\mathfrak{p}$.

Proof. For (i), we have that

$$\|\phi_i\|^2_{H^1(P)} = \int_P |\phi_i|^2 + |\nabla\phi_i|^2 \leq \left(1 + |\nabla\phi_i|^2_{L^\infty(P)}\right)|P| \leq (1 + C^2_{\nabla\phi})|P|,$$

where $|P|$ denotes the measure of P. Since $|P|$ must be contained inside a d-ball of unit radius, there is a constant C such that $|P| \leq C$ and the estimate follows.

For (ii), Theorem 2.1 ensures that any $u \in H^2(P)$ can be treated as a continuous function on $\overline{P}$, since $d = 2$ or 3. Thus, $u(\boldsymbol{v}_i)$ is well-defined and we have that

$$\|Iu\|_{H^1(P)} \leq \sum_{i=1}^{n} |u(\boldsymbol{v}_i)| \cdot \|\phi_i\|_{H^1(P)} \leq C_{H^1} C_{\mathrm{USE}}\, \|u\|_{H^2(P)}.$$

For (iii), first note that any choice of generalized barycentric coordinates ϕ ensures that

$$Ip = \sum_{i=1}^{n} p(\boldsymbol{v}_i)\phi_i = p \tag{2.11}$$

for any linear polynomial $p\colon P \to \mathbb{R}$. Now, given u, let p_u be the first-order polynomial satisfying (2.5). Then,

$$\begin{aligned}
\|u - Iu\|^2_{H^1(P)} &\le \|u - p_u\|^2_{H^1(P)} + \|p_u - Iu\|^2_{H^1(P)} && \text{by triangle inequality,}\\
&= \|u - p_u\|^2_{H^1(P)} + \|I(u - p_u)\|^2_{H^1(P)} && \text{by (2.11),}\\
&\le \|u - p_u\|^2_{H^1(P)} + C^2_{\mathrm{BI}}\,\|u - p_u\|^2_{H^2(P)} && \text{by (2.9),}\\
&\le (1 + C^2_{\mathrm{BI}})\|u - p_u\|^2_{H^1(P)} + C^2_{\mathrm{BI}}\,|u - p_u|^2_{H^2(P)} &&\\
&\le \left(C^2_{\mathrm{BH}}(1 + C^2_{\mathrm{BI}}) + C^2_{\mathrm{BI}}\right)|u|^2_{H^2(P)} && \text{by (2.5).}
\end{aligned}$$

The last step also uses the fact that the H^2 semi-norm of the first order polynomial p_u is zero. □

Fix any $h > 0$. The estimate on $\|u - Iu\|_{H^1(P)}$ just derived can be strengthened to an estimate on $\|u - Iu\|_{H^1(\Omega)}$, where Ω is a mesh made of elements from $\mathfrak{p}$ scaled by at most h. The resulting estimate is

$$\underbrace{\|u - u_h\|_{H^1(\Omega)}}_{\text{finite element error}} \le \underbrace{C_{\mathrm{C}}}_{\substack{\text{from}\\ (2.1)}} \underbrace{\|u - Iu\|_{H^1(\Omega)}}_{\text{interpolation error}} \le C_{\mathrm{C}}\, C_{\mathrm{IE}}\, h \underbrace{|u|_{H^2(\Omega)}}_{\substack{\text{2nd order}\\ \text{oscillation}}}, \tag{2.12}$$

where

$$C_{\mathrm{IE}} := \left(C^2_{\mathrm{BH}}(1 + C^2_{\mathrm{BI}}) + C^2_{\mathrm{BI}}\right)^{1/2},$$

with C_{BH} from (2.5) and C_{BI} from (2.9). As seen from the proof of Proposition 2.3, C_{BI} may depend in turn on the constants C_{H^1} from (2.8) and C_{USE} from (2.6) and C_{H^1} may depend on the constant $C_{\nabla\phi}$ from (2.7).

Equation (2.12) is the standard *optimal convergence estimate* for linear order, scalar-valued finite element methods. We can now see that the choice of generalized barycentric coordinates ϕ and class of polytopes $\mathfrak{p}$ is relevant at two key places in the analysis: the uniform Sobolev embedding estimate (2.6) and, more crucially, the gradient bound (2.7). We next investigate shape quality metrics that can ensure these two essential estimates on simplicial and polytopal meshes.

2.3 SHAPE QUALITY METRICS FOR SIMPLICES

The simplest case for our analysis is the class of triangles of diameter 1. Consider the following two types of subclasses:

$\mathfrak{t}_{\mathrm{mina}}$ — All angles are bounded away from zero: $\alpha_i > \alpha_* > 0$.

$\mathfrak{t}_{\mathrm{maxa}}$ — All angles are bounded away from 180°: $\alpha_i < \alpha^* < 180°$.

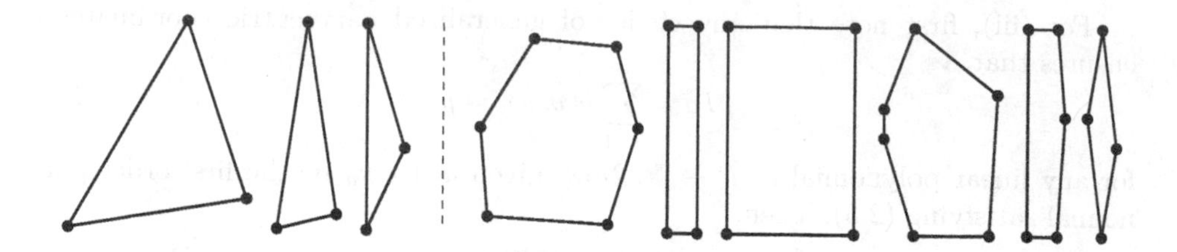

Figure 2.2 Left: Representative triangles satisfying one of three common quality metrics are shown: modest aspect ratio, one small angle, and one large angle. Right: Polygons are more challenging to classify in this manner as the presence of large or small angles neither implies nor excludes a good aspect ratio.

The terms *minimum angle condition* and *maximum angle condition*, which are common in the finite element literature, refer to classes of type $\mathfrak{t}_{\text{mina}}$ or $\mathfrak{t}_{\text{maxa}}$, respectively. In each case, estimate (2.7) can be established.

Lemma 2.4. *The inequality in (2.7) holds uniformly on* $\mathfrak{t}_{\text{mina}}$.

The proof of Lemma 2.4 is straightforward, since the minimum angle bound implies a minimum altitude, which in turn implies a maximum gradient on barycentric coordinates as required for (2.7).

Theorem 2.5. *The inequality in (2.10) holds uniformly on* $\mathfrak{t}_{\text{maxa}}$.

The proof of Theorem 2.5 is more involved since large angle triangles admit small altitudes and thus potentially large gradients locally. Instead, the proof involves mapping large angle triangles to those with bounded aspect ratio and following the error estimates through this mapping. Interpolation error estimates over these kinds of triangle classes originated independently in work by Babuška and Aziz [22] and Jamet [209].

Another way to describe the triangle quality is illustrated in Figure 2.2. The three triangles shown on the left have a modest aspect ratio, a single small angle, and a single large angle, respectively. Recall that the *aspect ratio* of a triangle is the ratio of its longest edge to the radius of its smallest inscribed circle. A bound on the aspect ratio is equivalent to a bound on the smallest angle in the triangle, hence any class of triangles with bounded aspect ratio is some class of the form of $\mathfrak{t}_{\text{mina}}$. Further, arbitrarily large angles only occur in conjunction with very small angles, hence, as a subclass of diameter 1 triangles, any class of type $\mathfrak{t}_{\text{mina}}$ is contained in a class of type $\mathfrak{t}_{\text{maxa}}$. The converse is false: each class of type $\mathfrak{t}_{\text{maxa}}$ includes some triangles with arbitrarily small angles, which is not contained in any class of type $\mathfrak{t}_{\text{mina}}$.

Simple examples demonstrate unbounded interpolation error on classes of triangles admitting arbitrarily large interior angles [22]. More precisely, it has been shown that the rate at which the interpolation error grows is proportional to the circumradius of the triangle [225, 317]. Thus, while a bound on the circumradius implies a bound on the maximum angle and vice versa, the former has a more direct connection to the interpolation error.

While a class of type $\mathfrak{t}_{\text{maxa}}$ is sufficient for the interpolation analysis outlined in Section 2.2, it is not strictly necessary for the convergence of finite element methods, as demonstrated in [185]. However, the examples of convergent finite element methods on meshes with arbitrarily large angles employ shape function spaces that contain subspaces corresponding to shape-regular meshes. Thus, some form of shape regularity still appears to underlie successful finite element methods.

These ideas have been generalized to tetrahedra and higher-dimensional simplices [7, 209, 231, 318]. On tetrahedra, the most widely known and used shape quality metric is aspect ratio, which on polyhedra is the ratio of the diameter of the element to the radius of the smallest inscribed sphere. Error analysis based on aspect ratio is straightforward to carry out, but is overly restrictive on the geometry, in the same way that a bound on the aspect ratio on triangles excludes permissible single-small-angle triangles. While a bound on the circumradius is closely tied to the interpolation error on triangles, the same is not true on tetrahedra due to the possibility of slivers. Slivers are very flat tetrahedra with evenly spaced vertices near a common circle such that the circumradius is modest. A metric called *coplanarity* [318] balances a weakened aspect ratio bound with an exclusion of sliver tetrahedra; it measures how far the edges of a tetrahedron lie to a common plane, thereby allowing narrow elements without allowing slivers.

2.4 SHAPE QUALITY METRICS FOR POLYGONS AND POLYHEDRA

For the class of diameter 1 triangles, we have just discussed how many common shape quality metrics are effectively equivalent. For the class of diameter 1 polygons, however, these equivalences do not hold and the relations among metrics are more subtle. For example, a polygon can have high aspect ratio without any small interior angles or can have a modest aspect ratio with interior angles near 180°. Some illustrative examples are shown in Figure 2.2.

To analyze the polygonal setting, let P denote a generic polygon of diameter 1 with edges $\{e_i\}_{i=1}^n$, vertices $\{\boldsymbol{v}_i\}_{i=1}^n$, interior angles $\{\alpha_i\}_{i=1}^n$, and maximum radius of an inscribed circle $r(P)$. We consider the following subclasses of polygons.*

$\mathfrak{p}_{\text{cvx}}$ — P is convex.

$\mathfrak{p}_{\text{ar}}$ — The aspect ratio is bounded: $\dfrac{\text{diam}(P)}{r(P)} \leq \gamma^*$.

$\mathfrak{p}_{\text{nv}}$ — The number of vertices is bounded: $n < n^*$.

$\mathfrak{p}_{\text{edge}}$ — The ratio of diameter to the shortest edge is bounded: $\dfrac{\text{diam}(P)}{|e_i|} < e^*$.

*We have included diam(P) in some of the definitions even though we have fixed it to be 1 for the subsequent analysis; this is intended to highlight the scaling effect of the diameter.

$\mathfrak{p}_{\text{mina}}$ — All interior angles are bounded away from zero: $\alpha_i > \alpha_* > 0$.

$\mathfrak{p}_{\text{maxa}}$ — All interior angles are bounded away from 180°: $\alpha_i < \alpha^* < 180°$.

The definitions of $\mathfrak{p}_{\text{cvx}}$, $\mathfrak{p}_{\text{ar}}$, $\mathfrak{p}_{\text{nv}}$, and $\mathfrak{p}_{\text{edge}}$ also define subclasses of diameter 1 polyhedra and of higher-dimensional polytopes. The definitions of $\mathfrak{p}_{\text{mina}}$ and $\mathfrak{p}_{\text{maxa}}$ are specific to polygons as stated, but can be generalized to higher dimensions by considering dihedral angles between faces with codimension 1. On polygons, we have previously shown some containments between these classes, as indicated in the following lemma; a proof is given in [164, Proposition 4].

Lemma 2.6. *For the class of diameter 1 polygons:*

(i) $\mathfrak{p}_{\text{cvx}} \cap \mathfrak{p}_{\text{ar}} \subset \mathfrak{p}_{\text{mina}}$.

(ii) $\mathfrak{p}_{\text{cvx}} \cap \mathfrak{p}_{\text{edge}} \subset \mathfrak{p}_{\text{nv}}$.

(iii) $\mathfrak{p}_{\text{cvx}} \cap \mathfrak{p}_{\text{maxa}} \subset \mathfrak{p}_{\text{nv}}$.

We have also shown that for P, a convex polygon of diameter 1, an aspect ratio and edge bound imply that a ball of some uniform radius r_* centered at any $\boldsymbol{x} \in P$ will intersect ∂P in one of three ways: at two adjacent edges, at a single edge, or not at all. This non-degeneracy result is helpful for our subsequent analysis and can be stated more precisely as follows; a proof is given in [164, Proposition 9].

Lemma 2.7. *For the class of diameter 1 polygons, there exists $r_* > 0$ such that for all $P \in \mathfrak{p}_{\text{cvx}} \cap \mathfrak{p}_{\text{ar}} \cap \mathfrak{p}_{\text{edge}}$ and for any $\boldsymbol{x} \in P$, $B(\boldsymbol{x}, r_*)$ does not intersect two non-adjacent edges of P.*

Beyond direct relationships between these geometric criteria, uniformity of the Sobolev embedding constant is an essential ingredient of our analysis involving some (relatively mild) geometric restrictions. This component is unique to polygonal analysis, since simplices can be mapped affinely to a single reference element.

Lemma 2.8. *Set $d = 2$. The uniform Sobolev embedding estimate (2.6) holds on $\mathfrak{p}_{\text{cvx}} \cap \mathfrak{p}_{\text{ar}}$.*

We omit a full proof of Lemma 2.8 and instead provide a sketch of the key ideas; the interested reader can find relevant technical details in the proof of [164, Proposition 10]. In the mathematical analysis literature, the constant in the Sobolev embedding theorem is typically derived from the "cone condition" [3]. This condition requires the existence of a cone such that for any point $\boldsymbol{x} \in P$, the cone can be rotated and translated so that it lies entirely inside P with its apex at $\boldsymbol{x}$. We can prove Lemma 2.8 by finding a *single* cone, depending only on the aspect ratio bound, such that the cone condition holds for any domain in $\mathfrak{p}_{\text{cvx}} \cap \mathfrak{p}_{\text{ar}}$. In the polygonal case, a minimum interior angle bound is required so that the specific cone can be placed near the vertices, and Lemma 2.6 (i) provides this guarantee.

Various *polytopal element methods* have recently appeared in the literature on numerical methods for partial differential equations that can accommodate polygonal and polyhedral meshes without the use of generalized barycentric coordinates. These include mimetic methods, virtual elements (see Chapter 15), weak Galerkin methods, gradient schemes, and many more that are beyond the scope of discussion in this chapter. All these approaches still require assumptions on shape quality of the mesh elements in order to carry out their desired error analyses. The typical assumption is that each mesh element is "star-shaped with respect to a ball," that is, there exists a d-ball $B_P \subset P$ of radius $r(B_P)$ such that the convex hull of $\{x\} \cup B$ is a subset of P [70]. Then, on a sequence of meshes, it is assumed that $\text{diam}(P)/r(B_P)$ is bounded above uniformly. Since $r(B_P) \leq r(P)$, this assumption is stronger than a bound on the aspect ratio as in $\mathfrak{p}_{\text{ar}}$. Since many of those methods are, in effect, using harmonic coordinates for their error analysis, it seems likely that their assumptions could be weakened accordingly.

Shape regularity conditions for polygonal and polyhedral meshes have also been considered recently by Mu, Wang, and Wang in the context of discontinuous Galerkin methods [288]. They provide four shape regularity conditions, denoted A1–A4, which bear many similarities to the metrics presented here. Condition A1 ensures that mesh elements have a good quality decomposition into simplices, which is related to our findings for harmonic coordinates; this is discussed further in Section 2.5 and [163]. Condition A2 requires the existence of one "large" edge per element, but this is later recast as condition A2″, which is exactly the same as a bound on the aspect ratio. Condition A3 allows a more general interpretation of how two polygons or polyhedra may meet in a mesh and A4 ensures that the ratio of diameters of adjacent mesh elements is bounded above; that is, the mesh is "graded" reasonably well.

In work by Floater et al. [147], the shape quality of a convex polytope P is characterized by a single parameter h_*, defined as the minimum distance between a vertex of P and any hyperplane containing a non-incident face. More formally,

$$h_*(P) := \min_{f \in F} \min_{\boldsymbol{v} \in V \setminus V_f} h_f(\boldsymbol{v}),$$

where F, V are the sets of faces and vertices of P, respectively, $V_f \subset V$ is the set of vertices belonging to face f, and $h_f(\boldsymbol{v})$ is the distance from vertex $\boldsymbol{v}$ to the $d-1$ dimensional hyperplane containing f. Consider the associated class of polytopes:

$\mathfrak{p}_{\text{vf}}$ — The value of h_* is bounded away from zero: $h_*(P) \geq h_* > 0$.

In the case of polygons, fixing a counterclockwise ordering of the vertices and $f := [\boldsymbol{v}_{i-1}, \boldsymbol{v}_i]$, we have $h_f(\boldsymbol{v}_{i+1}) = \|\boldsymbol{v}_{i+1} - \boldsymbol{v}_i\| \sin \alpha_i$, where $\alpha_i := \angle(\boldsymbol{v}_{i-1}, \boldsymbol{v}_i, \boldsymbol{v}_{i+1})$. Thus, a lower bound on h_* ensures a balance between short edges in P and the angles adjacent to them. A precise result is the following.

Lemma 2.9. *On polygons,* $(\mathfrak{p}_{\text{cvx}} \cap \mathfrak{p}_{\text{edge}} \cap \mathfrak{p}_{\text{mina}} \cap \mathfrak{p}_{\text{maxa}}) \subset \mathfrak{p}_{\text{vf}}$.

A proof of Lemma 2.9 can be found within the proof of [147, Corollary 2.4].

ϵ	HM*	WP	MV	DH	LS-1	LS-2	LS-3	ME-T	ME-U
0.1600	7.5e-1	1.1e-1	4.0e-2	2.3e-1	2.2e-1	3.4e-2	6.5e-2	2.5e-1	7.3e-1
0.0400	8.5e-1	3.9e-2	1.2e-2	8.3e-2	9.0e-2	1.3e-2	2.9e-2	9.6e-2	4.6e-1
0.0100	8.9e-1	1.2e-2	3.2e-3	1.6e-1	2.9e-2	4.3e-3	1.0e-2	3.0e-2	2.5e-1
0.0025	9.1e-1	3.3e-3	6.7e-4	3.7e-3	7.9e-3	1.2e-3	3.2e-3	8.9e-3	9.9e-2

Table 2.2 Comparison of some generalized barycentric coordinates over the family of domains depicted in Figure 2.1b. Coordinates and data are the same as in Table 2.1; that is, the H^1-seminorm of the interpolant is reported for harmonic coordinates and the H^1-seminorm difference from the harmonic interpolant is reported for other coordinates. The data shows that generalized barycentric coordinates *can* have controlled interpolation error in the presence of a small edge.

2.5 INTERPOLATION ERROR ESTIMATES ON POLYGONS

Let ϕ be a specific kind of barycentric coordinates; for example, Wachspress coordinates. What is a sufficient but not overly restrictive subclass of diameter 1 polygons for which one of the estimates (2.7)–(2.9) is guaranteed to hold for ϕ? In light of the analysis in Section 2.2, an answer to this question, coupled with a uniform Sobolev embedding estimate (2.6), implies a standard *a priori* error estimate of the form (2.12). We have answered this question for a variety of generalized barycentric coordinates, as we now discuss.

We first consider some simple examples demonstrating the robustness or failure of generalized barycentric interpolation over families of polygons admitting degenerate geometries. In Section 2.1, we discussed the family from Figure 2.1a, which included arbitrarily large angles. Now consider Figure 2.1b, which depicts a family of quadrilaterals containing arbitrarily small edge lengths, approaching a triangular geometry. Interpolation errors computed in Table 2.2 demonstrate that this degeneracy does not pose any practical problems. The data shows that while $\mathfrak{p}_{\text{edge}}$ is useful for analysis, it is not strictly necessary for controlled interpolation.

Moreover, the interpolation error can withstand the presence of many short edges, even over a class of polygons that is not of type $\mathfrak{p}_{\text{nv}}$. Table 2.3 contains interpolation errors over regular polygons with increasingly large numbers of vertices; this family of polygons is of type $\mathfrak{p}_{\text{cvx}}$, $\mathfrak{p}_{\text{ar}}$, and $\mathfrak{p}_{\text{mina}}$, but not of type $\mathfrak{p}_{\text{edge}}$, $\mathfrak{p}_{\text{maxa}}$, or $\mathfrak{p}_{\text{nv}}$. For every kind of coordinates tested, the interpolation error is practically bounded, although some errors could be growing very slowly. Another challenge also arises: large vertex counts make the computation of the coordinates increasingly expensive.

2.5.1 Triangulation coordinates

The simplest generalized barycentric coordinates to construct on a polygon P are the *triangulation coordinates*, which require the triangulation of P followed by piece-

n	HM*	WP	MV	DH	LS-1	LS-2	LS-3	ME-T	ME-U
4	1.5e0	0.0e0	4.5e-2	0.0e0	1.1e-1	8.3e-3	1.2e-2	0.0e0	0.0e0
8	1.9e0	5.0e-2	1.0e-2	5.0e-2	1.3e-1	9.9e-3	2.3e-2	7.5e-2	1.3e-1
16	1.9e0	7.9e-2	4.2e-2	7.9e-2	1.7e-1	5.5e-3	4.0e-2	8.2e-2	1.7e-1
32	2.0e0	8.8e-2	4.3e-2	8.8e-2	1.7e-1	2.3e-3	4.3e-2	6.8e-2	1.8e-1
64	2.0e0	9.1e-2	4.3e-2	9.1e-2	1.7e-1	9.7e-4	4.3e-2	5.2e-2	1.8e-1
128	2.0e0	9.2e-2	4.3e-2	9.2e-2	1.8e-1	5.8e-4	4.4e-2	3.9e-2	1.8e-1
256	2.0e0	9.2e-2	4.3e-2	9.2e-2	1.8e-1	5.0e-4	4.4e-2	2.9e-2	1.8e-1

Table 2.3 Comparison of some generalized barycentric coordinates over regular, fixed diameter polygons with n vertices. Coordinates and data are the same as in Tables 2.1 and 2.2.

wise linear barycentric interpolation over each triangle. Of course, the choice of triangulation matters significantly to the quality of interpolation. For example, the polygon in Figure 2.1a has three possible triangulations: one in which the center top vertex is connected to all four vertices and two where the top three vertices form an obtuse triangle. The first of these options provides bounded interpolation error independent of ϵ while the latter two yield large errors as ϵ shrinks.

As discussed in Section 2.3, the interpolation error can be large for triangles with large angles or, equivalently, triangles with large circumradii. Given a set of points (in our case, the vertices of P), the *Delaunay triangulation* is defined as the set of triangles with empty circumballs. The Delaunay triangulation provides a valid triangulation for any convex polygon and tends to avoid triangles with large circumradii as that would require the existence of a large empty ball passing through three of the vertices of P. More precisely, the Delaunay triangulation has a number of optimality properties: it maximizes the minimum angle [233, 359], minimizes the L^p error for the function $\|\boldsymbol{x}\|^2$ [92, 112, 327], and minimizes the H^1-seminorm of the interpolant [312, 326].

If P is non-convex, the *constrained Delaunay* triangulation [97] can be used to construct triangulation coordinates. This modified definition ensures that the edges of P exist in the triangulation while still avoiding large circumradii whenever possible. In applications that only require a piecewise linear interpolant, such as low-order finite element methods, triangulation coordinates using the (constrained) Delaunay triangulation perform quite well compared to more elaborate generalized barycentric coordinate types. To understand why this is the case, we turn next to the smoothest possible interpolant: harmonic coordinates.

2.5.2 Harmonic coordinates

Since harmonic coordinates are the solution to a simple elliptic partial differential equation, the standard, simplicial finite element theory provides a connection between triangulation coordinates (i.e., the discrete solution) and harmonic coordinates (i.e., the continuous solution). Specifically, the theory guarantees that the

H^1-seminorm of the triangulation coordinates must be larger, up to a constant factor depending on the diameter of the polygon, than that of the harmonic coordinates. Thus, finding a class of polygons for which the triangulation coordinates have bounded interpolation error estimates is sufficient to ensure bounded error in the harmonic coordinates. Starting with triangulation coordinates provides a distinct advantage: since interpolation error is very well understood for triangles of all types, there is no need to restrict our attention to a class of type $\mathfrak{p}_{\text{edge}}$ or $\mathfrak{p}_{\text{nv}}$.

This line of analysis can be broadened to allow any triangulation of the polygon, including any number of interior vertices, as long as the boundary of the triangulation exactly matches the edges of the polygon; that is, no vertices of the triangulation lie on edges of the polygon. The result is a straightforward approach to analyzing harmonic coordinates on a class of type $\mathfrak{p}_{\text{ar}}$: triangulate using a single additional vertex placed at the center of the largest inscribed circle. A proof is given in [164, Lemma 4].

Lemma 2.10. *The inequality in (2.9) holds uniformly for the harmonic coordinates on* $\mathfrak{p}_{\text{cvx}} \cap \mathfrak{p}_{\text{ar}}$.

Attempting to relax the bounded aspect ratio restriction yields an even stronger connection between triangulation and harmonic coordinates. Specifically, we show in [163] that whenever the Delaunay triangulation of a convex polygon contains a large circumradius triangle (which implies a poor interpolant via triangulation coordinates), the harmonic coordinates are also large (in H^1-norm). Selection of the Delaunay triangulation is essential for this analysis. For example, the polygon in Figure 2.1a can be triangulated in essentially two ways: one with an obtuse triangle formed by the top three vertices and another in which the top vertex is connected to all the other vertices. The former produces unbounded interpolation error as the angle approaches 180°, similar to Wachspress coordinates in Table 2.1, while the latter, which is the Delaunay triangulation, has controlled interpolation error.

We study harmonic interpolation on non-convex polygons first via numerical experiment on a simple example depicted in Figure 2.1c. The interpolation error for this case, with ϵ shrinking, is given in Table 2.4. However, this numerical experiment is somewhat inconclusive: the H^1 error is slightly increasing as ϵ is reduced but it is not clear if the trend is approaching a fixed value or growing unbounded.

The analysis given in [163] resolves this ambiguity. Whenever a family of polygons contains this kind of configuration where a polygon vertex can be arbitrarily close to the interior of a non-adjacent edge, the interpolation error under harmonic coordinates cannot be bounded. Hence, theoretically, the error *must* be unbounded as $\epsilon \to 0$ for all coordinates in Table 2.4. The details of this analysis are delicate and involve estimates at the critical case of $k = d/2$ in the Sobolev embedding theorem. Thus, it is not surprising that the divergence occurs very slowly.

The takeaway message is the following: poor quality triangles in the constrained Delaunay triangulation of a polygon will cause interpolation error estimates to fail. However, such situations pose few practical limitations as the error grows very slowly, at least for harmonic coordinates.

ϵ	HM*	MV	LS-1	LS-2	LS-3	ME-T
0.1600	2.4e0	8.2e-1	2.3e-0	4.2e-1	8.9e-1	1.4e-0
0.0400	2.2e0	7.4e-1	2.3e-0	3.0e-1	8.7e-1	1.4e-0
0.0100	2.4e0	7.6e-1	2.4e-0	2.6e-1	8.6e-1	1.6e-0
0.0025	2.6e0	8.2e-1	2.6e-0	2.4e-1	1.1e0	1.7e-0

Table 2.4 A comparison of some generalized barycentric coordinates over the family of non-convex domains depicted in Figure 2.1c. Coordinates and data are the same as in Tables 2.1 and 2.2, although only coordinates that are well-defined on non-convex polygons are included.

2.5.3 Wachspress coordinates

When analyzing explicitly defined coordinates (beginning with Wachspress), we follow an approach that hinges on deriving a uniform upper bound on the gradient of the coordinates. Selecting a class of polygons inside $\mathfrak{p}_{\rm cvx} \cap \mathfrak{p}_{\rm ar}$, the uniform Sobolev embedding estimate holds by Lemma 2.8. Then, Proposition 2.3 ensures that the desired interpolation error estimate holds. Thus the missing component of the analysis is addressed in the following result. A proof is given in [164, Lemma 6].

Lemma 2.11. *The inequality in (2.7) holds uniformly for the Wachspress coordinates on* $\mathfrak{p}_{\rm cvx} \cap \mathfrak{p}_{\rm ar} \cap \mathfrak{p}_{\rm edge} \cap \mathfrak{p}_{\rm maxa}$.

Rather than provide the complete proof, we briefly sketch why the geometric conditions are sufficient for ensuring the gradient bound. Without loss of generality, we consider a general diameter 1 polygon where the Wachspress coordinates are constructed from weights of the form

$$w_i = A(\boldsymbol{v}_{i-1}, \boldsymbol{v}_i, \boldsymbol{v}_{i+1}) \prod_{j \neq i, i+1} A(\boldsymbol{v}_{j-1}, \boldsymbol{v}_j, \boldsymbol{x}). \tag{2.13}$$

The relevant triangles $[\boldsymbol{v}_{i-1}, \boldsymbol{v}_i, \boldsymbol{v}_{i+1}]$ and $[\boldsymbol{v}_{i-1}, \boldsymbol{v}_i, \boldsymbol{x}]$ for this construction are depicted in the top row of Figure 2.3.

In the class $\mathfrak{p}_{\rm cvx} \cap \mathfrak{p}_{\rm ar} \cap \mathfrak{p}_{\rm edge} \cap \mathfrak{p}_{\rm maxa}$, weights can be bounded from above, since Lemma 2.6 ensures an upper bound on the number of terms in the expression and every triangle is a subset of the diameter 1 polygon. Further, gradients of the weights can be written as

$$\nabla w_i = A(\boldsymbol{v}_{i-1}, \boldsymbol{v}_i, \boldsymbol{v}_{i+1}) \sum_{k \neq i, i+1} \nabla A(\boldsymbol{v}_{k-1}, \boldsymbol{v}_k, \boldsymbol{x}) \prod_{j \neq i, i+1, k} A(\boldsymbol{v}_{j-1}, \boldsymbol{v}_j, \boldsymbol{x}).$$

These summands can be bounded from above via a similar argument, namely, that the norm of the gradient of a triangle area with respect to one of its vertices is half the length of the opposite edge, which is certainly bounded in a diameter 1 polygon.

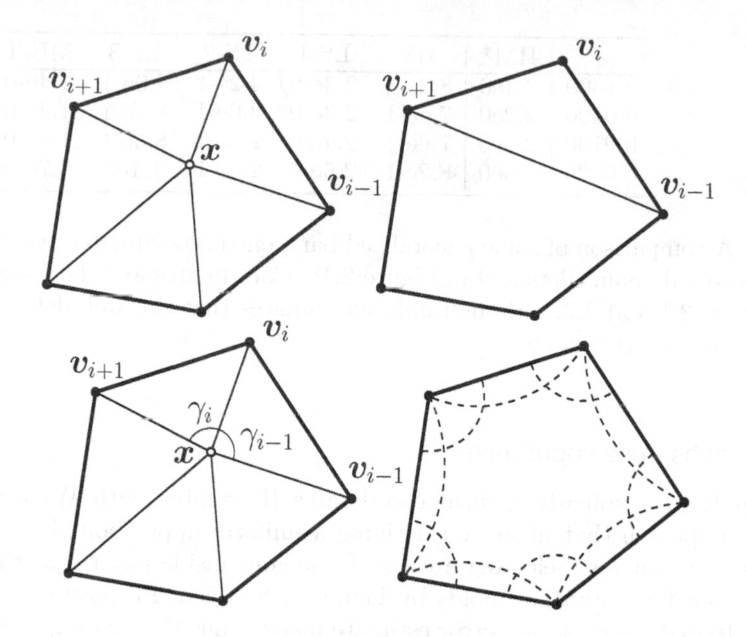

Figure 2.3 Notation and context for the interpolation error analysis of Wachspress (top two figures) and mean value (bottom two figures) coordinates.

The Wachspress coordinates are constructed by normalizing the weights from (2.13). Hence, to ensure the desired bound, we also must show that the sum of the weights is bounded from below. First, we observe that the maximum interior angle and edge length restrictions ensure that $A(\boldsymbol{v}_{j-1}, \boldsymbol{v}_j, \boldsymbol{x})$ can be small for at most 2 consecutive vertices, since $\boldsymbol{x}$ is very close to at most one vertex of the polygon. So, for any $\boldsymbol{x}$, there is one weight involving only non-small areas of the form $A(\boldsymbol{v}_{j-1}, \boldsymbol{v}_j, \boldsymbol{x})$ and the remaining term $A(\boldsymbol{v}_{i-1}, \boldsymbol{v}_i, \boldsymbol{v}_{i+1})$ also cannot be small, again thanks to the geometric restrictions. The full proof is given in [164], but the discussion above outlines the key points at which the geometric restrictions come into play.

A slightly different take on this result is given in [147, Theorem 2.3, $d = 2$], as restated here.

Lemma 2.12. *The inequality in (2.7) holds uniformly for the Wachspress coordinates on* $\mathfrak{p}_{\mathrm{cvx}} \cap \mathfrak{p}_{\mathrm{vf}}$.

Note that by Lemma 2.6 (i), a class of the form $\mathfrak{p}_{\mathrm{cvx}} \cap \mathfrak{p}_{\mathrm{ar}} \cap \mathfrak{p}_{\mathrm{edge}} \cap \mathfrak{p}_{\mathrm{maxa}}$ can be recast as a class of the form $\mathfrak{p}_{\mathrm{cvx}} \cap \mathfrak{p}_{\mathrm{edge}} \cap \mathfrak{p}_{\mathrm{mina}} \cap \mathfrak{p}_{\mathrm{maxa}}$, which by Lemma 2.9 can be recast as a class of the form $\mathfrak{p}_{\mathrm{cvx}} \cap \mathfrak{p}_{\mathrm{vf}}$. Thus, Lemma 2.12 implies Lemma 2.11, although it seems likely that the reverse implication is true as well. The generalization of Lemma 2.12 to dimensions $d \geq 3$ is discussed in Section 2.6.

2.5.4 Mean value coordinates

The analysis of mean value coordinates follows the same lines as that of Wachspress coordinates: we bound the L^∞ norm of the gradients of the individual coordinates as in (2.7), which is sufficient to imply the interpolation error estimate. The key differences are a new expression for the weights used to construct the coordinates and the fact that it is no longer necessary to exclude interior angles close to 180°. A proof of the following result is given in [319, Theorem 1].

Lemma 2.13. *The inequality in (2.7) holds uniformly for the mean value coordinates on* $\mathfrak{p}_{\text{cvx}} \cap \mathfrak{p}_{\text{ar}} \cap \mathfrak{p}_{\text{edge}}$.

Mean value coordinates are constructed from weights of the form,

$$w_i(\boldsymbol{x}) = \frac{\tan\left(\frac{\gamma_i(\boldsymbol{x})}{2}\right) + \tan\left(\frac{\gamma_{i-1}(\boldsymbol{x})}{2}\right)}{\|\boldsymbol{v}_i - \boldsymbol{x}\|},$$

where the angles $\gamma_i(\boldsymbol{x})$ are depicted in the bottom left image of Figure 2.3. The subsequent analysis in [319] hinges on the observation that these weights get large if either $\boldsymbol{x}$ is very close to a vertex of the polygon or if $\boldsymbol{x}$ is near one of the edges of the polygon. In the latter case, one of the angles γ_i is near 180°, causing the associated tangent term to be large. The analysis is divided into a number of cases, depending on which of these terms is nearly degenerate. The bottom right image in Figure 2.3 shows a polygon subdivided into different regions where these different cases apply. Lemma 2.7 helps to simplify this classification by ensuring that a vertex can only be near a limited number of polygon edges and vertices. Since the coordinates involve normalization of the weight functions, large weight functions do not prevent bounds on the coordinates and their gradients. In order to bound the gradient in each case, these large terms are shown to be offset by larger terms in the denominator. The analysis requires the convexity of the polygon to ensure all possible cases are covered; the non-convex case has not yet been considered.

2.6 INTERPOLATION ERROR ESTIMATES ON POLYHEDRA AND POLYTOPES

Moving from polygons to polyhedra and general polytopes, the challenge of quantifying shape quality becomes significantly more complex. Metrics involving polygonal interior angles (i.e., $\mathfrak{p}_{\text{mina}}$ and $\mathfrak{p}_{\text{maxa}}$) are essential in two-dimensional analysis, but do not have natural extensions to higher dimensions; dihedral angles are perhaps the closest analogue. Moreover, there are fewer definitions of generalized barycentric coordinates in higher dimensions, due in part to the fact that the restriction of such coordinates to a possibly non-simplicial boundary facet should produce a lower dimensional generalized barycentric coordinate in some pre-determined fashion. For simplicity, some constructions (e.g., mean value coordinates) restrict their definitions to polytopes with simplicial boundary faces, although others (e.g., harmonic and maximum entropy coordinates) are defined without restriction. The interpolation error has been analyzed under some strong geometric restrictions for harmonic and Wachspress coordinates, as discussed below.

2.6.1 Harmonic coordinates in 3D and higher

Based on the very general theory of elliptic partial differential equations, much of the analysis of harmonic coordinates applies to arbitrary dimensions. Only two key differences impact the final conclusions and statement of our results from the polygonal setting: the circumradius is no longer a good measure of simplex quality and the Sobolev embedding theorem allows more kinds of discontinuous functions in higher dimensions. We discuss each in turn.

Tetrahedra known as "slivers," defined by four vertices lying near a common circle, provide a simple example of the challenges of interpolation error analysis in three dimensions and higher. Slivers have a modest circumradius relative to their diameter but have a poor aspect ratio and exhibit large errors in linear interpolation. Accordingly, the 3D analogue of the constrained Delaunay triangulation, which minimizes large circumaradii, cannot be expected to control interpolation on par with harmonic coordinates.

Broadly speaking, the Sobolev embedding theorem describes which Sobolev spaces admit discontinuous functions and which contain only continuous ones. Two critical parameters in this regard are the spatial dimension, d, and the exponent used in integration, p. Discontinuous functions are admitted in the associated Sobolev space $W^{1,p}(\Omega)$ when $d > p$, and, as d increases with p fixed, more functions are included in the space.

What do these facts imply about harmonic coordinates? While harmonic coordinates can still be estimated by comparison to triangulation coordinates via a valid tetrahedralization of the domain, there is not a clear procedure for identifying the "best" tetrahedralization. Further, while it is true in $d = 2$ that the presence of a vertex near a non-incident edge of the polygon causes poor interpolation, analogous configurations in 3D still allow good interpolation errors. The shape quality metric of bounded aspect ratio can and is used to control the interpolation error, despite the fact that this is an overly restrictive geometric assumption. A sharp metric for harmonic coordinates in 3D is still unknown.

2.6.2 Wachspress coordinates in 3D and higher

One of the few types of generalized barycentric coordinates on polyhedra that is both relatively simple to implement and amenable to error analysis is the generalization of Wachspress coordinates by Warren et al. [411, 413]. Floater et al. [147] provide MATLAB™ code for the computation of these coordinates on generic convex polyhedra and derive an interpolation error bound over a subclass of simple, convex polyhedra.

A d-dimensional polytope is called *simple* if every vertex is incident to exactly d faces of dimension $d - 1$. In particular, every polygon is simple, every simplex is simple, and polyhedra generated from a Voronoi diagram are simple if the seed points are in general position. We introduce the associated class of polytopes:

$\mathfrak{p}_{\text{simp}}$ — P is simple.

The following result is a generalization of Lemma 2.12 to d dimensions. Recall the definition of $\mathfrak{p}_{\text{vf}}$ from the end of Section 2.4; the following result is from [147, Theorem 2.3].

Lemma 2.14. *Fix $d \geq 2$. Then the inequality in (2.7) holds uniformly for the Wachspress coordinates on $\mathfrak{p}_{\text{cvx}} \cap \mathfrak{p}_{\text{vf}} \cap \mathfrak{p}_{\text{simp}}$.*

A sketch of the proof is as follows. First, the quantity $\|\nabla\phi_i(\boldsymbol{x})\|$ is bounded by terms that are $O(h_f(\boldsymbol{x})^{-1})$; recall that $h_f(\boldsymbol{x})$ is the distance from point $\boldsymbol{x}$ to face f. These estimates are summed over all the vertices of P and regrouped. Since the function $h_f(\boldsymbol{x})$ is affine in $\boldsymbol{x}$, it can be written as a linear combination of $\{\phi_i(\boldsymbol{x})\}$, allowing further simplification of the estimate. Since $P \in \mathfrak{p}_{\text{vf}}$, we have $h_f(\boldsymbol{x}) \geq h_*(P) > 0$ and since $P \in \mathfrak{p}_{\text{simp}}$, we can factor out $2d$ to produce the result.

As highlighted by the above proof and other results in [147], the sum of gradients of Wachspress coordinates scales roughly like h_*^{-1}. Thus, if h_* is small, we would expect the Wachspress coordinates to have poor interpolation properties. Any polyhedron with a large dihedral angle has a small value of h_*, meaning any meshing scheme designed to allow Wachspress interpolation on the output mesh should aim to avoid large dihedral angles. Further analysis along these lines remains an important direction for future work.

2.7 EXTENSIONS AND FUTURE DIRECTIONS

The results from this chapter focus on interpolation error estimates for generalized barycentric coordinates in the context of linear finite element methods on polygonal and polyhedral meshes for scalar-valued elliptic partial differential equations. In [320], we have shown how to construct basis functions for quadratic order finite elements on polygons by taking linear combinations of pairwise products of generalized barycentric coordinates. The gradients of such functions have terms of the form $\phi_i\nabla\phi_j$ and thus our estimates on $\|\nabla\phi_j\|$ remain a crucial ingredient in error estimates for such higher order methods.

In [165], we have shown how to construct basis functions for linear order, vector-valued, $H(\text{curl})$- and $H(\text{div})$-conforming elements on polygons and polyhedra. Our construction uses generalized barycentric functions and their gradients in forms like $\phi_i\nabla\phi_j - \phi_j\nabla\phi_i$, mimicking the approach of Whitney functions for barycentric coordinates on simplices. Again, the bounds on $\|\nabla\phi_j\|$ explained in this chapter are necessary in the error analysis of such methods.

At the beginning of this chapter we raised the question of whether a change in mesh or an alternative selection of generalized barycentric coordinate type could fix a simulation code that would otherwise crash. The answer is certainly yes, both approaches *can* improve numerical methods for interpolation. In particular, the Wachspress, discrete harmonic, and maximum entropy coordinates with the uniform prior were shown to provide poor interpolation on polygons with large angles and are not defined on non-convex polygons. Changing the coordinate type or remeshing to avoid large angles and non-convex polygons avoids this failure.

On the other hand, our numerical experiments also demonstrate the surprising robustness of many coordinate types against extremely degenerate polygonal geometry. For instance, the element family of a quadrilateral collapsing to a triangle (Figure 2.1b) does not belong to the class $\mathfrak{p}_{\text{edge}}$ nor the class $\mathfrak{p}_{\text{vf}}$, excluding the Wachspress and mean value coordinates from the gradient estimates guaranteed by Lemma 2.11, Lemma 2.12, and Lemma 2.13. Nevertheless, the interpolation error remains small for *all* the coordinates that we tested on this family. Additional analysis and experiments will help sharpen existing gradient estimates and shape quality metrics in ways that increase their relevance to practical application settings.

CHAPTER 3

Transfinite Barycentric Coordinates

Alexander G. Belyaev

Heriot-Watt University, Edinburgh, UK

Pierre-Alain Fayolle

University of Aizu, Aizuwakamatsu, Japan

CONTENTS

IN THIS CHAPTER we study properties of transfinite barycentric interpolation schemes, which can be considered as continuous counterparts of generalized barycentric coordinates and are currently a subject of intensive research due to their fascinating mathematical properties and numerous applications in computational mechanics [170, 344], computer graphics [247, 248], and geometric modeling [93, 145, 227, 413] (see also references therein).

We start from a general construction of transfinite barycentric coordinates, which is obtained as a simple and natural generalization of Floater's mean value coordinates [144, 217], and investigate the properties of the continuous analogues of three-point coordinates [149]. We discuss the Gordon–Wixom interpolation approach [171] and consider its generalizations and modifications [37, 38]. Finally we introduce generalized mean value potentials and demonstrate how they can be used for distance function approximation purposes [40].

3.1 INTRODUCTION

While traditional generalized barycentric coordinates interpolate function values given at the vertices of a polygon or polyhedron, transfinite barycentric interpolation is based on smooth domains. Transfinite barycentric interpolation schemes establish bridges between barycentric coordinates and methods used to interpolate continuous data [38, 77, 93, 133, 151, 153, 171]. In addition, studying continuous versions of generalized barycentric coordinates helps us to achieve a better understanding of their discrete counterparts [37, 93, 248, 346, 413].

Similar to the discrete case (see Section 1.1.2), let us consider a convex domain Ω in $\mathbb{R}^d$ and a smooth function $\phi(\boldsymbol{x}, \boldsymbol{y})$, $\boldsymbol{x} \in \Omega$ and $\boldsymbol{y} \in \partial\Omega$, satisfying the two properties

$$\text{Partition of unity:} \qquad \int_{\partial\Omega} \phi(\boldsymbol{x}, \boldsymbol{y})\, ds_{\boldsymbol{y}} = 1, \tag{3.1}$$

$$\text{Linear precision:} \qquad \int_{\partial\Omega} \boldsymbol{y}\, \phi(\boldsymbol{x}, \boldsymbol{y})\, ds_{\boldsymbol{y}} = \boldsymbol{x}, \tag{3.2}$$

which are the continuous analogues of Conditions (1.8) and (1.9) in Definition 1.1. Now the transfinite interpolant of some function $u(\boldsymbol{y})$ defined on $\partial\Omega$ into Ω is given by

$$u(\boldsymbol{x}) = \int_{\partial\Omega} u(\boldsymbol{y}) \phi(\boldsymbol{x}, \boldsymbol{y})\, ds_{\boldsymbol{y}}. \tag{3.3}$$

In mathematical terminology, $\phi(\boldsymbol{x}, \boldsymbol{y})$ is the *kernel* of the integral operator defined by the right-hand side of (3.3).

Our main task is to establish links between various barycentric coordinates and the general transfinite interpolation construction described by (3.3) with some kernel $\phi(\boldsymbol{x}, \boldsymbol{y})$ satisfying (3.1) and (3.2).

In this chapter, we assume that the domain Ω is convex and bounded. In practice, of course, we would like to work with non-convex domains as well. An efficient and simple way to extend barycentric interpolation schemes to simply connected non-convex domains is proposed in [200]. An even simpler approach to deal with non-convex domains is suggested in [251]. We briefly discuss both approaches at the end of Section 3.2.

3.2 WEIGHTED MEAN VALUE INTERPOLATION

3.2.1 General construction

Let us start from the transfinite mean value interpolation scheme proposed in [217]. Let Ω be a bounded convex domain and $\boldsymbol{x}$ be a point inside Ω. Consider the unit sphere $S_{\boldsymbol{x}}$ centered at $\boldsymbol{x}$ and assume that $S_{\boldsymbol{x}}$ is parameterized by its outer unit normal $\boldsymbol{e}_\theta$. Let $\boldsymbol{y}$ be a point on $\partial\Omega$ and $\rho = \|\boldsymbol{x}-\boldsymbol{y}\|$. Denote by $\boldsymbol{z}$ the intersection point between the ray $[\boldsymbol{x},\boldsymbol{y})$ and $S_{\boldsymbol{x}}$ (see Figure 3.1, left). If we assume that we know the values of the function u on $\partial\Omega$ and at $\boldsymbol{x}$, then we can estimate $u(\boldsymbol{z})$ using linear interpolation,

$$u(\boldsymbol{z}) \approx \frac{(\rho-1)u(\boldsymbol{x}) + u(\boldsymbol{y})}{\rho}.$$

Now, applying S-averaging (circular averaging in the 2D case and spherical averaging in the 3D case) to the left and right sides of the above equation yields

$$\int_{S_{\boldsymbol{x}}} u(\boldsymbol{z})\, d\boldsymbol{\theta} = |S_{\boldsymbol{x}}|u(\boldsymbol{x}) - u(\boldsymbol{x})\int_{S_{\boldsymbol{x}}} \frac{d\boldsymbol{\theta}}{\|\boldsymbol{x}-\boldsymbol{y}\|} + \int_{S_{\boldsymbol{x}}} \frac{u(\boldsymbol{y})\, d\boldsymbol{\theta}}{\|\boldsymbol{x}-\boldsymbol{y}\|},$$

where $d\boldsymbol{\theta}$ is the area element of the unit sphere $S_{\boldsymbol{x}}$ at $\boldsymbol{z} \in S_{\boldsymbol{x}}$ (to distinguish the 2D case we denote the angular element of the unit circle by $d\theta$), and $|S_{\boldsymbol{x}}|$ is the unit sphere area. Assuming that u is harmonic, we arrive at the transfinite mean value interpolation scheme

$$u(\boldsymbol{x}) = \int_{S_{\boldsymbol{x}}} \frac{u(\boldsymbol{y})\, d\boldsymbol{\theta}}{\|\boldsymbol{x}-\boldsymbol{y}\|} \Bigg/ \int_{S_{\boldsymbol{x}}} \frac{d\boldsymbol{\theta}}{\|\boldsymbol{x}-\boldsymbol{y}\|}. \tag{3.4}$$

It is interesting that instead of mimicking the mean value property of harmonic functions we can derive (3.4) from the simple observation that the spherical (circular

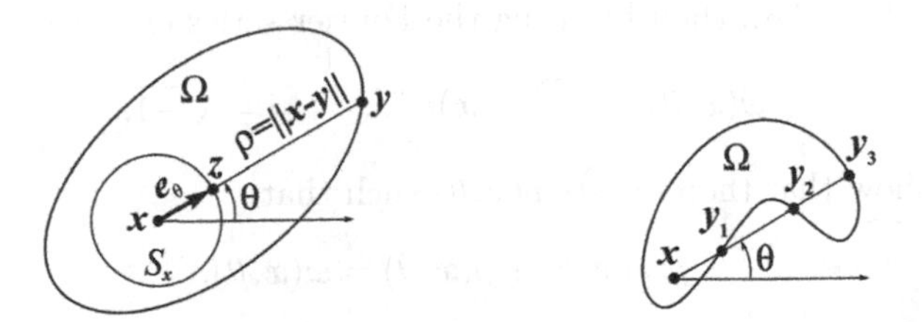

Figure 3.1 Left: notation used to define weighted mean value interpolation. Right: if Ω is non-convex, then the rays emitted from $\boldsymbol{x} \in \Omega$ may intersect $\partial\Omega$ in multiple points.

in the 2D case) mean of all the directional derivatives of a given smooth function is equal to zero. Now, approximating the directional derivative of u in the direction of the ray $[\boldsymbol{x}, \boldsymbol{y})$ by $(u(\boldsymbol{y}) - u(\boldsymbol{x}))/\|\boldsymbol{x} - \boldsymbol{y}\|$ and averaging with respect to all directions yields

$$\int_{S_{\boldsymbol{x}}} \frac{u(\boldsymbol{y}) - u(\boldsymbol{x})}{\|\boldsymbol{x} - \boldsymbol{y}\|}\, d\boldsymbol{\theta} = 0,$$

which is equivalent to (3.4).

The interpolation scheme (3.4) allows for a natural generalization

$$u(\boldsymbol{x}) = \int_{S_{\boldsymbol{x}}} \frac{u(\boldsymbol{y})\omega(\boldsymbol{x}, \boldsymbol{e}_\theta)}{\|\boldsymbol{x} - \boldsymbol{y}\|}\, d\boldsymbol{\theta} \Bigg/ \int_{S_{\boldsymbol{x}}} \frac{\omega(\boldsymbol{x}, \boldsymbol{e}_\theta)}{\|\boldsymbol{x} - \boldsymbol{y}\|}\, d\boldsymbol{\theta}, \tag{3.5}$$

where $\omega(\boldsymbol{x}, \boldsymbol{e}_\theta)$ is a weighting function that may depend on the interpolated location $\boldsymbol{x}$ and the direction $\boldsymbol{e}_\theta$. It is easy to see that (3.5) is just another form of (3.3) and we can also consider (3.5) as a transfinite version of the basic Shepard interpolation [354].

Let us check when (3.5) satisfies the linear precision property (3.2). Setting $u(\boldsymbol{x}) \equiv \boldsymbol{x}$ yields $u(\boldsymbol{y}) \equiv \boldsymbol{y} = \boldsymbol{x} + \rho\, \boldsymbol{e}_\theta$. Substituting the latter into (3.5) gives

$$0 = \int_{S_{\boldsymbol{x}}} \boldsymbol{e}_\theta\, \omega(\boldsymbol{x}, \boldsymbol{e}_\theta)\, d\boldsymbol{\theta} \qquad \forall \boldsymbol{x} \in \Omega, \tag{3.6}$$

which is necessary and sufficient for linear precision. One can see that (3.6) is satisfied, if the weighting function is centrally symmetric and coincides at each pair of antipodal points,

$$\omega(\boldsymbol{x}, \boldsymbol{e}_\theta) = \omega(\boldsymbol{x}, -\boldsymbol{e}_\theta).$$

Consider now the case of planar interpolation with $d = 2$ and $\boldsymbol{e}_\theta = (\cos\theta, \sin\theta)$,

$$u(\boldsymbol{x}) = \int_0^{2\pi} \frac{u(\boldsymbol{y})\omega(\boldsymbol{x}, \theta)}{\|\boldsymbol{x} - \boldsymbol{y}\|}\, d\theta \Bigg/ \int_0^{2\pi} \frac{\omega(\boldsymbol{x}, \theta)}{\|\boldsymbol{x} - \boldsymbol{y}\|}\, d\theta. \tag{3.7}$$

Then, (3.6) becomes

$$\int_0^{2\pi} \omega(\boldsymbol{x}, \theta) \cos\theta\, d\theta = 0 = \int_0^{2\pi} \omega(\boldsymbol{x}, \theta) \sin\theta\, d\theta \qquad \forall \boldsymbol{x} \in \Omega. \tag{3.8}$$

If $\omega(\boldsymbol{x}, \theta)$ satisfies (3.8), then by using the Fourier series expansion

$$\omega(\boldsymbol{x}, \theta) = \sum c_n(\boldsymbol{x}) e^{\mathrm{i}n\theta}, \qquad \mathrm{i} = \sqrt{-1},$$

it is easy to show that there exists $p(\boldsymbol{x}, \theta)$ such that

$$p''_{\theta\theta}(\boldsymbol{x}, \theta) + p(\boldsymbol{x}, \theta) = \omega(\boldsymbol{x}, \theta). \tag{3.9}$$

Here and below we often use prime-notations for the first- and second-order derivatives with respect to θ:

$$(\cdot)'_\theta \quad \text{stands for} \quad \frac{d}{d\theta}(\cdot) \qquad \text{and} \qquad (\cdot)''_{\theta\theta} \quad \text{stands for} \quad \frac{d^2}{d\theta^2}(\cdot).$$

The orthogonality condition (3.8) can be verified by simple integration by parts,

$$\int_0^{2\pi} \omega(\boldsymbol{x},\theta)e^{i\theta}d\theta = \int_0^{2\pi}\left(\frac{\partial^2}{\partial\theta^2}p + p\right)e^{i\theta}d\theta = \int_0^{2\pi} p(\boldsymbol{x},\theta)\left(\frac{\partial^2}{\partial\theta^2}e^{i\theta} + e^{i\theta}\right)d\theta = 0.$$

Representation (3.9) has an interesting geometric interpretation. Given a curve Σ and a point $\boldsymbol{x}$, the geometry of Σ is determined by its support function $p(\boldsymbol{x},\theta)$, the signed distance from $\boldsymbol{x}$ to the tangents of Σ. The support function satisfies the second-order differential equation

$$p''_{\theta\theta} + p = R(\theta) \equiv 1/k(\theta),$$

where $k(\theta)$ is the curvature of Σ and $R(\theta)$ is the radius of curvature of Σ. See, for example, [345] for details.

Now we can interpret (3.9) as follows. Each $\boldsymbol{x} \in \Omega$ defines a closed curve $\Sigma_{\boldsymbol{x}}$ with radius of curvature $R = \omega(\boldsymbol{x},\theta)$. The orthogonality condition (3.8) then corresponds to

$$\int_0^{2\pi} \boldsymbol{n}R\,d\theta \equiv \int_{\Sigma_{\boldsymbol{x}}} \boldsymbol{n}\,dl = 0,$$

where $\boldsymbol{n} = (\cos\theta, \sin\theta)$ is the outer unit normal of $\Sigma_{\boldsymbol{x}}$ and l denotes the arc-length parametrization of $\Sigma_{\boldsymbol{x}}$. For example, $\Sigma_{\boldsymbol{x}}$ is a unit circle for the 2D transfinite mean value coordinates and it is a unit sphere in $\mathbb{R}^3$ for the three-dimensional transfinite mean value coordinates.

If Ω is not convex, then a ray emitted from $\boldsymbol{x} \in \Omega$ may intersect $\partial\Omega$ in multiple points (see Figure 3.1, right). One way to cope with this situation is proposed in [200] and consists of using alternating signs for intersection points. For instance, the example shown in the right image of Figure 3.1 leads to

$$\frac{u(\boldsymbol{y}_1)}{\|\boldsymbol{x}-\boldsymbol{y}_1\|} - \frac{u(\boldsymbol{y}_2)}{\|\boldsymbol{x}-\boldsymbol{y}_2\|} + \frac{u(\boldsymbol{y}_3)}{\|\boldsymbol{x}-\boldsymbol{y}_3\|}$$

instead of $u(\boldsymbol{y})/\|\boldsymbol{x}-\boldsymbol{y}\|$ in (3.5). However, in practice it is often better to follow a simpler approach suggested in [251] and consider only the first intersections: $u(\boldsymbol{y}_1)/\|\boldsymbol{x}-\boldsymbol{y}_1\|$ instead of $u(\boldsymbol{y})/\|\boldsymbol{x}-\boldsymbol{y}\|$ in (3.5). For example, we use such a simplification in our numerical experiments with the so-called Gordon–Wixom coordinates and their extensions (see Section 3.3 and Figure 3.4).

3.2.2 Transfinite three-point coordinates

A general construction for discrete three-point coordinates (see Section 1.2.4) was introduced in [149]. Let $A(\boldsymbol{x},\boldsymbol{y},\boldsymbol{z})$ denote the signed area of the triangle formed by points $\boldsymbol{x}$, $\boldsymbol{y}$, and $\boldsymbol{z}$, Given a convex polygon with vertices $\boldsymbol{v}_1, \boldsymbol{v}_2, \ldots, \boldsymbol{v}_n$ and some point $\boldsymbol{x}$ inside the polygon, consider the signed triangle areas $A_i(\boldsymbol{x}) = A(\boldsymbol{x},\boldsymbol{v}_i,\boldsymbol{v}_{i+1})$ and $B_i(\boldsymbol{x}) = A(\boldsymbol{x},\boldsymbol{v}_{i-1},\boldsymbol{v}_{i+1})$. Then, according to [149], the three-point coordinates are given by the weights

$$w_i = \frac{F(r_{i+1})A_{i-1} - F(r_i)B_i + F(r_{i-1})A_i}{A_{i-1}A_i}, \tag{3.10}$$

where $r_i = \|\boldsymbol{x} - \boldsymbol{v}_i\|$ and $F(r)$ is an arbitrary function. Following [149], one can also rewrite (3.10) as

$$w_i = \frac{2}{r_i}\left(\frac{f(r_{i+1}) - f(r_i)\cos\theta_i}{\sin\theta_i} + \frac{f(r_{i-1}) - f(r_i)\cos\theta_{i-1}}{\sin\theta_{i-1}}\right), \tag{3.11}$$

where $f(r) = F(r)/r$ and θ_i is the angle between the rays $[\boldsymbol{x}, \boldsymbol{v}_i)$ and $[\boldsymbol{x}, \boldsymbol{v}_{i+1})$.

Assuming that $f(r)$ is sufficiently smooth, the number of vertices of the polygon tends to infinity, and all θ_i uniformly tend to zero, $\theta_i \approx d\theta \to 0$, we then have

$$\frac{f(r_{i+1}) - f(r_i)\cos\theta_i}{\sin\theta_i} \approx \left[f(\rho(\theta))'_\theta + f(r_i)\frac{\theta_i}{2}\right]_{\rho=r_{i+1}},$$

$$\frac{f(r_{i-1}) - f(r_i)\cos\theta_{i-1}}{\sin\theta_{i-1}} \approx \left[-f(\rho(\theta))'_\theta + f(r_i)\frac{\theta_{i-1}}{2}\right]_{\rho=r_i},$$

hence (3.11) is approximately equal to

$$\left[\frac{2}{\rho}\left(f(\rho(\theta))''_{\theta\theta} + f(\rho(\theta))\right)\right]_{\rho=r_i} d\theta.$$

This implies that the continuous or transfinite version of the three-point coordinates is described by (3.7) with

$$\omega(\boldsymbol{x}, \theta) = f(\rho(\theta))''_{\theta\theta} + f(\rho(\theta)). \tag{3.12}$$

3.2.3 Transfinite Laplace coordinates

Following [37, 94], let us define transfinite Laplace coordinates as the continuous version of discrete harmonic coordinates [309] (see Section 1.2.2), which are used widely in computational mechanics [375]. As discrete harmonic coordinates correspond to (3.10) with $F(r) = r^2$, their continuous version is given by (3.7) with the weight (3.12), where $f(r) = r$. In other words,

$$\omega(\boldsymbol{x}, \theta) = \rho(\theta)''_{\theta\theta} + \rho(\theta). \tag{3.13}$$

It turns out that (3.13) can also be derived by mimicking the Dirichlet energy minimization property of harmonic functions [37, 346]. Given $u(\boldsymbol{y})$, defined for each $\boldsymbol{y} \in \partial\Omega$, let us choose $\boldsymbol{x} \in \Omega$ and assume that $u(\boldsymbol{x})$ is known. Consider a ruled surface patch generated by straight segments connecting the inner point $(\boldsymbol{x}, u(\boldsymbol{x}))$ with the boundary points $(\boldsymbol{y}, u(\boldsymbol{y}))$. Now the value $u(\boldsymbol{x})$ is defined such that the Dirichlet energy of the constructed ruled surface attains its minimal value.

Let (r, θ) be the polar coordinates centered at $\boldsymbol{x}$. Then $\partial\Omega$ is described by $r = \rho(\theta)$. The ruled surface associated with $\boldsymbol{x}$ is given by

$$U_{\boldsymbol{x}}(\boldsymbol{z}) = \frac{(\rho - r)u(\boldsymbol{x}) + ru(\boldsymbol{y})}{\rho},$$

where $r = \|\boldsymbol{x} - \boldsymbol{z}\|$ and $\boldsymbol{y} \in \partial\Omega$ denotes the intersection point between $\partial\Omega$ and the ray from $\boldsymbol{x}$ through $\boldsymbol{z}$. We then arrive at the minimization problem

$$\begin{aligned}
\min \leftarrow & \int_{\Omega} |\nabla U_{\boldsymbol{x}}|^2 d\boldsymbol{z} \\
&= \int_0^{2\pi} d\theta \int_0^{\rho} r\,dr \left(\left[\frac{u(\boldsymbol{x}) - u(\boldsymbol{y})}{\rho} \right]^2 + \left[(u(\boldsymbol{y}) - u(\boldsymbol{x})) \left(\frac{1}{\rho}\right)'_{\theta} + \frac{1}{\rho} u'_{\theta}(\boldsymbol{y}) \right]^2 \right) \\
&= \frac{1}{2} \int_0^{2\pi} d\theta \left([u(\boldsymbol{x}) - u(\boldsymbol{y})]^2 + \left[u'_{\theta}(\boldsymbol{y}) + (u(\boldsymbol{y}) - u(\boldsymbol{x})) \frac{\rho'_{\theta}}{\rho} \right]^2 \right),
\end{aligned}$$

where the last integral is a quadratic function with respect to $u(\boldsymbol{x})$. Thus the optimal value of $u(\boldsymbol{x})$ is given by

$$u(\boldsymbol{x}) = \int_0^{2\pi} \left(u(\boldsymbol{y}) - u'_{\theta}(\boldsymbol{y})[\rho'_{\theta}/\rho] + u(\boldsymbol{y})[\rho'_{\theta}/\rho]^2 \right) d\theta \Big/ \int_0^{2\pi} \left(1 + [\rho'_{\theta}/\rho]^2\right) d\theta. \quad (3.14)$$

Integration by parts yields

$$-\int_0^{2\pi} u'_{\theta}(\boldsymbol{y})[\rho'_{\theta}/\rho]\,d\theta = \int_0^{2\pi} u(\boldsymbol{y})[\rho'_{\theta}/\rho]'_{\theta}\,d\theta = \int_0^{2\pi} u(\boldsymbol{y})\left[\rho''_{\theta\theta}/\rho - (\rho'_{\theta}/\rho)^2\right] d\theta,$$

$$\int_0^{2\pi} [\rho'_{\theta}/\rho]^2 d\theta = \int_0^{2\pi} \left(\rho''_{\theta\theta}/\rho - [\rho'_{\theta}/\rho]'_{\theta}\right) d\theta = \int_0^{2\pi} (\rho''_{\theta\theta}/\rho)\,d\theta,$$

and thus (3.14) is equal to

$$\int_0^{2\pi} u(\boldsymbol{y}) \frac{\rho''_{\theta\theta} + \rho}{\rho}\,d\theta \Big/ \int_0^{2\pi} \frac{\rho''_{\theta\theta} + \rho}{\rho}\,d\theta. \quad (3.15)$$

One can now see that (3.15) corresponds to (3.7) with (3.13).

3.2.4 Transfinite Wachspress coordinates

As demonstrated in [149], Wachspress coordinates (see Section 1.2.1) can be described by (3.10) with $F(r) \equiv 1$. This yields (3.7) with

$$\omega(\boldsymbol{x}, \theta) = (1/\rho)''_{\theta\theta} + 1/\rho \quad (3.16)$$

and corresponds to a curve $\Sigma_{\boldsymbol{x}}$ whose radius of curvature is given by (3.16). The domain enclosed by $\Sigma_{\boldsymbol{x}}$ is the polar dual (polar reciprocal) of Ω with respect to $\boldsymbol{x} \in \Omega$ [179] (see also [307] for modern proofs of properties of the polar duals). Interesting relationships between Wachspress coordinates and polar duals are established in [218, 346]. Below, for the 2D case, we link together (3.16), polar duals, and a curvature formula derived in [413].

Let $\partial\Omega$ be described by the radius vector $\boldsymbol{r}$ whose tail is situated at $\boldsymbol{x} \in \Omega$. Denote by $\boldsymbol{t}$ and $\boldsymbol{n}$ the unit tangent and normal vectors forming the Frenet frame

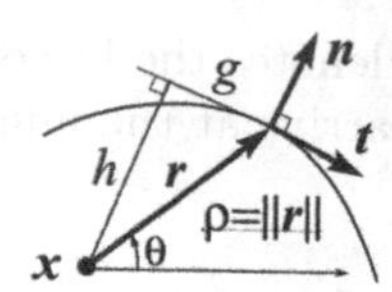

Figure 3.2 The Frenet frame $(\boldsymbol{t}, \boldsymbol{n})$ for $\partial\Omega$ is one of our main tools for studying geometric quantities of $\partial\Omega$ with respect to $\boldsymbol{x} \in \Omega$.

for $\partial\Omega$, as shown in Figure 3.2. Let s be the arc-length parameterization of $\partial\Omega$. We then have

$$\boldsymbol{t}' = k\boldsymbol{n}, \qquad \boldsymbol{n}' = -k\boldsymbol{t}, \qquad \boldsymbol{r} = g\boldsymbol{t} + h\boldsymbol{n} \quad \text{with} \quad g = \boldsymbol{r}\cdot\boldsymbol{t} \quad \text{and} \quad h = \boldsymbol{r}\cdot\boldsymbol{n},$$

where the derivatives are taken with respect to the arc-length s and k denotes the curvature of $\partial\Omega$.

It is easy to show [307] that the polar reciprocal curve $\Sigma_{\boldsymbol{x}}$ is given by the radius vector $\boldsymbol{n}/h$. Then a tangent vector for $\Sigma_{\boldsymbol{x}}$ can be obtained by

$$\left(\frac{\boldsymbol{n}}{h}\right)' = \frac{\boldsymbol{n}'h - \boldsymbol{n}h'}{h^2} = \frac{-k\boldsymbol{t}h + \boldsymbol{n}kg}{h^2} = \frac{g\boldsymbol{n} - h\boldsymbol{t}}{\sqrt{g^2+h^2}}\frac{k\rho}{h^2}, \tag{3.17}$$

where the simple observations that

$$h' = (\boldsymbol{r}\cdot\boldsymbol{n})' = \boldsymbol{r}'\cdot\boldsymbol{n} + \boldsymbol{r}\cdot\boldsymbol{n}' = \boldsymbol{r}\cdot\boldsymbol{n}' = -k\,\boldsymbol{r}\cdot\boldsymbol{t} = -kg \qquad \text{and} \qquad \sqrt{g^2+h^2} = |\boldsymbol{r}| = \rho$$

are used. Normalizing the tangent vector given by the right-hand side of (3.17) gives

$$dl/ds = k\rho/h^2,$$

where l is the arc-length of $\Sigma_{\boldsymbol{x}}$. The unit normal for $\Sigma_{\boldsymbol{x}}$ is given by $\boldsymbol{n}_\Sigma = \boldsymbol{r}/\rho$. Now we can find the curvature k_Σ of $\Sigma_{\boldsymbol{x}}$ by differentiating its unit normal $\boldsymbol{n}_\Sigma$ with respect to l,

$$\frac{d}{dl}\boldsymbol{n}_\Sigma = -k_\Sigma \boldsymbol{t}_\Sigma \quad \text{with} \quad \boldsymbol{n}_\Sigma = \frac{\boldsymbol{r}}{\rho} = \frac{g\boldsymbol{t} + h\boldsymbol{n}}{\sqrt{g^2+h^2}} \quad \text{and} \quad \boldsymbol{t}_\Sigma = -\boldsymbol{n}_\Sigma^\perp = \frac{h\boldsymbol{t} - g\boldsymbol{n}}{\sqrt{g^2+h^2}},$$

where $\perp$ denotes the $\pi/2$ rotation in a clockwise direction. Further,

$$\frac{d}{dl} = \frac{ds}{dl}\frac{d}{ds}, \qquad \frac{d}{ds}\left[\frac{\boldsymbol{r}}{\rho}\right] = \frac{\boldsymbol{r}'\rho - \boldsymbol{r}\rho'}{\rho^2} = \frac{\boldsymbol{t}\rho - \boldsymbol{r}g/\rho}{\rho^2} = \frac{h^2\boldsymbol{t} - gh\boldsymbol{n}}{\rho^3} = \frac{h}{\rho^2}\frac{h\boldsymbol{t} - g\boldsymbol{n}}{\sqrt{g^2+h^2}},$$

$$\frac{d}{dl}\boldsymbol{n}_\Sigma = \frac{ds}{dl}\frac{d}{ds}\left[\frac{\boldsymbol{r}}{\rho}\right] = \frac{h^2}{k\rho}\frac{d}{ds}\left[\frac{\boldsymbol{r}}{\rho}\right] = \frac{h^3}{k\rho^3}\boldsymbol{t}_\Sigma = -k_\Sigma \boldsymbol{t}_\Sigma.$$

Thus the curvature k_Σ of $\Sigma_{\boldsymbol{x}}$ is given by

$$k_\Sigma = -\frac{h^3}{k\rho^3} = -\frac{[\boldsymbol{n}(\boldsymbol{y})\cdot(\boldsymbol{y}-\boldsymbol{x})]^3}{k(\boldsymbol{y})\|\boldsymbol{x}-\boldsymbol{y}\|^3}$$

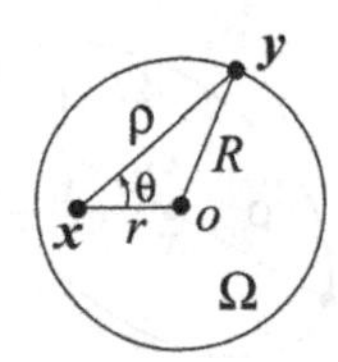

Figure 3.3 Notation used to demonstrate that transfinite Laplace and Wachspress coordinates coincide if Ω is a disk.

and therefore

$$-\frac{d\theta}{k_\Sigma} = \frac{k\rho^3}{h^3}\,d\theta = \frac{k\rho}{h^2}\,ds, \qquad \int_{\partial\Omega} \boldsymbol{r}\frac{k\,ds}{h^2} = \int_{\partial\Omega} \rho \boldsymbol{e}_\theta \frac{k\,ds}{h^2} = -\int_0^{2\pi} \frac{d\theta}{k_\Sigma} = 0,$$

which is in agreement with [413].

3.2.5 Transfinite Laplace and Wachspress coordinates coincide for a disk

As shown in [149], discrete harmonic and Wachspress coordinates are the same for a circumscribable polygon. The continuous or transfinite version of this remarkable result is established in [93]. Below we give a different and elementary proof that transfinite Laplace and Wachspress coordinates coincide if Ω is a disk.

Using the notations of Figure 3.3 we have

$$\cos\theta = \frac{\rho^2 + r^2 - R^2}{2\rho r} = A\rho - B\frac{1}{\rho}$$

for some values of A and B, which do not depend on θ. Thus,

$$0 = (\cos\theta)'' + \cos\theta = A\big[\rho''_{\theta\theta} + \rho\big] - B\big[(1/\rho)''_{\theta\theta} + 1/\rho\big].$$

It remains to note that the kernels $\rho''_{\theta\theta} + \rho$ and $(1/\rho)''_{\theta\theta} + 1/\rho$ correspond to the transfinite Laplace and Wachspress coordinates, respectively.

3.3 GORDON–WIXOM INTERPOLATION

3.3.1 Lagrange-type Gordon–Wixom interpolation

Harmonic coordinates and harmonic mappings deliver extremely useful tools for high quality interpolation and shape deformation [95, 101, 241] (see also references therein). However, harmonic mappings and coordinates do not allow for closed-form solutions in general and sophisticated numerical schemes are required for computing accurate approximations. One possible way to overcome this difficulty consists of using mappings and coordinates, which can be calculated effortlessly and mimic some properties of harmonic mappings and coordinates. In particular, the so-called pseudo-harmonic barycentric coordinates [94, 247] can be considered as

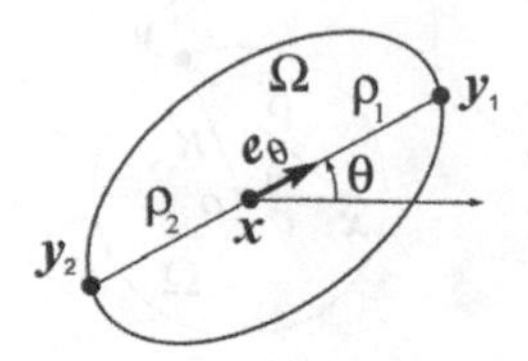

Figure 3.4 The main idea behind Gordon–Wixom interpolation consists of combining linear interpolation for each chord passing through $\boldsymbol{x}$ with circular averaging with respect to all directions $\boldsymbol{e}_\theta$.

potentially good substitutes for harmonic coordinates [215]. Below we consider a pseudo harmonic transfinite barycentric interpolation scheme introduced initially by Gordon and Wixom [171] and then studied and extended by a number of researchers [37, 38, 145, 146, 266].

Following [171], let us call a transfinite barycentric interpolation scheme *pseudo-harmonic* if it reproduces harmonic functions in a ball (a disk in the 2D case). It turns out that transfinite mean value (3.4), Laplace (3.13), and Wachspress coordinates (3.16) are not pseudo-harmonic [37]. In contrast, the Gordon–Wixom interpolation scheme [171] and some of its generalizations and modifications [37, 38] are pseudo-harmonic.

The basic idea behind the approach of Gordon and Wixom is simple and elegant. Given a point $\boldsymbol{x}$ inside a bounded convex domain $\Omega \subset \mathbb{R}^2$, consider a straight line passing through $\boldsymbol{x}$, forming an angle θ with some fixed direction, and intersecting $\partial\Omega$ in two points $\boldsymbol{y}_1$ and $\boldsymbol{y}_2$ (see Figure 3.4). We then use linear interpolation between $u(\boldsymbol{y}_1)$ and $u(\boldsymbol{y}_2)$ to obtain an estimate $u(\boldsymbol{x},\theta)$ associated with the direction θ,

$$\left[\frac{u(\boldsymbol{y}_1)}{\rho_1}+\frac{u(\boldsymbol{y}_2)}{\rho_2}\right] \Big/ \left[\frac{1}{\rho_1}+\frac{1}{\rho_2}\right].$$

Circular averaging of $u(\boldsymbol{x},\theta)$ with respect to θ then yields the Gordon–Wixom interpolation of some function u from $\partial\Omega$ to $\boldsymbol{x}\in\Omega$,

$$u(\boldsymbol{x}) = \frac{1}{2\pi}\int_0^{2\pi}\left(\left[\frac{u(\boldsymbol{y}_1)}{\rho_1}+\frac{u(\boldsymbol{y}_2)}{\rho_2}\right] \Big/ \left[\frac{1}{\rho_1}+\frac{1}{\rho_2}\right]\right)d\theta. \tag{3.18}$$

To demonstrate that the Gordon–Wixom interpolation scheme is pseudo-harmonic, we rewrite (3.18) as

$$u(\boldsymbol{x}) = \frac{1}{\pi}\int_0^{2\pi}\frac{\rho_2}{\rho_1+\rho_2}u(\boldsymbol{y}_1)\,d\theta \tag{3.19}$$

and assume that Ω is a disk of radius R. We now recall that, according to the intersecting chords theorem,

$$\rho_1\rho_2 \equiv c(\boldsymbol{x}),$$

where $c(\boldsymbol{x})$ depends only on $\boldsymbol{x}$. In the notation of Figure 3.5 we further have

$$h_1 = \rho_1\sin\varphi, \qquad h_2 = \rho_2\sin\varphi, \qquad \rho_1+\rho_2 = 2R\sin\varphi,$$

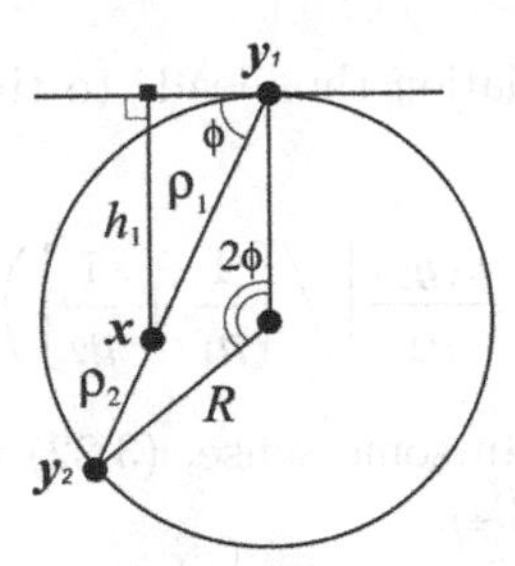

Figure 3.5 Geometric notation used to demonstrate that the Gordon–Wixom interpolation scheme (3.18) and its modification (3.22) reproduce harmonic functions if Ω is a disk.

$$\frac{h_1 h_2}{\sin^2\varphi} = c(\boldsymbol{x}), \qquad \frac{\rho_2}{\rho_1+\rho_2} = \frac{h_2}{2R\sin^2\varphi} = \frac{c(\boldsymbol{x})}{2R}\frac{1}{h_1}.$$

Thus, (3.19) reduces to

$$u(\boldsymbol{x}) = \int_0^{2\pi} \frac{u(\boldsymbol{y})\,d\theta}{h(\boldsymbol{x},\boldsymbol{y})} \bigg/ \int_0^{2\pi} \frac{d\theta}{h(\boldsymbol{x},\boldsymbol{y})}, \tag{3.20}$$

where $h(\boldsymbol{x},\boldsymbol{y})$ denotes the distance from $\boldsymbol{x}\in\Omega$ to the straight line tangent to $\partial\Omega$ at $\boldsymbol{y}\in\partial\Omega$. It is easy to see that the length element $ds_{\boldsymbol{y}}$ of $\partial\Omega$ at $\boldsymbol{y}\in\partial\Omega$ is given by

$$ds_{\boldsymbol{y}} = \rho^2 d\theta/h \quad \text{with } \rho = \|\boldsymbol{x}-\boldsymbol{y}\| \quad \text{and} \quad h = h(\boldsymbol{x},\boldsymbol{y}).$$

Thus, (3.20) can be rewritten as

$$u(\boldsymbol{x}) = \int_{\partial\Omega} \frac{u(\boldsymbol{y})\,ds_{\boldsymbol{y}}}{\|\boldsymbol{x}-\boldsymbol{y}\|^2} \bigg/ \int_{\partial\Omega} \frac{ds_{\boldsymbol{y}}}{\|\boldsymbol{x}-\boldsymbol{y}\|^2},$$

which is equivalent to the classical Poisson integral formula for recovering the values of a harmonic function inside a disk from its boundary values.

Another connection between the Gordon–Wixom approach and harmonic interpolation can be derived from the following observation. If ρ_1 and ρ_2 are small, then we can write

$$\frac{\partial^2 u}{\partial \boldsymbol{e}_\theta^2}(\boldsymbol{x}) \approx \frac{2}{\rho_1\rho_2}\left(\left[\frac{u(\boldsymbol{y}_1)}{\rho_1} + \frac{u(\boldsymbol{y}_2)}{\rho_2}\right] \bigg/ \left[\frac{1}{\rho_1}+\frac{1}{\rho_2}\right] - u(\boldsymbol{x})\right). \tag{3.21}$$

Note that

$$\frac{1}{\pi}\int_0^{2\pi} \frac{\partial^2}{\partial \boldsymbol{e}_\theta^2}\,d\theta = \frac{1}{\pi}\int_0^{2\pi}\left(\cos\theta\frac{\partial}{\partial x_1} + \sin\theta\frac{\partial}{\partial x_2}\right)^2 d\theta = \Delta,$$

which together with (3.21) implies

$$\Delta u \approx \frac{2}{\pi}\int_0^{2\pi}\left(\left[\frac{u(\boldsymbol{y}_1)}{\rho_1} + \frac{u(\boldsymbol{y}_2)}{\rho_2}\right] \bigg/ \left[\frac{1}{\rho_1}+\frac{1}{\rho_2}\right] - u(\boldsymbol{x})\right)\frac{d\theta}{\rho_1\rho_2}.$$

Mimicking harmonic interpolation thus leads to the modification of the Gordon–Wixom scheme (3.18)

$$u(\boldsymbol{x}) = \int_0^{2\pi} \left(\left[\frac{u(\boldsymbol{y}_1)}{\rho_1} + \frac{u(\boldsymbol{y}_2)}{\rho_2} \right] \Big/ \left[\frac{1}{\rho_1} + \frac{1}{\rho_2} \right] \right) \frac{d\theta}{\rho_1 \rho_2} \Big/ \int_0^{2\pi} \frac{d\theta}{\rho_1 \rho_2}, \tag{3.22}$$

which was obtained in [38]. In some sense, (3.22) resembles the transfinite mean value interpolation scheme (3.4).

The modified Gordon–Wixom interpolation scheme (3.22) is also pseudo-harmonic. Indeed, if Ω is a circle, then $\rho_1 \rho_2$ in (3.22) does not depend on the integration variable θ and (3.22) reduces to (3.18), which is pseudo-harmonic. The modified Gordon–Wixom scheme also has linear precision.

Both (3.18) and (3.22) can be considered as particular cases of weighted mean value transfinite interpolation (3.7). Indeed, both can be rewritten as

$$u(\boldsymbol{x}) = \int_0^{2\pi} \left(\left[\frac{u(\boldsymbol{y}_1)}{\rho_1} + \frac{u(\boldsymbol{y}_2)}{\rho_2} \right] \Big/ \left[\frac{1}{\rho_1} + \frac{1}{\rho_2} \right] \right) \tilde{\omega}(\boldsymbol{x}, \theta) \, d\theta \Big/ \int_0^{2\pi} \tilde{\omega}(\boldsymbol{x}, \theta) \, d\theta, \tag{3.23}$$

where the weighting functions $\tilde{\omega}(\boldsymbol{x}, \theta)$ in (3.23) and $\omega(\boldsymbol{x}, \theta)$ in (3.7) are related by

$$\tilde{\omega}(\boldsymbol{x}, \theta) = (1/\rho_1 + 1/\rho_2) \omega(\boldsymbol{x}, \theta).$$

Setting $\tilde{\omega}(\boldsymbol{x}, \theta) = (1/\rho_1 + 1/\rho_2)$ in (3.23) corresponds to the transfinite mean value interpolation scheme (3.4), choosing $\tilde{\omega}(\boldsymbol{x}, \theta) = 1$ yields the Gordon–Wixom scheme (3.18), and $\tilde{\omega}(\boldsymbol{x}, \theta) = 1/(\rho_1 \rho_2)$ gives the modified Gordon–Wixom scheme (3.22).

In practice (3.22) demonstrates a higher quality of interpolation compared with (3.18). In our numerical experiments (3.18) and (3.22) are evaluated using the composite trapezoidal rule (typically 100 equispaced directions are used). For non-convex domains, the ray $[\boldsymbol{x}, \boldsymbol{y})$ defined by the unit vector $\boldsymbol{e}_\theta$ can intersect the boundary at more than one point. For the two opposite directions $\boldsymbol{e}_\theta$ and $-\boldsymbol{e}_\theta$ we consider the intersections $\boldsymbol{y}_1$ and $\boldsymbol{y}_2$ that are closest to $\boldsymbol{x}$, respectively. Figure 3.6 compares the approximations obtained with the Gordon–Wixom scheme (3.18) and its modification (3.22) for a harmonic function $x^3 - 3xy^2$ in a geometrically complex domain. The absolute errors between the given function and its interpolations is used to illustrate the advantages of (3.22) over (3.18).

3.3.2 Hermite-type Gordon–Wixom interpolation

Gordon and Wixom extended their scheme in [171] to handle Hermite-type interpolation. Replacing linear interpolation with cubic interpolation of $(u(\boldsymbol{y}_1), u'(\boldsymbol{y}_1))$ and $(u(\boldsymbol{y}_2), u'(\boldsymbol{y}_2))$ gives

$$\begin{aligned} u(\boldsymbol{x}, \theta) = {} & \frac{(3\rho_1 + \rho_2)\rho_2^2}{(\rho_1 + \rho_2)^3} u(\boldsymbol{y}_1) + \frac{(\rho_1 + 3\rho_2)\rho_1^2}{(\rho_1 + \rho_2)^3} u(\boldsymbol{y}_2) \\ & - \frac{\rho_1 \rho_2^2}{(\rho_1 + \rho_2)^2} \frac{\partial u}{\partial \boldsymbol{e}_\theta}(\boldsymbol{y}_1) + \frac{\rho_1^2 \rho_2}{(\rho_1 + \rho_2)^2} \frac{\partial u}{\partial \boldsymbol{e}_\theta}(\boldsymbol{y}_2). \end{aligned} \tag{3.24}$$

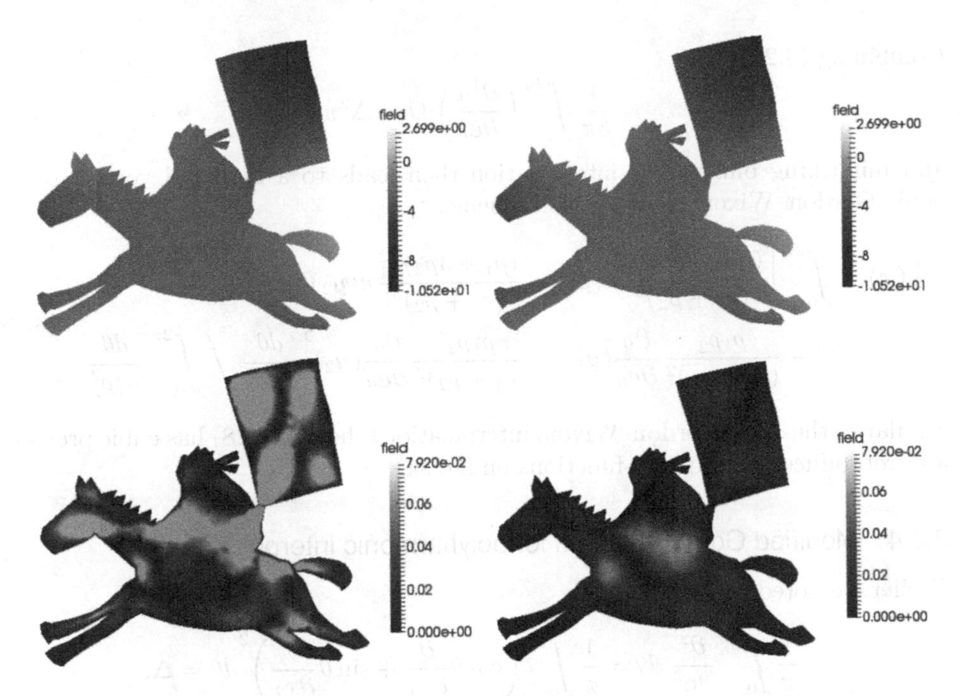

Figure 3.6 The original (top left) and modified (top right) Gordon–Wixom interpolation schemes (3.18) and (3.22) are used to interpolate a polynomial in a geometrically complex domain. The modified scheme yields a lower approximation error (bottom right) than the original scheme (bottom left).

Averaging (3.24) with respect to θ leads to the Hermite-type Gordon–Wixom interpolation scheme

$$u(\boldsymbol{x}) = \frac{1}{2\pi} \int_0^{2\pi} u(\boldsymbol{x}, \theta)\, d\theta. \tag{3.25}$$

The transfinite interpolation scheme (3.25) has cubic precision and reproduces biharmonic functions if Ω is a disk. The biharmonic property was conjectured in [171] and rigorously proved in [145].

3.3.3 Modified Hermite-type Gordon–Wixom interpolation

If ρ_1 and ρ_2 are small, then an approximation of the 4th-order directional derivative of $u(\boldsymbol{x})$ in the direction of $\boldsymbol{e}_\theta$ is given by

$$\begin{aligned} \frac{\partial^4 u}{\partial \boldsymbol{e}_\theta^4}(\boldsymbol{x}) \approx \frac{24}{\rho_1^2 \rho_2^2} \Bigg[& u(\boldsymbol{x}) - \frac{(3\rho_1 + \rho_2)\rho_2^2}{(\rho_1 + \rho_2)^3} u(\boldsymbol{y}_1) - \frac{(\rho_1 + 3\rho_2)\rho_1^2}{(\rho_1 + \rho_2)^3} u(\boldsymbol{y}_2) \\ & + \frac{\rho_1 \rho_2^2}{(\rho_1 + \rho_2)^2} \frac{\partial u}{\partial \boldsymbol{e}_\theta}(\boldsymbol{y}_1) - \frac{\rho_1^2 \rho_2}{(\rho_1 + \rho_2)^2} \frac{\partial u}{\partial \boldsymbol{e}_\theta}(\boldsymbol{y}_2) \Bigg]. \end{aligned} \tag{3.26}$$

Combining (3.26) with

$$\frac{4}{3\pi}\int_0^{2\pi}\left(\frac{\partial^4 u}{\partial \boldsymbol{e}_\theta^4}\right)d\theta = \Delta^2 u \tag{3.27}$$

and mimicking biharmonic interpolation then leads to a modified version of the cubic Gordon–Wixom interpolation scheme,

$$\begin{aligned} u(\boldsymbol{x}) = \int_0^{2\pi}\Bigg[&\frac{(3\rho_1+\rho_2)\rho_2^2}{(\rho_1+\rho_2)^3}u(\boldsymbol{y}_1) + \frac{(\rho_1+3\rho_2)\rho_1^2}{(\rho_1+\rho_2)^3}u(\boldsymbol{y}_2) \\ &- \frac{\rho_1\rho_2^2}{(\rho_1+\rho_2)^2}\frac{\partial u}{\partial \boldsymbol{e}_\theta}(\boldsymbol{y}_1) + \frac{\rho_1^2\rho_2}{(\rho_1+\rho_2)^2}\frac{\partial u}{\partial \boldsymbol{e}_\theta}(\boldsymbol{y}_2)\Bigg]\frac{d\theta}{\rho_1^2\rho_2^2}\Bigg/\int_0^{2\pi}\frac{d\theta}{\rho_1^2\rho_2^2}. \end{aligned} \tag{3.28}$$

Similar to the cubic Gordon–Wixom interpolation scheme, (3.28) has cubic precision and reproduces biharmonic functions on a disk.

3.3.4 Modified Gordon–Wixom for polyharmonic interpolation

Earlier we noted that

$$\frac{1}{\pi}\int_0^{2\pi}\frac{\partial^2}{\partial \boldsymbol{e}_\theta^2}\,d\theta = \frac{1}{\pi}\int_0^{2\pi}\left(\cos\theta\frac{\partial}{\partial x_1} + \sin\theta\frac{\partial}{\partial x_2}\right)^2 d\theta = \Delta.$$

Together with (3.27), this suggests that similar formulas can be obtained for the poly-Laplacian. Indeed, direct computations yield

$$\frac{8}{5\pi}\int_0^{2\pi}\frac{\partial^6 u}{\partial \boldsymbol{e}_\theta^6}\,d\theta = \Delta^3 u$$

and more generally

$$c_n\int_0^{2\pi}\frac{\partial^{2n} u}{\partial \boldsymbol{e}_\theta^{2n}}\,d\theta = \Delta^n u, \quad \text{where} \quad c_n = 1\Bigg/\int_0^{2\pi}(\cos\theta)^{2n}\,d\theta. \tag{3.29}$$

For example, $c_1 = 1/\pi$, $c_2 = 4/(3\pi)$, and $c_3 = 8/(5\pi)$.

In order to get a modified Gordon–Wixom scheme for polyharmonic interpolation, we then need to compute an approximation of $\partial^{2n}u/\partial\boldsymbol{e}_\theta^{2n}$ in terms of $u(\boldsymbol{x}), u(\boldsymbol{y}_1), u(\boldsymbol{y}_2), \ldots, \partial^{n-1}u(\boldsymbol{y}_1)/\partial\boldsymbol{e}_\theta^{n-1}, \partial^{n-1}u(\boldsymbol{y}_2)/\partial\boldsymbol{e}_\theta^{n-1}$.

If ρ_1 and ρ_2 are small, then an approximation of the $(2n)$-th order directional derivative of $u(\boldsymbol{x})$ in the direction of $\boldsymbol{e}_\theta$ can be obtained using the Taylor series

$$u(\boldsymbol{x}+\rho_1\boldsymbol{e}_\theta) = u(\boldsymbol{x}) + \rho_1\frac{\partial u}{\partial \boldsymbol{e}_\theta}(\boldsymbol{x}) + \cdots + \frac{\rho_1^{2n}}{(2n)!}\frac{\partial^{2n} u}{\partial \boldsymbol{e}_\theta^{2n}}(\boldsymbol{x}),$$

$$u(\boldsymbol{x}-\rho_2\boldsymbol{e}_\theta) = u(\boldsymbol{x}) - \rho_2\frac{\partial u}{\partial \boldsymbol{e}_\theta}(\boldsymbol{x}) + \cdots + \frac{(-\rho_2)^{2n}}{(2n)!}\frac{\partial^{2n} u}{\partial \boldsymbol{e}_\theta^{2n}}(\boldsymbol{x}),$$

$$\vdots$$

$$\frac{\partial^{n-1}u}{\partial e_\theta^{n-1}}(\boldsymbol{x}+\rho_1 e_\theta) = \frac{\partial^{n-1}u}{\partial e_\theta^{n-1}}(\boldsymbol{x}) + \rho_1 u^{(n)}(\boldsymbol{x}) + \cdots + \frac{\rho_1^{2n}}{(2n)!}\frac{\partial^{2n}u}{\partial e_\theta^{2n}}(\boldsymbol{x}),$$
$$\frac{\partial^{n-1}u}{\partial e_\theta^{n-1}}(\boldsymbol{x}-\rho_2 e_\theta) = \frac{\partial^{n-1}u}{\partial e_\theta^{n-1}}(\boldsymbol{x}) - \rho_2 u^{(n)}(\boldsymbol{x}) + \cdots + \frac{(-\rho_2)^{2n}}{(2n)!}\frac{\partial^{2n}u}{\partial e_\theta^{2n}}(\boldsymbol{x}).$$

Eliminating terms in $\partial u(\boldsymbol{x})/\partial e_\theta, \ldots, \partial^{2n-1}u(\boldsymbol{x})/\partial e_\theta^{2n-1}$ we arrive at

$$\frac{\partial^{2n}u}{\partial e_\theta^{2n}}(\boldsymbol{x}) = \frac{(2n)!}{\rho_1^n \rho_2^n}\Big[u(\boldsymbol{x}) - \alpha_1 u(\boldsymbol{y}_1) - \alpha_2 u(\boldsymbol{y}_2) - \cdots$$
$$- \alpha_{2n-1}\frac{\partial^{n-1}u}{\partial e_\theta^{n-1}}(\boldsymbol{y}_1) - \alpha_{2n}\frac{\partial^{n-1}u}{\partial e_\theta^{n-1}}(\boldsymbol{y}_2)\Big],$$

where $\alpha_1, \ldots, \alpha_{2n}$ are some weights. Combining with (3.29) and mimicking polyharmonic interpolation gives a high-order generalization of the modified Gordon–Wixom interpolation scheme.

3.4 GENERALIZED MEAN VALUE POTENTIALS AND DISTANCE FUNCTION APPROXIMATIONS

In this section we study properties of generalized mean value potentials, which are generalizations of the normalization function for the transfinite mean value coordinates; that is, the denominator in (3.4). We show that these generalized mean value potentials deliver accurate approximations of the distance function $\mathrm{dist}(\boldsymbol{x}, \partial\Omega)$ near $\partial\Omega$ and can be seen as smooth distance functions. Efficient computation of smooth distance functions is important for a number of applications, including pattern recognition [435], computational mechanics [157], medical imaging [338], and computational fluid dynamics [332, 399].

3.4.1 Generalized mean value potentials for smooth domains

Floater et al. [77, 133] observed some remarkable properties of the mean value normalization function

$$\Phi(\boldsymbol{x}) = \int_{S_{\boldsymbol{x}}} \frac{d\boldsymbol{\theta}}{\|\boldsymbol{x}-\boldsymbol{y}\|},$$

which is the denominator in (3.4). In particular, for $\Psi(\boldsymbol{x}) = 1/\Phi(\boldsymbol{x})$, they show that

$$\Psi(\boldsymbol{y}) = 0 \quad \text{and} \quad \frac{\partial \Psi(\boldsymbol{y})}{\partial \boldsymbol{n}_{\boldsymbol{y}}} = \frac{1}{V_{d-1}} \qquad \forall \boldsymbol{y} \in \partial\Omega, \tag{3.30}$$

where $\boldsymbol{n}_{\boldsymbol{y}}$ is the outer unit normal for the boundary of $\Omega \subset \mathbb{R}^d$ at $\boldsymbol{y} \in \partial\Omega$ and V_{d-1} is the volume of the unit ball in $\mathbb{R}^{d-1}$ ($V_1 = 2$ and $V_2 = \pi$). In view of (3.30), the function $V_{d-1}\Psi(\boldsymbol{x})$ delivers an accurate approximation of the distance function $\mathrm{dist}(\boldsymbol{x}, \partial\Omega)$ near $\partial\Omega$.

This property can be used for Hermite interpolation purposes as in [77, 133]. More generally, the distance function property also plays an important role in computer graphics, visualization, and modeling. For example, it is used as a parameter for material modeling in [51], for the animation of shapes in [106], and in level-set methods [297]. See also [214] and the references therein for further applications in computer graphics and visualization.

It seems natural to consider the L^p generalizations of $\Phi(\boldsymbol{x})$ and $\Psi(\boldsymbol{x})$. Following [39, 40], let us consider

$$\Phi_p(\boldsymbol{x}) = \int_{S_{\boldsymbol{x}}} \frac{d\boldsymbol{\theta}}{\|\boldsymbol{x}-\boldsymbol{y}\|^p} \qquad \text{and} \qquad \Psi_p(\boldsymbol{x}) = 1/\Phi_p(\boldsymbol{x}) \tag{3.31}$$

with $p > 0$. Below, for the sake of simplicity, we deal with the two-dimensional version of (3.31). A similar study of the generalized mean value potential $\Phi_p(\boldsymbol{x})$ in the 3D case can be found in [39].

In the 2D case, $\Phi_p(\boldsymbol{x})$ can be written as

$$\Phi_p(\boldsymbol{x}) = \int_0^{2\pi} \frac{d\theta}{\rho(\theta)^p},$$

where, as before, θ is the angle between the vector $\boldsymbol{y} - \boldsymbol{x}$ and a fixed direction and $\rho(\theta) = \|\boldsymbol{x} - \boldsymbol{y}\|$. The generalized mean value potential $\Phi_p(\boldsymbol{x})$ satisfies some interesting properties, including [40, Proposition 3]

$$\Delta\Phi_p(\boldsymbol{x}) = p(p+2)\Phi_{p+2}(\boldsymbol{x}). \tag{3.32}$$

It is clear that $\Phi_p(\boldsymbol{x}) > 0$ and $\Phi_p(\boldsymbol{x}) \to \infty$ as $\boldsymbol{x}$ approaches $\partial\Omega$. Further, (3.32) implies that $\Phi_p(\boldsymbol{x})$ has no local maxima inside Ω, because otherwise $\Delta\Phi_p > 0$ at a local maximum and (3.32) cannot be satisfied. Hence, $\Psi_p(\boldsymbol{x})$ is positive inside Ω, vanishes on $\partial\Omega$, and has no local minima inside Ω. Below we study the asymptotic properties of $\Phi_p(\boldsymbol{x})$ near $\partial\Omega$ (or, equivalently, the asymptotic properties of $\Psi_p(\boldsymbol{x})$ near $\partial\Omega$) and show that, after a proper constant rescaling, $\Psi_p(\boldsymbol{x})$ delivers an accurate approximation of $\operatorname{dist}(\boldsymbol{x}, \partial\Omega)$ near $\partial\Omega$.

Let $\boldsymbol{x}$ be near $\partial\Omega$ and denote by $h = \operatorname{dist}(\boldsymbol{x}, \partial\Omega)$ the distance from $\boldsymbol{x}$ to $\partial\Omega$. We keep following [39] and now show that in the planar case ($d = 2$) for $p > 1$ we have

$$\Phi_p(\boldsymbol{x}) = \int_0^{2\pi} \frac{d\theta}{\rho(\theta)^p} = \frac{c_p}{h^p} + k\frac{c_{p-2}}{2h^{p-1}} + O\left(\frac{1}{h^{p-2}}\right) \tag{3.33}$$

with

$$c_p = \int_{-\pi/2}^{\pi/2} \cos^p \theta \, d\theta,$$

where k is the curvature of $\partial\Omega$ at $\boldsymbol{y} \in \partial\Omega$. The case $p = 1$, which corresponds to the mean value normalization function, yields

$$\Phi_1(\boldsymbol{x}) \sim \frac{2}{h} + \frac{k}{2}\ln\frac{1}{h} + O(1) \quad \text{as} \quad h \to 0, \tag{3.34}$$

which is an extension of one of the main results in [133].

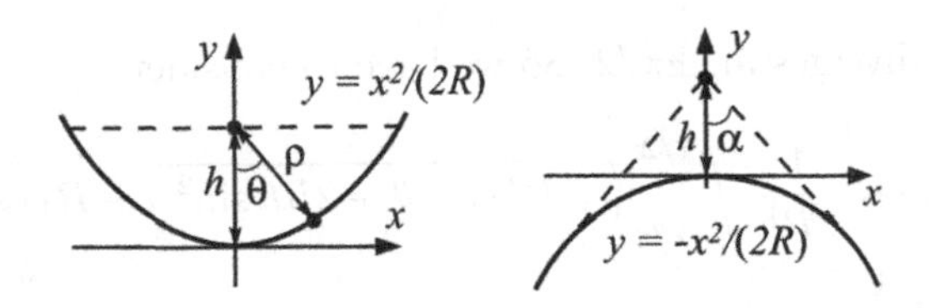

Figure 3.7 To analyze the asymptotic behavior of $\Phi_p(\boldsymbol{x})$ when $\boldsymbol{x}$ approaches $\partial\Omega$ $(h \to 0)$ we locally approximate $\partial\Omega$ by osculating parabolas.

Let us demonstrate (3.33) and (3.34). Given a point $\boldsymbol{x} \in \Omega$, situated at the distance $h \ll 1$ from $\partial\Omega$, let us introduce Euclidean and polar coordinates, as shown in Figure 3.7, where the origin of coordinates is located at the closest point to $\boldsymbol{x}$ on $\partial\Omega$ and the y-axis coincides with the direction of the orientation normal.

Let us start with the case of positive curvature k of $\partial\Omega$ at the origin of coordinates. Locally, $\partial\Omega$ is approximated by the osculating parabola

$$y = x^2/(2R),$$

where $R = 1/k$ is the curvature radius, as seen in Figure 3.7 (left). In polar coordinates

$$x = \rho \sin\theta, \qquad y = h - \rho\cos\theta$$

the parabola becomes

$$\rho^2 \sin^2\theta + 2R\rho\cos\theta - 2Rh = 0.$$

Solving this quadratic equation for $\rho > 0$ yields

$$\rho = \frac{-R\cos\theta + \sqrt{D}}{\sin^2\theta}, \qquad D = R^2\cos^2\theta + 2Rh\sin^2\theta$$

and we can write

$$\frac{1}{\rho} = \frac{1}{2Rh}\left(R\cos\theta\sqrt{1 + \frac{2h}{R}\tan^2\theta} + R\cos\theta\right) = \frac{\cos\theta}{h} + \frac{1}{2R}\frac{\sin^2\theta}{\cos\theta} + O(h).$$

Therefore,

$$\Phi_p(\boldsymbol{x}) \sim \int_{-\pi/2}^{\pi/2} \frac{d\theta}{\rho(\theta)^p} = \frac{c_p}{h^p} + \frac{1}{R}\frac{\tilde{c}_p}{h^{p-1}} + O\left(\frac{1}{h^{p-2}}\right),$$

where

$$c_p = \int_{-\pi/2}^{\pi/2} \cos^p\theta\, d\theta \quad \text{and} \quad \tilde{c}_p = \frac{p}{2}\int_{-\pi/2}^{\pi/2} \cos^{p-2}\theta \sin^2\theta\, d\theta. \tag{3.35}$$

Simple calculations show that $\tilde{c}_p = c_{p-2}/2$.

Above we assumed that $p > 1$. If $p = 1$ (the case of mean value coordinates), then we have

$$\Phi_1(\boldsymbol{x}) = \int_{-\pi/2}^{\pi/2} \frac{d\theta}{\rho(\theta)} = \frac{2}{h} + \frac{1}{2R}\int_{-\pi/2}^{\pi/2} \frac{\sin^2\theta}{\cos\theta}\, d\theta + O(h)$$

and the integral diverges at $\pm\pi/2$. So we have to consider

$$\Phi_1(\boldsymbol{x}) \sim \frac{1}{2Rh}\int_{-\pi/2}^{\pi/2}\left(\sqrt{R^2\cos^2\theta + 2Rh\sin^2\theta} + R\cos\theta\right)d\theta \tag{3.36}$$

as $h \to 0$. Note that

$$\begin{aligned}\int_{-\pi/2}^{\pi/2}\sqrt{R^2\cos^2\theta + 2Rh\sin^2\theta}\,d\theta &= 2R\int_0^{\pi/2}\sqrt{1-\left(1-\frac{2h}{R}\right)\sin^2\theta}\,d\theta\\ &= 2R\,\mathcal{E}\left(1-\frac{2h}{R}\right),\end{aligned}$$

where

$$\mathcal{E}(t) = \int_0^{\pi/2}\sqrt{1-t\sin^2\theta}\,d\theta$$

is the complete elliptic integral of the second kind. It can be shown that

$$\mathcal{E}(1-\varepsilon) = 1 - \frac{1}{4}\varepsilon\ln\varepsilon + \cdots \quad \text{as} \quad \varepsilon \to 0.$$

Thus,

$$\Phi_1(\boldsymbol{x}) \sim \frac{2}{h} + \frac{1}{2R}\ln\frac{1}{h} + O(1) \quad \text{as} \quad h \to 0,$$

which completes our proof of (3.34) in the case of positive curvature.

Now let us consider the negative curvature case, shown in Figure 3.7 (right). We have

$$x = \rho\sin\theta, \qquad y = h - \rho\cos\theta, \qquad y = -x^2/(2R),$$

$$\rho^2\sin^2\theta - 2R\rho\cos\theta + 2Rh = 0, \qquad \rho = \left(R\cos\theta - \sqrt{R^2\cos^2\theta - 2Rh\sin^2\theta}\right)/\sin^2\theta,$$

$$\frac{1}{\rho} = \frac{1}{2Rh}\left(\sqrt{R^2\cos^2\theta - 2Rh\sin^2\theta} + R\cos\theta\right) = \frac{\cos\theta}{h} - \frac{1}{2R}\frac{\sin^2\theta}{\cos\theta} + O(h).$$

Thus we arrive at

$$\frac{1}{\rho^p} = \frac{\cos^p\theta}{h^p} - \frac{p}{2R}\frac{\cos^{p-2}\theta\sin^2\theta}{h^{p-1}} + O\left(\frac{1}{h^{p-2}}\right). \tag{3.37}$$

Now let us study the asymptotic behavior of

$$\int_{-\alpha(h)}^{\alpha(h)}\frac{d\theta}{\rho(\theta)^p},$$

where the integration limits $\alpha(h)$ and $-\alpha(h)$ correspond to the two rays originating from $\boldsymbol{x}$ and the tangent to the osculating parabola $y = -x^2/(2R)$, as shown in

Figure 3.7 (right). Note that $\alpha(h) = \pi/2 + O(h)$ and $\cos\theta = O(h)$ for $|\theta|$ between $\alpha(h)$ and $\pi/2$, as $h \to 0$. Thus, in view of (3.37), we have

$$\begin{aligned}
\Phi_p(\boldsymbol{x}) &\sim \int_{-\alpha}^{\alpha} \frac{d\theta}{\rho(\theta)^p} \\
&= \frac{1}{h^p}\int_{-\alpha}^{\alpha} \cos^p\theta\, d\theta - \frac{p}{2Rh^{p-1}}\int_{-\alpha}^{\alpha}\cos^{p-2}\theta \sin^2\theta\, d\theta + O\Big(\frac{1}{h^{p-2}}\Big) \\
&= \frac{1}{h^p}\int_{-\pi/2}^{\pi/2} \cos^p\theta\, d\theta - \frac{p}{2Rh^{p-1}}\int_{-\pi/2}^{\pi/2}\cos^{p-2}\theta \sin^2\theta\, d\theta + O\Big(\frac{1}{h^{p-2}}\Big) \\
&= \frac{c_p}{h^p} - \frac{1}{R}\frac{\tilde{c}_p}{h^{p-1}} + O\Big(\frac{1}{h^{p-2}}\Big)
\end{aligned}$$

with the same c_p and $\tilde{c}_p$ as in (3.35) (the case of positive curvature). This completes our proof of (3.33).

Similarly, for $p = 1$, instead of (3.36) we have

$$\begin{aligned}
\Phi_1(\boldsymbol{x}) &= \frac{1}{2Rh}\int_{-\alpha(h)}^{\alpha(h)} \Big(\sqrt{R^2\cos^2\theta - 2Rh\sin^2\theta} + R\cos\theta\Big)d\theta + O(1) \\
&= \frac{1}{2Rh}\int_{-\pi/2}^{\pi/2} \Big(\sqrt{R^2\cos^2\theta - 2Rh\sin^2\theta} + R\cos\theta\Big)d\theta + O(1) \quad \text{as} \quad h \to 0.
\end{aligned}$$

Further,

$$\int_{-\pi/2}^{\pi/2} \sqrt{R^2\cos^2\theta - 2Rh\sin^2\theta}\, d\theta = 2R\,\mathcal{E}\Big(1 + \frac{2h}{R}\Big),$$

$$\Phi_1(\boldsymbol{x}) \sim \frac{2}{h} - \frac{1}{2R}\ln\frac{1}{h} + O(1) \quad \text{as} \quad h \to 0,$$

which completes the proof of (3.34) for the negative curvature case.

3.4.2 Generalized potentials for polygons

When the domain boundary $\partial\Omega$ is approximated by a polygon in 2D, then the potential $\Phi_p(\boldsymbol{x})$ at a point $\boldsymbol{x}$ is obtained by computing the contribution of each edge of the polygon and summing these contributions.

In the 2D case, given a point $\boldsymbol{x}$, an edge $[\boldsymbol{v}_i, \boldsymbol{v}_{i+1}]$, and the angle α between the vectors $\boldsymbol{v}_i - \boldsymbol{x}$ and $\boldsymbol{v}_{i+1} - \boldsymbol{x}$, we need to express the distance $\rho(\theta)$ between $\boldsymbol{x}$ and the point $\boldsymbol{y} \in [\boldsymbol{v}_i, \boldsymbol{v}_{i+1}]$, where θ is the angle between $\boldsymbol{v}_i - \boldsymbol{x}$ and $\boldsymbol{y} - \boldsymbol{x}$. Let $r_i = \|\boldsymbol{x} - \boldsymbol{v}_i\|$ and $r_{i+1} = \|\boldsymbol{x} - \boldsymbol{v}_{i+1}\|$ (see Figure 3.8). The area of the triangle $[\boldsymbol{x}, \boldsymbol{v}_i, \boldsymbol{v}_{i+1}]$ is equal to the sum of the triangles $[\boldsymbol{x}, \boldsymbol{v}_i, \boldsymbol{y}]$ and $[\boldsymbol{x}, \boldsymbol{y}, \boldsymbol{v}_{i+1}]$,

$$r_i r_{i+1}\sin\alpha = r_i\rho(\theta)\sin\theta + r_{i+1}\rho(\theta)\sin(\alpha - \theta),$$

giving

$$\rho(\theta) = \frac{r_i r_{i+1}\sin\alpha}{r_i\sin\theta + r_{i+1}\sin(\alpha - \theta)}.$$

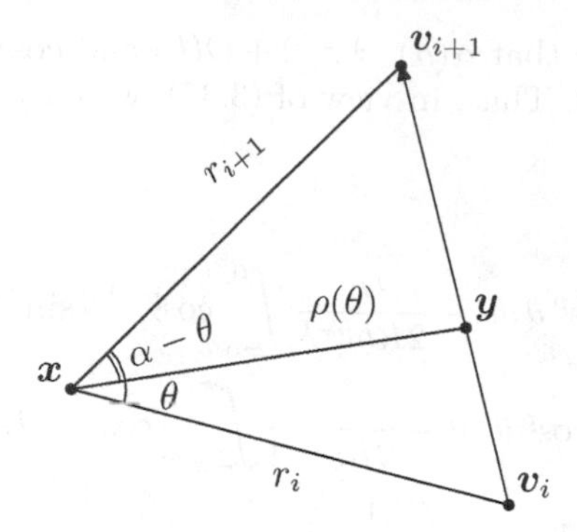

Figure 3.8 Notation used to define the generalized mean value potential induced by the edge $[\boldsymbol{v}_i, \boldsymbol{v}_{i+1}]$.

The potential $\Phi_p(\boldsymbol{x})$ for the edge $[\boldsymbol{v}_i, \boldsymbol{v}_{i+1}]$ is then given by

$$\Phi_p(\boldsymbol{x}) = \int_0^\alpha \frac{d\theta}{\rho(\theta)^p} = \frac{1}{(r_i r_{i+1} \sin\alpha)^p} \int_0^\alpha (r_i \sin\theta + r_{i+1}\sin(\alpha - \theta))^p d\theta.$$

For a given value of p the integral can be computed symbolically. In particular, when p is odd, it can be expressed as a polynomial of degree p of the variable $t = \tan(\alpha/2)$. The first expressions for $p = 1, 3$ are given by

$$\Phi_1(\boldsymbol{x}) = \left(\frac{1}{r_i} + \frac{1}{r_{i+1}}\right)t, \qquad \Phi_3(\boldsymbol{x}) = \frac{1}{3}\left(\frac{1}{r_i^3} + \frac{1}{r_{i+1}^3}\right)t + \frac{1}{6}\left(\frac{1}{r_i} + \frac{1}{r_{i+1}}\right)^3 t(1 + t^2).$$

The expression for $\Phi_1(\boldsymbol{x})$ corresponds to the expression for the mean value weight in [144, 200]. For even p, the resulting expressions are no longer simple anymore, except for the case $p = 0$, which corresponds to the angle α. The interested reader can find further details in [40].

In the 3D case we need to compute the contribution of the potential induced by a triangle $[\boldsymbol{v}_i, \boldsymbol{v}_j, \boldsymbol{v}_k]$ at $\boldsymbol{x}$. The case $p = 0$ corresponds to the solid angle at which the triangle $[\boldsymbol{v}_i, \boldsymbol{v}_j, \boldsymbol{v}_k]$ is seen from $\boldsymbol{x}$. The case $p = 1$ corresponds to the mean value weight for which an expression is given in [217]. For other values of p, a closed form expression for the potential $\Phi_p(\boldsymbol{x})$ induced by one triangle can be obtained as a particular case of the expressions for singular potential integrals derived in [85].

CHAPTER 4

Barycentric Mappings*

Teseo Schneider

Università della Svizzera italiana, Lugano, Switzerland

CONTENTS

BARYCENTRIC MAPPINGS allow us to naturally warp a source polygon to a corresponding target polygon, or, more generally, to create mappings between closed curves or polyhedra. Unfortunately, bijectivity of such barycentric mappings can only be guaranteed for the special case of warping between convex polygons. In fact, for any barycentric coordinates, it is always possible to construct a pair of polygons such that the barycentric mapping is not bijective. However, if the two polygons are sufficiently close, the mapping is close to the identity and hence bijective. This fact suggests "splitting" it into several intermediate mappings and creating a composite barycentric mapping, which is guaranteed to be bijective between arbitrary polygons, polyhedra, or closed planar curves.

*This work was supported by the SNSF under project number 200020_156178.

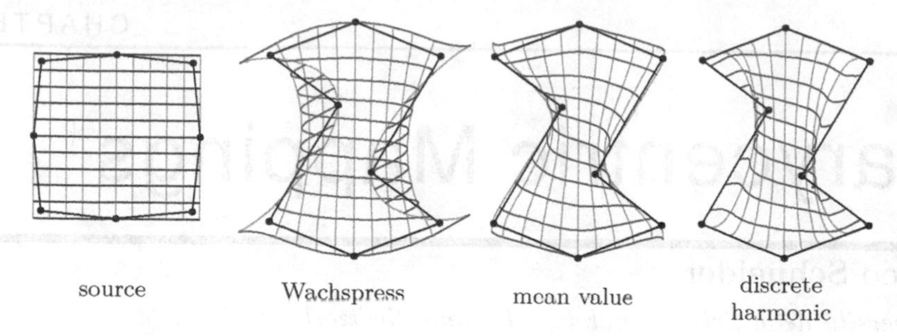

Figure 4.1 Example of a barycentric mapping based on different barycentric coordinates. For the discrete harmonic mapping we only show the image of the interior of $\bar{\Omega}^0$.

4.1 INTRODUCTION

A *barycentric mapping* between a *source polygon* $\bar{\Omega}^0 \subset \mathbb{R}^2$ with $n \geq 3$ *source vertices* $\boldsymbol{v}_i^0$, $i = 1, \ldots, n$, ordered anticlockwise, and a *target polygon* $\bar{\Omega}^1 \subset \mathbb{R}^2$ with the same number of *target vertices* $\boldsymbol{v}_i^1$, is a mapping

$$\boldsymbol{f} \colon \bar{\Omega}^0 \to \bar{\Omega}^1, \qquad \boldsymbol{f}(\boldsymbol{x}) = \sum_{i=1}^{n} \phi_i(\boldsymbol{x}) \boldsymbol{v}_i^1, \tag{4.1}$$

where the functions $\phi_i \colon \bar{\Omega}^0 \to \mathbb{R}$, $i = 1, \ldots, n$ are *barycentric coordinates* with respect to $\bar{\Omega}^0$ (see Chapter 1). Since $\boldsymbol{f}$ depends on the particular choice of coordinates, we denote the mapping $\boldsymbol{f}$ with the name of the coordinate generating it. For instance, a *Wachspress mapping* is a barycentric mapping based on Wachspress coordinates.

Because of the Lagrange property (1.10b), it is clear that $\boldsymbol{f}(\boldsymbol{v}_i^0) = \boldsymbol{v}_i^1$. Moreover, if a point $\boldsymbol{x} = (1-\mu)\boldsymbol{v}_i^0 + \mu \boldsymbol{v}_{i+1}^0$ lies on the edge $[\boldsymbol{v}_i^0, \boldsymbol{v}_{i+1}^0]$, the only non-zero coordinates are $\phi_i(\boldsymbol{x}) = 1 - \mu$ and $\phi_{i+1}(\boldsymbol{x}) = \mu$, because the coordinates are linear along the edges of the polygon (1.10c). Hence the mapping is $\boldsymbol{f}(\boldsymbol{x}) = (1-\mu)\boldsymbol{v}_i^1 + \mu \boldsymbol{v}_{i+1}^1$, which means that it is linear along the edges, too. Finally, the mapping $\boldsymbol{f}$ is always surjective, because the coordinates ϕ_i are continuous, which implies that $\boldsymbol{f}$ is continuous on $\bar{\Omega}^0$, and edges are mapped to edges.

Figure 4.1 shows an example of a barycentric mapping for different coordinates. We see that the mapping is linear along the edges, regardless of the choice of coordinates, and that it is not bijective; that is, the grid folds over in concave regions.

4.1.1 Convex polygons

Barycentric mappings between *convex* polygons are guaranteed to be bijective only for the special case of Wachspress mappings [152]. Figure 4.2 shows an extreme example of a barycentric mapping between two convex polygons. In the close-up we

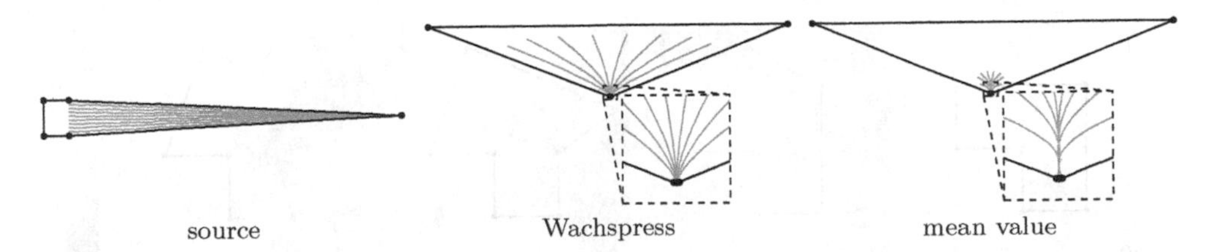

Figure 4.2 Example of a convex source and a convex target polygon with 5 vertices, taken from [152], for which the mean value mapping (right) is not bijective, whereas the Wachspress (middle) is.

see that the mean value mapping is not bijective, whereas the Wachspress mapping is.

4.1.2 Arbitrary polygons

For any choice of barycentric coordinates, it is possible to construct a source and a target polygon such that the barycentric mapping is not bijective [208]. In order to construct such a counterexample, let us consider the barycentric mapping between a square and a deformed square, as shown in Figure 4.3.

Because we only move $\boldsymbol{v}_3^0$, we know that $\boldsymbol{v}_i^1 - \boldsymbol{v}_i^0 = 0$ for $i = 1, 2, 4$, and we can rewrite the mapping $\boldsymbol{f}$, evaluated at the origin, as

$$\boldsymbol{f}(\mathbf{0}) = \sum_{i=1}^{4} \phi_i(\mathbf{0})(\boldsymbol{v}_i^1 - \boldsymbol{v}_i^0 + \boldsymbol{v}_i^0) = \sum_{i=1}^{4} \phi_i(\mathbf{0})(\boldsymbol{v}_i^1 - \boldsymbol{v}_i^0) = \phi_3(\mathbf{0})(\boldsymbol{v}_3^1 - \boldsymbol{v}_3^0) = \phi_3(\mathbf{0})(\boldsymbol{v}_3^1 + \boldsymbol{v}_1^0),$$

where we exploit the fact that $\boldsymbol{v}_3^0 = -\boldsymbol{v}_1^0$ in the last step.

We first assume that $\phi_3(\mathbf{0}) > 0.5$ and show that there exists a choice of $\boldsymbol{v}_3^1$ such that $\boldsymbol{f}(\mathbf{0}) = \boldsymbol{v}_1^1$, which contradicts the bijectivity of the mapping. To this end,

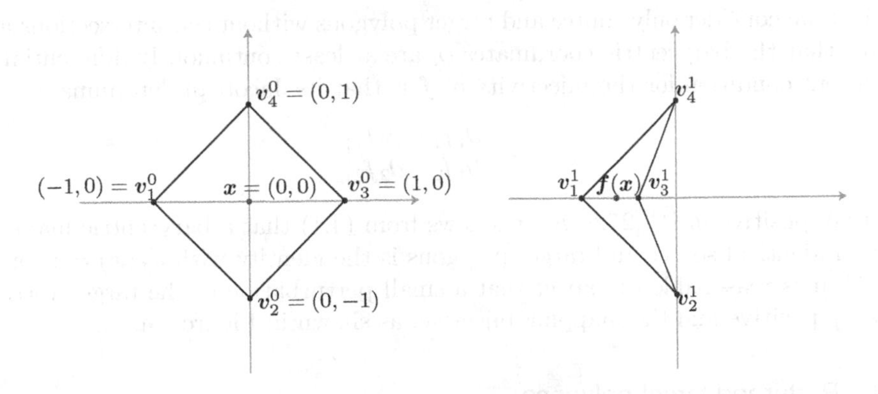

Figure 4.3 Source and target polygon for the construction of a non-bijective barycentric mapping.

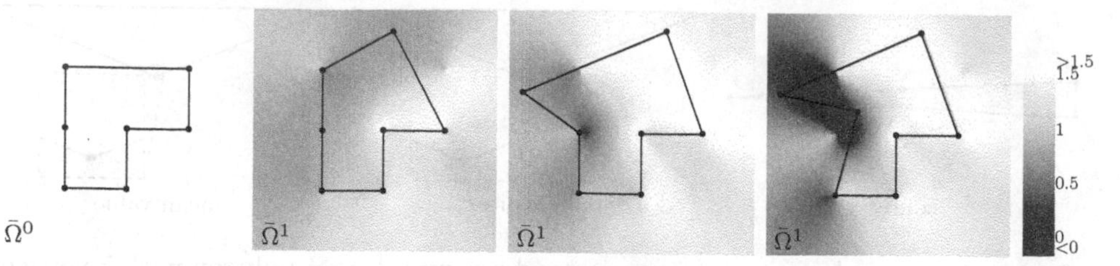

Figure 4.4 Color-coded plots of $J_{\boldsymbol{f}}$ for the mean value mapping $\boldsymbol{f}: \bar{\Omega}^0 \to \bar{\Omega}^1$ for different target polygons, which remains positive when the perturbation is small, but becomes negative for large deformations. See color insert.

observe that

$$x = \frac{\phi_3(\mathbf{0}) - 1}{\phi_3(\mathbf{0})} > -1$$

by the assumption and choose $\boldsymbol{v}_3^1 = (x, 0) = -x\boldsymbol{v}_1^0$. With this choice,

$$\boldsymbol{f}(\mathbf{0}) = \phi_3(\mathbf{0})(1 - x)\boldsymbol{v}_1^0 = \boldsymbol{v}_1^1.$$

Assuming next that $\phi_3(\mathbf{0}) < 0.5$, we conclude with similar considerations that $\boldsymbol{f}(\mathbf{0}) = \boldsymbol{v}_3^1$, again contradicting the bijectivity of $\boldsymbol{f}$.

Finally, consider the case when $\phi_3(\mathbf{0}) = 0.5$. Rotating the source polygon by 90 degrees, keeping in mind that $\boldsymbol{f}$ is invariant under rotations, and repeating the same reasoning as in the previous two cases, we conclude that $\phi_i(\mathbf{0}) = 0.5$ for all i, which contradicts the partition of unity property.

4.2 BIJECTIVE BARYCENTRIC MAPPING

Let us denote the *partial derivatives* of the barycentric mapping $\boldsymbol{f} = (f_1, f_2)$ in (4.1) at $\boldsymbol{x} = (x_1, x_2) \in \bar{\Omega}^0$ by $\partial_k \boldsymbol{f}(\boldsymbol{x}) = \partial \boldsymbol{f} / \partial x_k$, $k = 1, 2$, and the *gradients* of its two components f_i by $\nabla f_i = (\partial_1 f_i, \partial_2 f_i)$.

Since we consider only source and target polygons without self-intersections and assume that the barycentric coordinates ϕ_i are at least continuously differentiable, a sufficient condition for the injectivity of $\boldsymbol{f}$ is that its Jacobian determinant

$$J_{\boldsymbol{f}} = \begin{vmatrix} \partial_1 f_1 & \partial_2 f_1 \\ \partial_1 f_2 & \partial_2 f_2 \end{vmatrix}$$

is strictly positive in $\bar{\Omega}^0$ [272]. As it follows from (4.1) that a barycentric mapping between identical source and target polygons is the identity with $J_{\boldsymbol{f}}(\boldsymbol{x}) = 1$ for all $\boldsymbol{x} \in \bar{\Omega}^0$, it is reasonable to expect that a small perturbation of the target vertices keeps $J_{\boldsymbol{f}}$ positive and the mapping bijective, as shown in Figure 4.4.

4.2.1 Perturbed target polygons

Consider first a target polygon with a single perturbed vertex, as shown in Figure 4.5. Formally, the target vertices are $\boldsymbol{v}_i^1 = \boldsymbol{v}_i^0 + \boldsymbol{u}$ for some i and $\boldsymbol{v}_j^1 = \boldsymbol{v}_j^0$

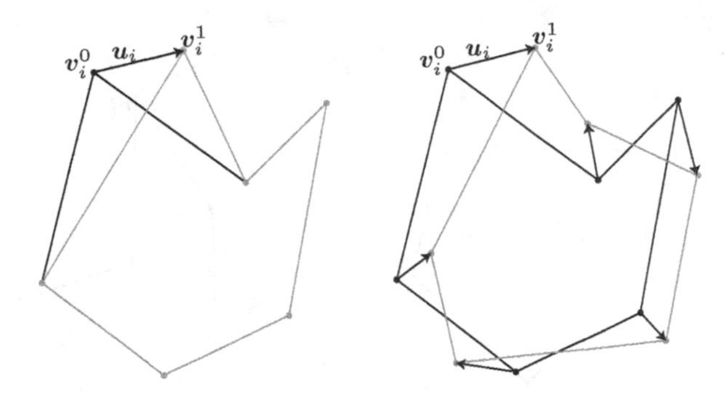

Figure 4.5 Perturbation of one vertex (left) and all vertices (right) of the target polygon.

for $j \neq i$. Substituting these target vertices $\boldsymbol{v}_i^1$ in (4.1) and recalling the linear reproduction property (1.9), we obtain

$$\boldsymbol{f}(\boldsymbol{x}) = \boldsymbol{x} + \phi_i(\boldsymbol{x})\boldsymbol{u}$$

for any $\boldsymbol{x} \in \bar{\Omega}^0$, and further,

$$J_{\boldsymbol{f}}(\boldsymbol{x}) = \begin{vmatrix} 1 + \partial_1\phi_i(\boldsymbol{x})\boldsymbol{u}_1 & \partial_2\phi_i(\boldsymbol{x})\boldsymbol{u}_1 \\ \partial_1\phi_i(\boldsymbol{x})\boldsymbol{u}_2 & 1 + \partial_2\phi_i(\boldsymbol{x})\boldsymbol{u}_2 \end{vmatrix} = 1 + \nabla\phi_i(\boldsymbol{x}) \cdot \boldsymbol{u}.$$

Therefore,

$$J_{\boldsymbol{f}}(\boldsymbol{x}) \geq 1 - |\nabla\phi_i(\boldsymbol{x}) \cdot \boldsymbol{u}| \geq 1 - \|\boldsymbol{u}\|\|\nabla\phi_i(\boldsymbol{x})\|,$$

which is strictly positive for $\|\boldsymbol{u}\| < 1/M_i$ with

$$M_i = \sup_{\boldsymbol{x} \in \bar{\Omega}^0} \|\nabla\phi_i(\boldsymbol{x})\|.$$

This result nicely extends to a perturbation of all vertices, where we consider a target polygon with vertices $\boldsymbol{v}_i^1 = \boldsymbol{v}_i^0 + \boldsymbol{u}_i$ for $i = 1, \ldots, n$ (Figure 4.5). Using the linear reproduction property again, we reformulate the mapping as

$$\boldsymbol{f}(\boldsymbol{x}) = \boldsymbol{x} + \sum_{i=1}^{n} \phi_i(\boldsymbol{x})\boldsymbol{u}_i$$

for any $\boldsymbol{x} \in \bar{\Omega}^0$, hence

$$\begin{aligned} J_{\boldsymbol{f}}(\boldsymbol{x}) &= \begin{vmatrix} 1 + \sum_i \partial_1\phi_i(\boldsymbol{x})u_{i,1} & \sum_i \partial_2\phi_i(\boldsymbol{x})u_{i,1} \\ \sum_i \partial_1\phi_i(\boldsymbol{x})u_{i,2} & 1 + \sum_i \partial_2\phi_i(\boldsymbol{x})u_{i,2} \end{vmatrix} \\ &= 1 + \sum_i \nabla\phi_i(\boldsymbol{x}) \cdot \boldsymbol{u}_i + \sum_i \sum_j \partial_1\phi_i(\boldsymbol{x})\partial_2\phi_j(\boldsymbol{x})(\boldsymbol{u}_i \times \boldsymbol{u}_j), \end{aligned}$$

where $\boldsymbol{u}_i = (u_{i,1}, u_{i,2})$ and the sums range from 1 to n. Therefore,

$$J_{\boldsymbol{f}}(\boldsymbol{x}) \geq 1 - Md - M^2d^2$$

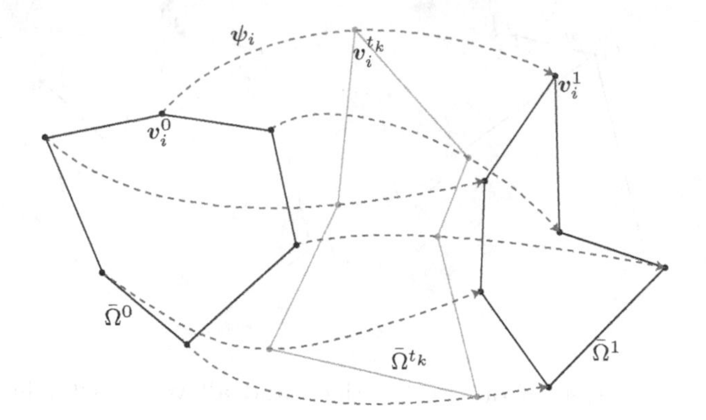

Figure 4.6 Construction of the intermediate polygon $\bar{\Omega}^{t_k}$ using the paths ψ_i.

with $M = M_1 + \cdots + M_n$ and $d = \max_{1 \le i \le n} \|u_i\|$. Overall, this implies that the mapping f is injective if

$$d < \frac{\sqrt{5}-1}{2M}.$$

4.3 BIJECTIVE COMPOSITE BARYCENTRIC MAPPING

Section 4.2.1 suggests that if source and target polygon are sufficiently close, the mapping is close to the identity and hence bijective. Therefore, by "splitting" the barycentric mapping from source to target polygon into a finite number of intermediate steps, where each step perturbs the vertices only slightly, it should be possible to obtain a bijective composite mapping.

To this end, suppose that $\psi_i\colon [0,1] \to \mathbb{R}^2$, $i = 1, \ldots, n$ are a set of continuous *vertex paths* between each source vertex $v_i^0 = \psi_i(0)$ and its corresponding target vertex $v_i^1 = \psi_i(1)$, as shown in Figure 4.6. Let $\tau = (t_0, t_1, \ldots, t_m)$ with $t_0 = 0$, $t_m = 1$, let $t_k < t_{k+1}$ for $k = 0, \ldots, m-1$ be a *partition* of $[0,1]$ and let f_k be the barycentric mapping from $\bar{\Omega}^{t_k}$ to $\bar{\Omega}^{t_{k+1}}$, based on the barycentric coordinates $\phi_i^{t_k}$.

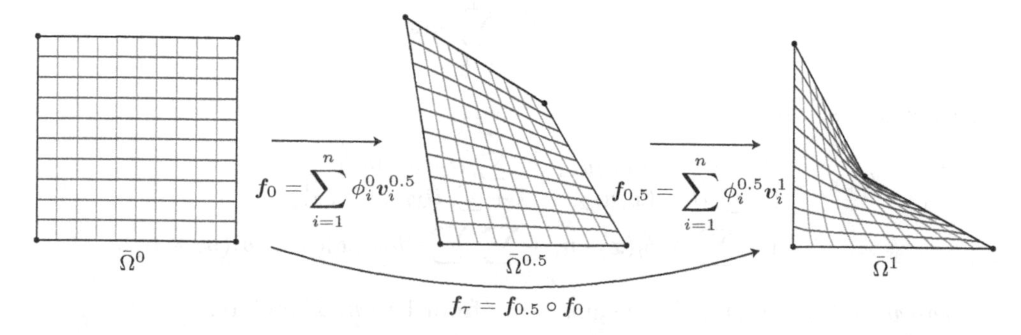

Figure 4.7 Construction of a *composite barycentric mapping* for $\tau = [0, 0.5, 1]$.

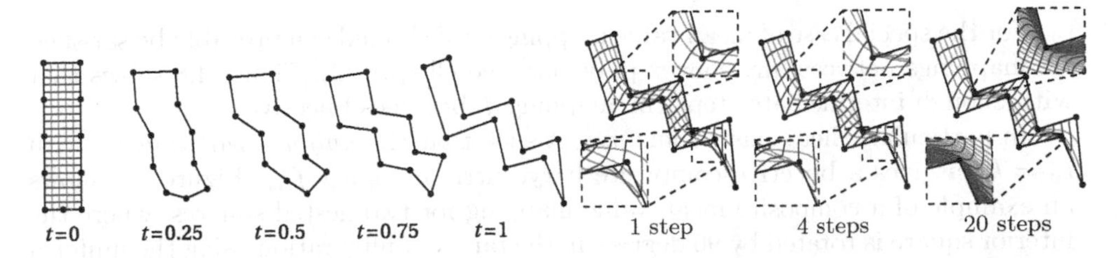

Figure 4.8 Examples of uniform composite mean value mappings for different numbers of uniform steps.

The mapping

$$f_\mathcal{T} = f_{m-1} \circ f_{m-2} \circ \cdots \circ f_0$$

is called a *composite barycentric mapping* from $\bar{\Omega}^0$ to $\bar{\Omega}^1$ [350]. An example of a composite barycentric mapping between a square and a concave quadrilateral is shown in Figure 4.7.

Denoting the maximum displacement distance between $\bar{\Omega}^{t_k}$ and $\bar{\Omega}^{t_{k+1}}$ by

$$d_k = \max_{1 \le i \le n} \left\| \boldsymbol{v}_i^{t_k} - \boldsymbol{v}_i^{t_{k+1}} \right\|,$$

it follows from the previous results that $f_\mathcal{T}$ is bijective if

$$d_\mathcal{T} = \max_{0 \le k < m} d_k < \frac{\sqrt{5} - 1}{2nM^*},$$

where

$$M^* = \max_{1 \le i \le n} \sup_{t \in [0,1]} \sup_{\boldsymbol{x} \in \bar{\Omega}^t} \left\| \nabla \phi_i^t(\boldsymbol{x}) \right\|.$$

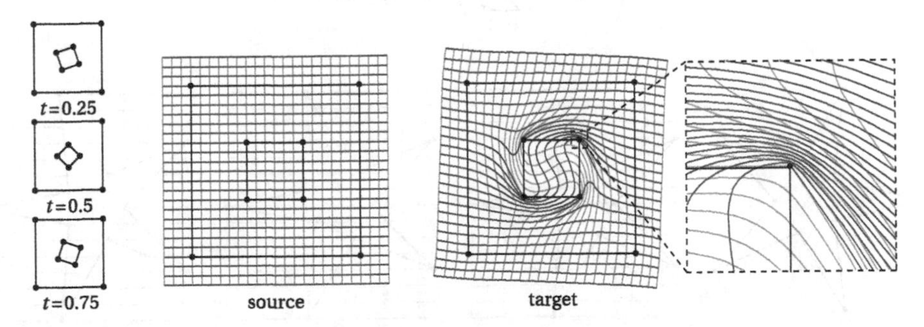

Figure 4.9 Example of a composite mean value mapping with 1000 uniform steps. The resolution of the grid is increased by a factor of 4 in the close-up.

For the special case of mean value mappings, this bound can provably be satisfied for mappings between any convex polygons (see Chapter 2). Figure 4.8 shows that with enough intermediate steps the mapping $\boldsymbol{f}$ becomes bijective.

In particular, there exists some $m \in \mathbb{N}$ such that the *uniform partition* τ_m with $t_k = k/m$ gives a bijective composite barycentric mapping f_{τ_m}. Figure 4.9 shows an example of a composite mean value mapping for two nested squares, where the interior square is rotated by 90 degrees in the target configuration using the uniform partition τ_{1000}.

4.3.1 Limit of composite barycentric mappings

The idea of uniform composite barycentric mappings leads to the interesting question of the behavior in the limit. To this end, we consider the *infinite composite barycentric mapping* $\boldsymbol{f}_\infty = \lim_{m\to\infty} \boldsymbol{f}_{\tau_m}$ and its *backward mapping* $\boldsymbol{g}_\infty : \bar{\Omega}^1 \to \bar{\Omega}^0$. Figure 4.10 shows the result of mapping a star with a uniform composite mapping $\boldsymbol{f}_{\tau_m}$ composed with its backward mapping $\boldsymbol{g}_{\tau_m} : \bar{\Omega}^1 \to \bar{\Omega}^0$ with the same number of steps. Computing the maximum distance $\|\boldsymbol{x} - \boldsymbol{g}_{\tau_m}(\boldsymbol{f}_{\tau_m}(\boldsymbol{x}))\|$ for one million random points $\boldsymbol{x} \in \bar{\Omega}^0$ indicates that this distance converges to zero and so $\boldsymbol{g}_{\tau_m} = \boldsymbol{f}_{\tau_m}^{-1}$ as $m \to \infty$. Consequently, the inverse of an infinite barycentric mapping is likely to be an infinite barycentric mapping itself, which is not the case for standard barycentric mappings.

This observation can be proven by considering the difference between two successive steps and taking its limit [145]. For any point $\boldsymbol{x}^0 \in \bar{\Omega}^0$ we evaluate the first step $\boldsymbol{f}_0$ of the composite mapping $\boldsymbol{f}$,

$$\boldsymbol{x}^{t_1} = \boldsymbol{f}_0(\boldsymbol{x}^0) = \sum_{i=1}^{n} \phi_i^{t_0}(\boldsymbol{x}^0)\boldsymbol{\psi}_i(t_1).$$

Because of the linear reproduction property,

$$\boldsymbol{x}^0 = \sum_{i=1}^{n} \phi_i^{t_0}(\boldsymbol{x}^0)\boldsymbol{\psi}_i(t_0),$$

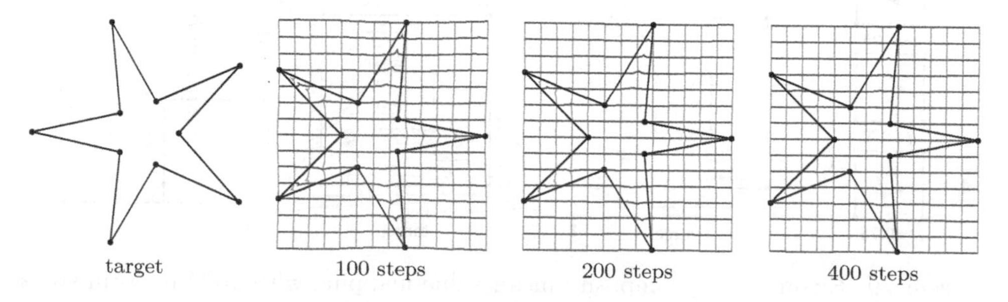

Figure 4.10 Composing the mapping $\boldsymbol{f}_{\tau_m}$ and the backward mapping $\boldsymbol{g}_{\tau_m}$ converges to the identity as the number of uniform steps m increases.

hence

$$x^{t_1} - x^0 = \sum_{i=1}^{n} \phi_i^{t_0}(x^0)(\psi_i(t_1) - \psi_i(t_0)).$$

Dividing by $t_1 - t_0$ and taking the limit for $t_1 \to t_0$, yields

$$x'(t) = \sum_{i=1}^{n} \phi_i(x(t), t)\psi_i'(t),$$

which is a first-order differential equation in $x(t)$ of the form

$$x'(t) = F(t, x(t)).$$

with initial condition $x(0) = x^0$.

If the barycentric coordinates ϕ_i are Lipschitz-continuous, then F is Lipschitz-continuous, too, which is a sufficient condition for the existence of a local unique solution, according to the Picard–Lindelöf Theorem [250, 308]. The solution of the differential equation is also a global solution, since $x(t)$ stays inside all intermediate polygons. This is the case because the mapping is bijective and edges are mapped to edges.

In order to show that $g^{-1} = f$, we consider, for any point $x^1 = x(1) \in \bar{\Omega}^1$,

$$y'(t) = \sum_{i=1}^{n} \phi_i(x(1-t), 1-t)\bar{\psi}_i'(t), \qquad \bar{\psi}_i(t) = \psi_i(1-t),$$

with $y(0) = x(1)$ as the mapping of x^1 from $\bar{\Omega}^1$ to $\bar{\Omega}^0$. Because $\bar{\psi}_i'(t) = -\psi_i'(1-t)$, we conclude that $y(t) = x(1-t)$. Therefore, $y(1) = x^0$ and g has an inverse generated by the previous equation with $g^{-1} = f$.

4.4 EXTENSIONS

The sufficient condition for guaranteeing bijectivity of a composite barycentric mapping between polygons can be naturally extended to mappings between closed curves and polyhedra.

4.4.1 Closed planar curves

We define the *barycentric mapping* between a closed planar source curve $\gamma^0(s)$ and a closed planar target curve $\gamma^1(s)$ with $s \in [0, 1]$, as

$$f(x) = \int_0^1 \phi(s, x)\gamma^1(s)ds,$$

where $\phi(s, x)$ are the transfinite barycentric coordinates with respect to γ^0 (see Chapter 3). Analogously to the polygonal case, we now consider a perturbed target

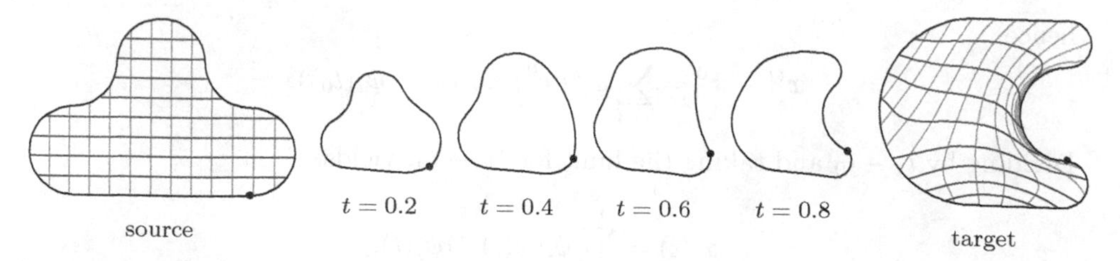

Figure 4.11 Example of a composite mean value mapping between two closed planar curves. The grid shows how the interior of the source curve is morphed to the interior of the target curve.

curve $\gamma^1(s) = \gamma^0(s) + \boldsymbol{u}(s)$ and assume that the maximum displacement distance satisfies

$$d = \sup_{s\in[0,1]} \|\boldsymbol{u}(s)\| < \frac{\sqrt{5}-1}{2M},$$

where

$$M = \sup_{s\in[0,1]} \sup_{\boldsymbol{x}\in\gamma^0} \|\nabla\phi(s,\boldsymbol{x})\|.$$

We exploit the linear reproduction property to rewrite

$$\boldsymbol{f}(\boldsymbol{x}) = \boldsymbol{x} + \int_0^1 \phi(s,\boldsymbol{x})\boldsymbol{u}(s)ds$$

and compute

$$\begin{aligned} J_{\boldsymbol{f}}(\boldsymbol{x}) &= \begin{vmatrix} 1 + \int \partial_1\phi(s,\boldsymbol{x})u_1(s)ds & \int \partial_2\phi(s,\boldsymbol{x})u_1(s)ds \\ \int \partial_1\phi(s,\boldsymbol{x})u_2(s)ds & 1 + \int \partial_2\phi(s,\boldsymbol{x})u_2(s)ds \end{vmatrix} \\ &= 1 + \int \nabla\phi(s,\boldsymbol{x}) \cdot \boldsymbol{u}(s)ds + \iint \partial_1\phi(s,\boldsymbol{x})\partial_2\phi(r,\boldsymbol{x})(\boldsymbol{u}(s) \times \boldsymbol{u}(r))ds\,dr, \end{aligned}$$

where the integrals range from zero to one. Therefore,

$$J_{\boldsymbol{f}}(\boldsymbol{x}) \geq 1 - Md - M^2d^2,$$

which is strictly positive for $Md < (\sqrt{5}-1)/2$. Hence the mapping is injective as long as $d < \frac{\sqrt{5}-1}{2M}$. It is interesting to note that this bound is the same as in the polygonal case.

Now suppose that $\boldsymbol{\psi}$ is a continuous *closed curve interpolation* function between the source curve $\gamma^0 = \boldsymbol{\psi}(0)$ and the target curve $\gamma^1 = \boldsymbol{\psi}(1)$. Again we let $\tau = (t_0, t_1, \ldots, t_m)$ with $t_0 = 0$, $t_m = 1$, and $t_k < t_{k+1}$ for $k = 0, \ldots, m-1$ be a *partition* of $[0,1]$, and let $\boldsymbol{f}_k$ be the barycentric mapping from γ^{t_k} to $\gamma^{t_{k+1}}$, based on the barycentric coordinates ϕ^{t_k}. The mapping

$$\boldsymbol{f}_\tau = \boldsymbol{f}_{m-1} \circ \boldsymbol{f}_{m-2} \circ \cdots \circ \boldsymbol{f}_0,$$

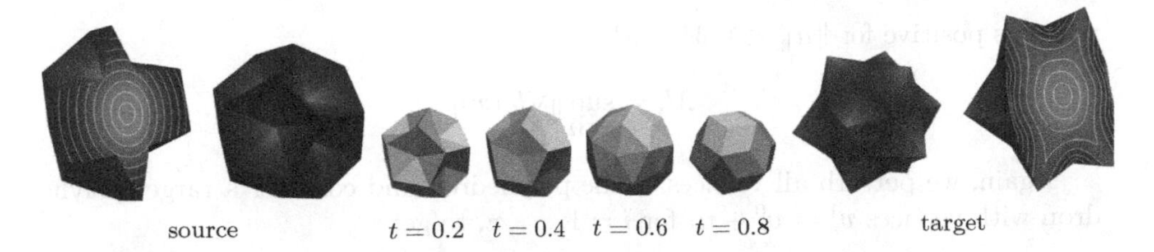

Figure 4.12 Example of a composite mean value mapping between two polyhedra. The color shows how the interior of the source polyhedron is morphed to the interior of the target polyhedron. See color insert.

is called a *composite barycentric mapping* from γ^0 to γ^1. An example of a composite barycentric mapping between two closed curves is shown in Figure 4.11.

Denoting the maximum displacement distance between γ^{t_k} and $\gamma^{t_{k+1}}$ by

$$d_k = \sup_{s\in[0,1]} \left\|\gamma^{t_k}(s) - \gamma^{t_{k+1}}(s)\right\|,$$

it follows from the previous result that $\boldsymbol{f}_\mathcal{T}$ is bijective if

$$d_\mathcal{T} = \max_{0\le k<m} d_k < \frac{\sqrt{5}-1}{2nM^*},$$

where

$$M^* = \sup_{s\in[0,1]} \sup_{t\in[0,1]} \sup_{\boldsymbol{x}\in\gamma^t} \left\|\nabla\phi^t(s,\boldsymbol{x})\right\|.$$

4.4.2 Polyhedra

A barycentric mapping between two polyhedra with the same number of vertices n and the same topology is defined as a function

$$\boldsymbol{f}\colon \bar{\Omega}^0 \to \bar{\Omega}^1, \qquad \boldsymbol{f}(\boldsymbol{x}) = \sum_{i=1}^{n} \phi_i(\boldsymbol{x})\boldsymbol{v}_i^1,$$

where $\boldsymbol{v}_i^1$ are the vertices of the target polyhedron and $\phi_i(\boldsymbol{x})$ are 3D barycentric coordinates (see Section 1.3). An example of such a mapping is illustrated in Figure 4.12.

As in the polygonal case we first perturb only one vertex and consider a target polyhedron with vertices $\boldsymbol{v}_i^1 = \boldsymbol{v}_i^0 + \boldsymbol{u}$ for some i and $\boldsymbol{v}_j^1 = \boldsymbol{v}_j^0$ for $j \neq i$. Then,

$$J_{\boldsymbol{f}}(\boldsymbol{x}) = \begin{vmatrix} 1+\partial_1\phi_i(\boldsymbol{x})u_1 & \partial_2\phi_i(\boldsymbol{x})u_1 & \partial_3\phi_i(\boldsymbol{x})u_1 \\ \partial_1\phi_i(\boldsymbol{x})u_2 & 1+\partial_2\phi_i(\boldsymbol{x})u_2 & \partial_3\phi_i(\boldsymbol{x})u_2 \\ \partial_1\phi_i(\boldsymbol{x})u_3 & \partial_2\phi_i(\boldsymbol{x})u_3 & 1+\partial_3\phi_i(\boldsymbol{x})u_3 \end{vmatrix} = 1 + \nabla\phi_i(\boldsymbol{x})\cdot\boldsymbol{u}$$

and

$$|J_{\boldsymbol{f}}(\boldsymbol{x})| \ge 1 - |\phi_i(\boldsymbol{x})\cdot\boldsymbol{u}| \ge 1 - \|\phi_i(\boldsymbol{x})\|\|\boldsymbol{u}\|,$$

which is positive for $\|\boldsymbol{u}\| < 1/M_i$ with

$$M_i = \sup_{\boldsymbol{x}\in\Omega} \|\nabla b_i(\boldsymbol{x})\|.$$

Again, we perturb all vertices of the polyhedron and consider a target polyhedron with vertices $\boldsymbol{v}_i^1 = \boldsymbol{v}_i^0 + \boldsymbol{u}_i$ for $i = 1, \ldots, n$, to get

$$\begin{aligned} J_{\boldsymbol{f}}(\boldsymbol{x}) &= \begin{vmatrix} 1+\sum_i \partial_1\phi_i(\boldsymbol{x})u_{i,1} & \sum_i \partial_2\phi_i(\boldsymbol{x})u_{i,1} & \sum_i \partial_3\phi_i(\boldsymbol{x})u_{i,1} \\ \sum_i \partial_1\phi_i(\boldsymbol{x})u_{i,2} & 1+\sum_i \partial_2\phi_i(\boldsymbol{x})u_{i,2} & \sum_i \partial_3\phi_i(\boldsymbol{x})u_{i,2} \\ \sum_i \partial_1\phi_i(\boldsymbol{x})u_{i,3} & \sum_i \partial_2\phi_i(\boldsymbol{x})u_{i,3} & 1+\sum_i \partial_3\phi_i(\boldsymbol{x})u_{i,3} \end{vmatrix} \\ &= 1 + \sum_i \nabla\phi_i(\boldsymbol{x}) \cdot \boldsymbol{u} + \sum_i \sum_j \sum_k \partial_1\phi_i(\boldsymbol{x})\,\partial_2\phi_j(\boldsymbol{x})\,\partial_3\phi_k(\boldsymbol{x})\,|(\boldsymbol{u}_i\ \boldsymbol{u}_j\ \boldsymbol{u}_k)| \\ &\quad + \sum_i \sum_j \Big(\partial_1\phi_i(\boldsymbol{x})\,\partial_2\phi_j(\boldsymbol{x})\,D_3 + \partial_1\phi_i(\boldsymbol{x})\,\partial_3\phi_j(\boldsymbol{x})\,D_2 + \partial_2\phi_i(\boldsymbol{x})\,\partial_3\phi_j(\boldsymbol{x})\,D_1\Big), \end{aligned}$$

where the sums range from 1 to n and D_k is the k-th component of $\boldsymbol{u}_i \times \boldsymbol{u}_j$. Therefore,

$$J_{\boldsymbol{f}}(\boldsymbol{x}) \geq 1 - Md - 3M^2d^2 - M^3d^3,$$

where

$$d = \max_{1\leq i\leq n} \|\boldsymbol{u}_i\| \qquad \text{and} \qquad M = M_1 + \cdots + M_n,$$

which is positive if $Md < \sqrt{2} - 1$, implying that $\boldsymbol{f}$ is injective for $d < \frac{\sqrt{2}-1}{M}$.

Now, suppose that $\boldsymbol{\psi}_i \colon [0,1] \to \mathbb{R}^3$, $i = 1, \ldots, n$ are a set of continuous *vertex paths* between each source vertex $\boldsymbol{v}_i^0 = \boldsymbol{\psi}_i(0)$ and its corresponding target vertex $\boldsymbol{v}_i^1 = \boldsymbol{\psi}_i(1)$. We define the *composite barycentric mapping* from $\bar{\Omega}^0$ to $\bar{\Omega}^1$ as

$$\boldsymbol{f}_\tau = \boldsymbol{f}_{m-1} \circ \boldsymbol{f}_{m-2} \circ \cdots \circ \boldsymbol{f}_0,$$

where $\boldsymbol{f}_k \colon \bar{\Omega}^{t_k} \to \bar{\Omega}^{t_{k+1}}$ are barycentric mappings based on the partition $\tau = (t_0, t_1, \ldots, t_m)$ of the interval $[0,1]$.

Denoting the maximum displacement distance between $\bar{\Omega}^{t_k}$ and $\bar{\Omega}^{t_{k+1}}$ by

$$d_k = \max_{1\leq i\leq n} \left\|\boldsymbol{v}_i^{t_k} - \boldsymbol{v}_i^{t_{k+1}}\right\|,$$

it follows that $\boldsymbol{f}_\tau$ is bijective if

$$d_\tau = \max_{0\leq k<m} d_k < \frac{\sqrt{2}-1}{2nM^*},$$

where

$$M^* = \max_{1\leq i\leq n} \sup_{t\in[0,1]} \sup_{\boldsymbol{x}\in\bar{\Omega}^t} \left\|\nabla\phi_i^t(\boldsymbol{x})\right\|.$$

The composite barycentric mapping $\boldsymbol{f}_\tau$ can also be extended to closed smooth surfaces, and by following a similar reasoning we can conclude that the mapping $\boldsymbol{f}_\tau$ is bijective if $d_\tau < (\sqrt{2} - 1)/(2nM^*)$.

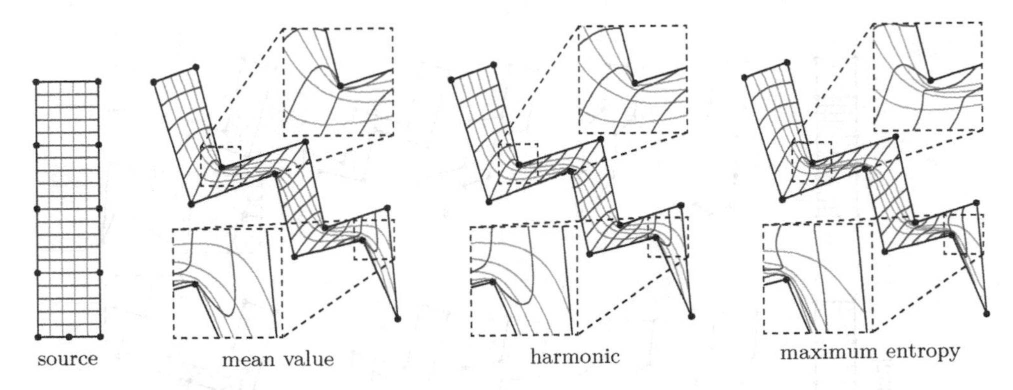

Figure 4.13 Composite barycentric map for different types of barycentric coordinates for 100 uniform steps.

4.5 PRACTICAL CONSIDERATIONS

Composite barycentric mappings depend on two main choices: the underlying barycentric coordinates ϕ and the vertex paths ψ. In this section we illustrate the influence of these choices on the composite mapping.

4.5.1 Choosing the coordinates

In general, the vertex paths ψ_i produce an arbitrary intermediate shape, which may be concave or weakly convex. For this reason we need barycentric coordinates that are well-defined for arbitrary simple polygons. We suggest using mean value (see Section 1.2.3), harmonic (see Section 1.2.8), or maximum entropy coordinates (see Section 1.2.9). Figure 4.13 shows an example of a composite barycentric mapping for these three different coordinates.

4.5.2 Choosing the vertex paths

To create intermediate shapes, the natural vertex path that works in any dimension and for any shape is to linearly interpolate the shapes from source to target. However, this interpolation is not invariant with respect to similarity transformations, while the mapping is. Moreover, linear interpolation of the vertices is more likely to produce self-intersecting polygons in the case of rotations, which invalidates the mapping, since barycentric coordinates are well-defined only for non-self-intersecting polygons. For these reasons, linear interpolation is not well-suited for creating the intermediate shapes for the composite barycentric mappings.

Instead, the intermediate polygons can be computed by linearly interpolating the turning angle and the edge lengths [352], which generates more natural results than linearly interpolating the vertices, because edges and angles are invariant with respect to similarity transformations. Figure 4.14 shows how the composite barycentric mapping changes when this method is used instead of linearly interpolating the

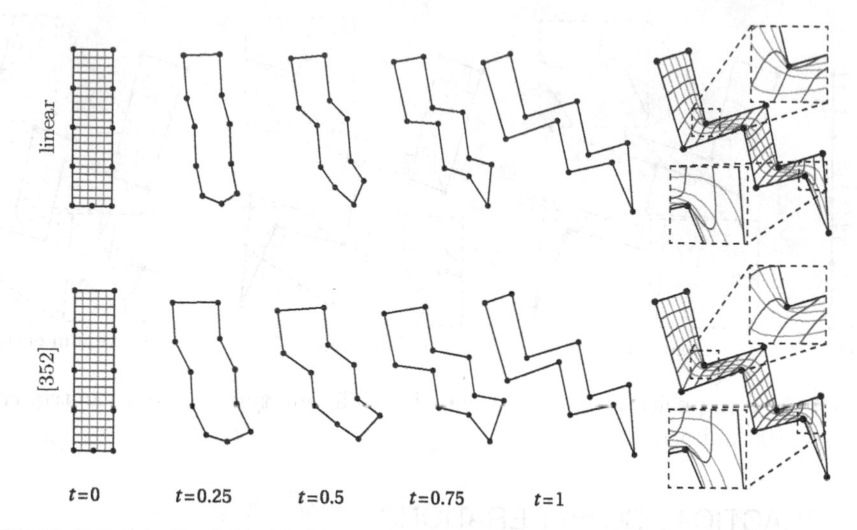

Figure 4.14 Example of a composite barycentric map for different vertex paths for 100 uniform steps.

vertices. A similar approach can be employed for closed planar curves where the curvature is interpolated between source and target curve [343]. An alternative approach consists of creating a multi-resolution representation of the two input curves (or polygons) and to interpolate between these multi-resolution representations to create the intermediate shape [168].

By combining both ideas in 3D we interpolate edges and dihedral angles and, to derive a globally coherent solution, we follow a hierarchical shape-matching approach [428]. An alternative and simpler approach involves decomposing the global interpolation into local affine transformations and splitting them into a rotational and a scale/shear part [381, 382]. While the scale/shear part can then be interpolated linearly, the rotational part should be treated in log-space. However, the problem is that this method cannot properly handle large global rotations.

Unfortunately, none of these methods guarantees that the intermediate shapes are intersection-free. To overcome this limitation in two dimensions, initial results deal with pairs of polygons that have corresponding parallel edges [180]. A more general method embeds the two polygons inside a convex region, generates a pair of compatible triangulations, and interpolates the stochastic matrices whose eigenvectors encode the geometry [173]. An alternative idea is to create the intermediate polygons by "unfolding" the source polygon and "refolding" it back to the target polygon [206].

CHAPTER 5

A Primer on Laplacians

Max Wardetzky

Georg-August-University Göttingen, Germany

CONTENTS

IN THIS CHAPTER we review some important properties of Laplacians, smooth and discrete. We place special emphasis on a unified framework for treating smooth Laplacians on Riemannian manifolds alongside discrete Laplacians on graphs and simplicial manifolds. We cast this framework into the language of linear algebra, with the intent to make this topic as accessible as possible. We combine perspectives from smooth geometry, discrete geometry, spectral analysis, machine learning, numerical analysis, and geometry processing within this unified framework. The connection to generalized barycentric coordinates is established through harmonic functions that interpolate given boundary conditions.

5.1 INTRODUCTION

The Laplacian is perhaps *the* prototypical differential operator for various physical phenomena. It describes, for example, heat diffusion, wave propagation, steady state fluid flow, and it is key to the Schrödinger equation in quantum mechanics. In Euclidean space, the Laplacian of a smooth function $u\colon \mathbb{R}^n \to \mathbb{R}$ is given as the sum of second partial derivatives along the coordinate axes,

$$\Delta u = -\left(\frac{\partial^2 u}{\partial x_1^2} + \frac{\partial^2 u}{\partial x_2^2} + \cdots + \frac{\partial^2 u}{\partial x_n^2}\right),$$

where we adopt the geometric perspective of using a minus sign.

5.1.1 Basic properties

The Laplacian has many intriguing properties. For the remainder of this exposition, consider an open and bounded domain $\Omega \subset \mathbb{R}^n$ and the L^2 inner product

$$(f, g) = \int_\Omega fg$$

on the linear space of square-integrable functions on Ω. Let $u, v\colon \Omega \to \mathbb{R}$ be two (sufficiently smooth) functions that vanish on the boundary of Ω. Then the Laplacian Δ is a *symmetric* (or, to be precise, a formally self-adjoint) linear operator with respect to this inner product, since integration by parts yields

$$\text{(SYM)} \qquad (u, \Delta v) = \int_\Omega \nabla u \cdot \nabla v = (\Delta u, v).$$

Here ∇ denotes the standard gradient operator and $\nabla u \cdot \nabla v$ denotes the standard inner product between vectors in $\mathbb{R}^n$. The choice of using a minus sign in the definition of the Laplacian makes this operator positive semidefinite, since

$$\text{(PSD)} \qquad (u, \Delta u) = \int_\Omega \nabla u \cdot \nabla u \geq 0.$$

If one considers only functions that vanish on the boundary of Ω, (PSD) implies that the only functions that lie in the kernel of the Laplacian ($\Delta u = 0$) are those functions that vanish on the entire domain. Moreover, properties (SYM) and (PSD) imply that the Laplacian can be diagonalized and its eigenvalues are nonnegative,

$$\Delta u = \lambda u \quad \Rightarrow \quad \lambda \geq 0.$$

Another prominent property of smooth Laplacians is the *maximum principle*. Let $u\colon \Omega \to \mathbb{R}$ be *harmonic*; that is, $\Delta u = 0$. The maximum principle asserts that

$$\text{(MAX)} \qquad u \text{ is harmonic} \quad \Rightarrow \quad u \text{ has no strict local maximum in } \Omega,$$

where we no longer assume that u vanishes on the boundary of Ω. Likewise, no harmonic function can have a local minimum in Ω.

The maximum principle can be derived from another important property of harmonic functions, the *mean value property*. Consider a point $\boldsymbol{x} \in \Omega$ and a closed ball $B(\boldsymbol{x}, r)$ of radius r centered at $\boldsymbol{x}$ that is entirely contained in Ω. Every harmonic function has the property that the value $u(\boldsymbol{x})$ can be recovered from the average of the values of u in the ball $B(\boldsymbol{x}, r)$:

$$u(\boldsymbol{x}) = \frac{1}{\text{vol}(B(\boldsymbol{x}, r))} \int_{B(\boldsymbol{x},r)} u(\boldsymbol{y}) d\boldsymbol{y}.$$

A simple argument by contradiction shows that the mean value property implies property (MAX).

The properties mentioned so far play an important role in applications; specifically, in the context of barycentric coordinates, they give rise to mean value coordinates (see Section 1.2.3) and harmonic coordinates (see Section 1.2.8). Below we discuss additional properties of Laplacians. For further reading we refer to the books [43, 140, 333] and the lecture notes [83, 109].

5.2 LAPLACIANS ON RIEMANNIAN MANIFOLDS

The standard Laplacian in $\mathbb{R}^n$ can be expressed as

$$\Delta u = -\text{div}\,\nabla u,$$

where div is the usual divergence operator acting on vector fields in $\mathbb{R}^n$. Written in integral form, the negative divergence operator is the (formal) adjoint of the gradient: If X is a vector field and $u \colon \Omega \to \mathbb{R}$ is a function that vanishes on the boundary of Ω, then

$$\int_\Omega \nabla u \cdot X = \int_\Omega u\,(-\text{div} X).$$

This perspective can be generalized to Riemannian manifolds, which incorporate the notion of curvature. The Laplacian plays an important role in the study of these curved spaces.

5.2.1 Exterior calculus

Although gradient and divergence can readily be defined on Riemannian manifolds, it is more convenient to work with the differential (or *exterior derivative*) d instead of the gradient and with the codifferential d^* instead of divergence.

The differential d is similar to (but not the same as) the gradient. Indeed, given a function $u \colon \Omega \to \mathbb{R}$, one has

$$du(X) = \nabla u \cdot X$$

for every vector field X. In particular, the differential does not require the notion of a metric, whereas the gradient does. The codifferential d^* is defined as the formal adjoint to d, in the same way as divergence is the adjoint of the gradient. In contrast to the divergence operator, which acts on vector fields, the codifferential d^* acts on 1-forms. A 1-form is a covector at every point of Ω; that is, if X is a vector field on

Ω and α is a 1-form, then $\alpha(X)$ is a real-valued function on Ω. In order to define the codifferential d^*, consider a 1-form α and a function $u\colon \Omega \to \mathbb{R}$ that vanishes on the boundary of Ω. Then

$$\int_\Omega du \cdot \alpha = \int_\Omega u\, d^*\alpha,$$

where the dot product is the inner product between covectors induced from the inner product between vectors. Notice that different from the differential d, the codifferential *does* require the notion of a metric. The Laplacian of a function u can be expressed as

$$\Delta u = d^* du,$$

which is equivalent to the representation $\Delta u = -\mathrm{div}\nabla u$ given above.

In order to carry over this framework to manifolds, let M be a smooth orientable manifold with smooth Riemannian metric g. Suppose for simplicity that M is compact and has an empty boundary. The Riemannian metric induces a pointwise inner product between tangent vectors on M, which, analogously to the above discussion, induces an inner product between 1-forms. More generally, one works with k-forms for $k \geq 0$. A 0-form, by convention, is a real-valued function on M. A 1-form can be thought of as an oriented 1-volume in the sense that applying a 1-form to a vector field returns a real value at every point. Likewise, a k-form for $k > 1$ can be thought of as an oriented k-volume in the sense of returning a real number at every point when applied to an ordered k-tuple (parallelepiped) of tangent vectors. As a consequence, k-forms can be integrated over (sub)manifolds of dimension k. In the sequel we let Λ^k denote the linear space of k-forms on M.

Analogous to the L^2 inner product between function in $\mathbb{R}^n$, let

$$(\alpha, \beta)_k = \int_M g(\alpha, \beta)\mathrm{vol}_g$$

denote the L^2 inner product between k-forms α and β on M, where, by slight abuse of notation, we let $g(\alpha, \beta)$ denote the (pointwise) inner product induced by the Riemannian metric.

The differential $d\colon \Lambda^k \to \Lambda^{k+1}$ maps k-forms to $(k+1)$-forms for $0 \leq k \leq \mathrm{dim}M$, where one sets $d\alpha = 0$ for any k-form with $k = \mathrm{dim}M$. One can define the differential acting on k-forms by postulating *Stokes's theorem*,

$$\int_U d\alpha = \int_{\partial U} \alpha,$$

for every k-form α and every (sufficiently smooth) submanifold $U \subset M$ of dimension $(k+1)$ with boundary ∂U. If one asserts this equality as the defining property of the differential d, then it immediately follows that $d \circ d = 0$ since the boundary of a boundary of a manifold is empty ($\partial(\partial U) = \emptyset$).

The codifferential d^*, taking $(k+1)$-forms back to k-forms, is the (formal) adjoint of d with respect to the L^2 inner products on k- and $(k+1)$-forms. It is defined by requiring that

$$(d\alpha, \beta)_{k+1} = (\alpha, d^*\beta)_k$$

for all k-forms α and all $(k+1)$-forms β. Finally, the *Laplace–Beltrami operator* $\Delta\colon \Lambda^k \to \Lambda^k$ acting on k-forms is defined as

$$\Delta\alpha = dd^*\alpha + d^*d\alpha.$$

Notice that this expression reduces to $\Delta u = d^*du$ for 0-forms (functions) on M. It follows almost immediately from the definition of the Laplacian that a k-form α is *harmonic* ($\Delta\alpha = 0$) if and only if α is closed ($d\alpha = 0$) and co-closed ($d^*\alpha = 0$).

From a structural perspective it is important to note that properties (SYM), (PSD), and (MAX) mentioned earlier remain true (among various other properties) in the setting of Riemannian manifolds. For further details on exterior calculus and the Laplace–Beltrami operator, we refer to [333].

5.2.2 Hodge decomposition

Every sufficiently smooth k-form α on M admits a unique decomposition

$$\alpha = d\mu + d^*\nu + h,$$

known as the *Hodge decomposition* (or Hodge–Helmholtz decomposition), where μ is a $(k-1)$-form, ν is a $(k+1)$-form, and h is a harmonic k-form. This decomposition is unique and orthogonal with respect to the L^2 inner product on k-forms,

$$0 = (d\mu, d^*\nu)_k = (h, d\mu)_k = (h, d^*\nu)_k,$$

which immediately follows from the fact that $d \circ d = 0$ and the fact that harmonic forms satisfy $dh = d^*h = 0$. The Hodge decomposition can be thought of as a (formal) application of the well-known fact from linear algebra that the orthogonal complement of the kernel of a linear operator is equal to the range of its adjoint (transpose) operator.

By duality between vector fields and 1-forms, the Hodge decomposition for 1-forms carries over to a corresponding decomposition for vector fields into curl-free and divergence-free components, which has applications for fluid mechanics [13] and Maxwell's equations for electromagnetism [155].

Geometrically, the Hodge decomposition establishes relations between the Laplacian and global properties of manifolds. Indeed, the linear space of harmonic k-forms is finite-dimensional for compact manifolds and isomorphic to $H^k(M;\mathbb{R})$, the k-th cohomology of M. As an application of this fact, consider a compact orientable surface without boundary. Then the dimension of the space of harmonic 1-forms is equal to twice the genus of the surface; that is, this dimension is zero for the 2-sphere, two for the two-dimensional torus, four for a genus-two surface (pretzel) and so on. Hence the Laplacian provides global information about the topology of the underlying space.

For a thorough treatment of Hodge decompositions, including the case of manifolds with boundary, we refer to [351].

5.2.3 The spectrum

One cannot speak about the Laplacian without discussing its spectrum. On a compact orientable manifold without a boundary, it follows from the inequality

$$(\Delta u, u)_0 = (du, du)_1 \geq 0$$

that the spectrum is nonnegative and that the only functions in the kernel of the Laplacian are constant functions. Thus zero is a trivial eigenvalue of the Laplacian with a one-dimensional space of eigenfunctions. The next (non-trivial) eigenvalue $\lambda_1 > 0$ is much more interesting. By the min-max principle, this eigenvalue satisfies

$$\lambda_1 = \min_{(u,1)_0=0} \frac{(du, du)_1}{(u, u)_0},$$

where one takes the minimum over all functions that are L^2-orthogonal to the constants. Higher eigenvalues can be obtained by successively applying the min-max principle to the orthogonal complements of the eigenspaces of lower eigenvalues.

The first non-trivial eigenvalue already tells a great deal about the geometry of the underlying Riemannian manifold. As an example, consider Cheeger's isoperimetric constant

$$\lambda_C = \inf_N \left\{ \frac{\mathrm{vol}_{n-1}(N)}{\min(\mathrm{vol}_n(M_1), \mathrm{vol}_n(M_2))} \right\},$$

where N runs over all compact codimension-1 submanifolds that partition M into two disjoint open sets M_1 and M_2 with $N = \partial M_1 = \partial M_2$. Intuitively, the optimal N for which λ_C is attained partitions M into two sets that have maximal volume and minimal perimeter. As an example, suppose that M has the shape of the surface of a smooth dumbbell. Then N is a curve going around the axis of the dumbbell at the location where the dumbbell is thinnest.

A relation of Cheeger's constant to the first non-trivial eigenvalue of the Laplacian is provided by the Cheeger inequalities

$$\frac{\lambda_C^2}{4} \leq \lambda_1 \leq c(K\lambda_C + \lambda_C^2),$$

where the constant c only depends on dimension and $K \geq 0$ provides a lower bound on the Ricci curvature of M in the sense that $Ric(M, g) \geq -K^2(n-1)$; see [79, 89]. Recall that for surfaces, Ricci curvature and Gauß curvature coincide. The first non-trivial eigenvalue of the Laplacian is thus related to the metric problem of minimal cuts—thus providing a relation between an analytical quantity (the first eigenvalue) and a purely geometric quantity (the Cheeger constant). Intuitively, if λ_1 is small, then M must have a small bottleneck; vice-versa, if λ_1 is large, then M is somewhat thick.

Equipped with the full set of eigenfunctions $\{\varphi_i\}$ of the Laplace–Beltrami operator, one can perform Fourier analysis on manifolds by decomposing any square-integrable function u into its modes,

$$u = \sum_i (u, \varphi_i)_0 \varphi_i,$$

provided that one chooses the eigenfunctions such that $(\varphi_i, \varphi_i) = \delta_{ij}$. (Notice that $(\varphi_i, \varphi_j)_0 = 0$ is automatic for eigenfunctions belonging to different eigenvalues.) The Fourier perspective is of great relevance in signal and geometry processing.

Maintaining a spectral eye on geometry, it is natural to ask the inverse question: How much *geometric* information can be reconstructed from information about the Laplacian? If the entire Laplacian is known on a smooth manifold, then one can reconstruct the metric, for example, by using the expression of Δ in local coordinates. If, however, "only" the spectrum is known, then less can be said in general. For example, Kac's famous question *Can one hear the shape of a drum?* [220], that is, whether the entire geometry can be inferred from the spectrum alone, has a negative answer: There exist isospectral but non-isometric manifolds [169, 383].

5.3 DISCRETE LAPLACIANS

Discrete Laplacians can be defined on simplicial manifolds or, more generally, on graphs. We treat the case of graphs first and discuss simplicial manifolds further below.

5.3.1 Laplacians on graphs

Consider an undirected graph $\Gamma = (V, E)$ with vertex set V and edge set E. For simplicity, we only consider finite graphs here. Suppose that every edge $c \in E$ between vertices $i \in V$ and $j \in V$ carries a real-valued weight $\omega_e = \omega_{ij} = \omega_{ji} \in \mathbb{R}$. We discuss below how weights can be chosen; suppose for now that such a choice has been made. A discrete Laplacian acting on a function $u\colon V \to \mathbb{R}$ is defined as

$$(Lu)_i = \sum_{j \sim i} \omega_{ij}(u_i - u_j), \tag{5.1}$$

where the sum ranges over all vertices j that are connected by an edge with vertex i.* This allows for representing the linear operator L as a matrix by

$$L_{ij} = \begin{cases} -\omega_{ij}, & \text{if there is an edge between } i \text{ and } j, \\ \sum_{k \sim i} \omega_{ik}, & \text{if } i = j, \\ 0, & \text{otherwise.} \end{cases}$$

The matrix L is called the discrete Laplace matrix. This definition may seem to come a bit out of the blue. In order to see how it relates to smooth Laplacians, consider again the smooth case and the quantity

$$E_D[u] = \frac{1}{2} \int_\Omega \|\nabla u\|^2,$$

*Some authors include a division by vertex weights in the definition of Laplacians on graphs. Such a division arises naturally when considering strongly defined Laplacians, instead of weakly defined Laplacians. We cover this distinction below.

which is known as the *Dirichlet energy* of u. In the discrete setup, one may discretize the gradient ∇u along an edge $e = (i, j)$ as the finite difference $(u_i - u_j)$. Accordingly, one defines discrete Dirichlet energy as

$$E_D[u] = \frac{1}{2} \sum_{e \in E} \omega_{ij} (u_i - u_j)^2,$$

where the sum ranges over all edges. One then has

$$\text{(discrete)}\ \ E_D[u] = \frac{1}{2} u^T L u \qquad \text{vs.} \qquad \text{(smooth)}\ \ E_D[u] = \frac{1}{2} (u, \Delta u)_0,$$

which justifies calling L a Laplace matrix. Due to this representation (and using the language of partial differential equations [140]) we call discrete Laplacians of the form (5.1) *weakly defined* instead of *strongly defined*. We return to this distinction below.

Weakly defined discrete Laplacians of the form (5.1) are always symmetric—they satisfy (SYM) due to the assumption that $\omega_{ij} = \omega_{ji}$. However, whether or not a discrete Laplacian satisfies (at least some of) the other properties of the smooth setting heavily depends on the choice of weights.

The simplest choice of weights is to set $\omega_{ij} = 1$ whenever there is an edge between vertices i and j. This results in the so-called *graph Laplacian*. The diagonal entries of the graph Laplacian are equal to the degree of the respective vertex, that is, the number of edges adjacent to that vertex. The graph Laplacian is just a special case of what we call a *Laplacian on graphs* here.

Positive edge weights are a natural choice if weights resemble transition probabilities of a random walker. Discrete Laplacians with positive weights are always positive semidefinite (PSD) and, just like in the smooth setting, they only have the constant functions in their kernel provided that the graph is connected. As a word of caution we remark that positivity of weights is *not* necessary to guarantee (PSD). Below we discuss Laplacians that allow for (some) negative edge weights but still satisfy (PSD).

Laplacians with positive edge weights always satisfy the mean value property since every harmonic function u (a function for which $Lu = 0$) satisfies

$$u_i = \sum_{j \sim i} l_{ij} u_j \qquad \text{with} \qquad l_{ij} = \frac{\omega_{ij}}{L_{ii}} > 0.$$

Therefore, discrete Laplacians with positive weights also satisfy the maximum principle (MAX) since $\sum_{j \sim i} l_{ij} = 1$ and thus u_i is a convex combination of its neighbors u_j for discrete harmonic functions. This convex combination property establishes a connection between Laplacians and barycentric coordinates via the partition of unity property (1.8).

5.3.2 The spectrum

As in the smooth case, one cannot discuss discrete Laplacians without mentioning their spectrum and their eigenfunctions, which provide a fingerprint of the structure of the underlying graph.

As an example, consider again Cheeger's isoperimetric constant. In order to define this constant in the discrete setting, consider a partitioning of Γ into two disjoint subgraphs Γ_1 and $\bar{\Gamma}_1$ such that the vertex set V of Γ is the disjoint union of the vertex sets V_1 of Γ_1 and $\bar{V}_1$ of $\bar{\Gamma}_1$. Here a subgraph of Γ denotes a graph whose vertex set is a subset of the vertex set of Γ such that two vertices in the subgraph are connected by an edge if and only if they are connected by an edge in Γ. For positive edge weights, the discrete Cheeger constant (sometimes called *conductance* of a weighted graph) is defined as

$$\lambda_C = \min\left\{\frac{\text{vol}(\Gamma_1, \bar{\Gamma}_1)}{\min(\text{vol}(\Gamma_1), \text{vol}(\bar{\Gamma}_1))}\right\},$$

where

$$\text{vol}(\Gamma_1, \bar{\Gamma}_1) = \sum_{i \in V_1, j \notin V_1} \omega_{ij} \quad \text{and} \quad \text{vol}(\Gamma_1) = \sum_{i \in V_1} \omega_{ii},$$

and similarly for $\text{vol}(\bar{\Gamma}_1)$. Here, $\omega_{ii} := L_{ii}$. Notice that for the case of the graph Laplacian, $\text{vol}(\Gamma_1, \bar{\Gamma}_1)$ equals the number of edges with one vertex in Γ_1 and another vertex in its complement.†

Similar to the smooth case, one then obtains the Cheeger inequalities

$$\frac{\lambda_C^2}{2} \leq \tilde{\lambda}_1 \leq 2\lambda_C,$$

where $\tilde{\lambda}_1$ is the first nontrivial eigenvalue of the rescaled Laplace matrix

$$\tilde{L} = SLS,$$

where S is a diagonal matrix with $S_{ii} = 1/\sqrt{\omega_{ii}}$. This rescaling is necessary since λ_C is invariant under a uniform rescaling of edge weights (and so is $\tilde{L}$), whereas L scales linearly with the edge weights. A proof of the discrete Cheeger inequalities for the case of the graph Laplacians ($\omega_{ij} = 1$) can be found in [104]; the proof for arbitrary positive edge weights is nearly identical.

The Cheeger constant—and alongside, the corresponding partitioning of Γ into the two disjoint subgraphs Γ_1 and $\bar{\Gamma}_1$—has applications in graph clustering, since the edges connecting Γ_1 and $\bar{\Gamma}_1$ tend to "cut" the graph along its bottleneck [222].

Any discrete Laplacian (having positive edge weights or not) that satisfies (SYM) and (PSD) can be used for Fourier analysis on graphs. Indeed, let $\{\varphi_i\}$ be the eigenfunctions of L, chosen such that $\varphi_i^T \varphi_j = \delta_{ij}$. Then one has

$$u = \sum_i (u^T \varphi_i)\varphi_i,$$

just like in the smooth setting. This decomposition of discrete functions on graphs

†Some authors use a different version of the respective volumes in the definition of the Cheeger constant, resulting in different versions of the Cheeger inequalities; see [384].

into their Fourier modes has a plethora of applications in geometry processing; see [242] and references therein.

As in the smooth setting, it is natural to ask the inverse question of how much geometric information can be inferred from information about the Laplacian. Recall that in the the smooth case, knowing the full Laplacian allows for recovering the Riemannian metric. Similarly, in the discrete case it can be shown that for simplicial surfaces the knowledge of the cotan weights for discrete Laplacians (covered below) allows for reconstructing edge lengths (i.e., the discrete metric) of the underlying mesh up to a global scale factor [432]. If, however, only the spectrum of the Laplacian is known, then there exist isospectral but non-isomorphic graphs [76]. In fact, for the case of the cotan Laplacian (see below), the exact same isospectral domains considered in [169] that were originally proposed for showing that "One cannot hear the shape of a drum" for smooth Laplacians work in the discrete setup.

A curious fact concerning the connection between discrete Laplacians and the underlying geometry is *Rippa's theorem* [326]: The Delaunay triangulation of a fixed point set in $\mathbb{R}^n$ minimizes the Dirichlet energy of any piecewise linear function over this point set. In [96], this result is taken a step further, where the authors show that the spectrum of the cotan Laplacian obtains its minimum on a Delaunay triangulation in the sense that the i-th eigenvalue of the cotan Laplacian of any triangulation of a fixed point set in the plane is bounded below by the i-th eigenvalue resulting from the cotan Laplacian associated with the Delaunay triangulation of the given point set.

5.3.3 Laplacians on simplicial manifolds

Recall that in the smooth case, Laplacians acting on k-forms take the form

$$\Delta = dd^* + d^*d.$$

In order to mimic this construction in the discrete setting, one requires a bit more structure than just an arbitrary graph. To this end, consider a simplicial manifold, such as a triangulated surface. We keep referring to this manifold as M. As in the smooth case, for simplicity, suppose that M is orientable and has no boundary.

Simplicial manifolds allow for a natural definition of discrete k-forms as duals of k-cells. Indeed, every 0-form is a function defined on vertices, a 1-form α is dual to edges, that is, $\alpha(e)$ is a real number for any oriented 1-cell (oriented edge), a 2-form is dual to oriented 2-cells, and so forth. In the sequel, let the linear space of discrete k-forms (better known as simplicial cochains) be denoted by C^k.

As in the smooth case, the discrete differential δ (better known as the coboundary operator) maps discrete k-forms to discrete $(k+1)$-forms, $\delta\colon C^k \to C^{k+1}$. Again as in the smooth case, the discrete differential can be defined by *postulating* Stokes's formula: Let α be a discrete k-form. Then one requires that

$$\delta\alpha(\sigma) = \alpha(\partial\sigma)$$

for all $(k+1)$-cells σ, where ∂ denotes the simplicial boundary operator. The simplicial boundary operator, when applied to a vertex returns zero (since vertices do not have a boundary). When applied to an oriented edge, the boundary operator returns the difference between the edge's vertices. Likewise, ∂ applied to an oriented 2-cell σ returns the sum of oriented edges of σ (with the orientation induced by that of σ). Since the boundary of a boundary is empty ($\partial \circ \partial = 0$) one has $\delta \circ \delta = 0$, just like in the smooth case. Notice that the definition of δ does not require the notion of inner products.

In order to define the discrete codifferential δ^* one additionally requires an inner product $(\cdot, \cdot)_k$ on the linear space of k-forms for each k. Below we discuss the construction of such inner products. Given a fixed choice of inner products on discrete k-forms, the codifferential is defined by requiring that

$$(\delta\alpha, \beta)_{k+1} = (\alpha, \delta^*\beta)_k$$

for all k-forms α and all $(k+1)$-forms β, and the discrete *strongly defined* Laplacian acting on k-forms takes the form

$$\mathbb{L} = \delta\delta^* + \delta^*\delta. \tag{5.2}$$

This perspective is that of *discrete exterior calculus* (DEC) [109, 118], where—by slight abuse of notation—inner products are referred to as "discrete Hodge stars."

Strongly defined Laplacians are self-adjoint with respect to the inner products $(\cdot, \cdot)_k$ on discrete k-forms, since

$$(\mathbb{L}\alpha, \beta)_k = (\delta\alpha, \delta\beta)_{k+1} + (\delta^*\alpha, \delta^*\beta)_{k-1} = (\alpha, \mathbb{L}\beta)_k.$$

Moreover, strongly defined Laplacians are always positive semidefinite (PSD), since

$$(\mathbb{L}\alpha, \alpha)_k = (\delta\alpha, \delta\alpha)_{k+1} + (\delta^*\alpha, \delta^*\alpha)_{k-1} \geq 0.$$

In particular, a discrete k-form is harmonic ($\mathbb{L}\alpha = 0$) if and only if $\delta\alpha = \delta^*\alpha = 0$, just like in the smooth setting.

5.3.4 Strongly and weakly defined Laplacians

Every strongly defined Laplacian acting on 0-forms has a weakly defined cousin L acting on discrete functions u. The weak version is obtained by requiring that at every vertex i the resulting function Lu is equal to

$$(Lu)_i = (\mathbb{L}u, 1_i)_0 = (\delta^*\delta u, 1_i)_0 = (\delta u, \delta 1_i)_1,$$

where 1_i is the indicator function of vertex i. In particular, let $\mathbb{M}_0$ and $\mathbb{M}_1$ be the symmetric positive definite matrices that encode the inner products between 0-forms and 1-forms, respectively,

$$(u, v)_0 = u^T \mathbb{M}_0 v \qquad \text{and} \qquad (\alpha, \beta)_1 = \alpha^T \mathbb{M}_1 \beta.$$

Then the weakly and strongly defined Laplacians satisfy, respectively,

$$L = \delta^T \mathbb{M}_1 \delta \qquad \text{and} \qquad \mathbb{L} = \mathbb{M}_0^{-1} L.$$

As an example, consider diagonal inner products on 0-forms and 1-forms,

$$(u,v)_0 = \sum_{i\in V} m_i u_i v_i \qquad \text{and} \qquad (\alpha,\beta)_1 = \sum_{e\in E} \omega_e \alpha(e)\beta(e),$$

with positive vertex weights $m_i > 0$ and positive edge weights $\omega_e > 0$. The resulting strongly defined Laplacian acting on 0-forms (functions) takes the form

$$(\mathbb{L}u)_i = \frac{1}{m_i}\sum_{j\sim i} \omega_{ij}(u_i - u_j)$$

and its associated weakly defined cousin is the Laplacian on graphs defined in (5.1).

5.3.5 Hodge decomposition

Given a choice of inner products for k-forms on simplicial manifolds, one always obtains a discrete Hodge decomposition. Indeed, for every discrete k-form α one has

$$\alpha = \delta\mu + \delta^*\nu + h,$$

where μ is a $(k-1)$-form, ν is a $(k+1)$-form and h is a harmonic k-form ($\mathbb{L}h = 0$). As in the smooth case, not only is this decomposition unique, it is also orthogonal with respect to the inner products on k-forms,

$$0 = (\delta\mu, \delta^*\nu)_k = (h, \delta\mu)_k = (h, \delta^*\nu)_k,$$

which immediately follows from $\delta\circ\delta = 0$ and the fact that harmonic forms satisfy $\delta h = \delta^* h = 0$.

Akin to the smooth case, the Hodge decomposition establishes relations to global properties of simplicial manifolds, since the linear space of harmonic k-forms is isomorphic to the k-th simplicial cohomology of the simplicial manifold M. Again, as an application of this fact, consider a compact simplicial surface without a boundary. Then the dimension of the space of harmonic 1-forms is equal to twice the genus of the surface—independent of the concrete choice of inner products on k-forms.

5.3.6 The cotan Laplacian and beyond

We conclude the discussion of Laplacians on simplicial manifolds by providing an important example of inner products. In [426], Whitney constructs a map from simplicial k-forms (k-cochains) to piecewise linear differential k-forms. In a nutshell, the idea is to linearly interpolate simplicial k-forms across full-dimensional cells. As the simplest example, consider linear interpolation of 0-forms (functions) on vertices. This kind of interpolation can be extended to arbitrary k-forms. The resulting map

$$W\colon C^k \to L^2\Lambda^k$$

takes simplicial k-forms to square-integrable k-forms on the simplicial manifold, where we assume each simplex carries the standard Euclidean structure. The Whitney map is the right inverse of the so-called de Rham map,

$$\alpha(\sigma) = \int_\sigma W(\alpha)$$

for all discrete k-forms α and all k-cells σ. For details we refer to [426]. The Whitney map W is a chain map—it commutes with the differential ($dW = W\delta$) and thus factors to cohomology.

Dodziuk and Patodi [124] use the Whitney map in order to define an inner product on discrete k-forms (k-cochains) by

$$(\alpha, \beta)_k = \int_M g(W\alpha, W\beta)\mathrm{vol}_g,$$

where (in our case) g denotes a piecewise Euclidean metric on the simplicial manifold. From the perspective of the *Finite Element Method* (FEM), Whitney's construction is a special case of constructing stable finite elements; see [12].

For triangle meshes the resulting strongly defined Laplacian acting on 0-forms (functions) is given by

$$\mathbb{L} = \mathbb{M}_0^{-1} L,$$

where $\mathbb{M}_0$ is the *mass matrix* given by

$$(\mathbb{M}_0)_{ij} = \begin{cases} \frac{A_{ij}}{12}, & \text{if } i \sim j, \\ \frac{A_i}{6}, & \text{if } i = j, \\ 0, & \text{otherwise.} \end{cases}$$

Here A_{ij} denotes the combined area of the two triangles incident to edge (i, j) and A_i is the combined area of all triangles incident to vertex i. The corresponding weakly defined Laplacian L is the *stiffness matrix* with entries

$$L_{ij} = \begin{cases} -\frac{1}{2}(\cot \alpha_{ij} + \cot \beta_{ij}), & \text{if } i \sim j, \\ -\sum_{j \sim i} L_{ij}, & \text{if } i = j, \\ 0, & \text{otherwise,} \end{cases}$$

where α_{ij} and β_{ij} are the two angles opposite to edge (i, j). The matrix L is often referred to as the *cotan Laplacian*.

The cotan Laplacian has been rediscovered many times in different contexts [130, 134, 309]; the earliest explicit mention seems to go back to MacNeal [262], but perhaps it was already known at the time of Courant. The cotan Laplacian has been enjoying a wide range of applications in geometry processing (see [109, 242, 341] and references therein), including barycentric coordinates (see in particular Section 1.2.2 in this book), mesh parameterization, mesh compression, fairing, denoising, spectral fingerprints, shape clustering, shape matching, physical simulation of thin structures, and geodesic distance computation.

The construction of discrete Laplacians based on inner products can be extended from simplicial surfaces to meshes with (not necessarily planar) polygonal faces; see [8] for such an extension that is similar to the approach considered in [73, 74] for planar polygons. The cotan Laplacian has furthermore been extended to semi-discrete surfaces [84] as well as to subdivision surfaces [114].

5.3.7 Discrete versus smooth Laplacians

Structural properties of discrete Laplacians play an important role for computations; for example, when solving the Poisson problem

$$\mathbb{L}u = f$$

for a given right-hand side f and an unknown function u with Dirichlet boundary conditions. For solving this problem, one often prefers to work with the weak formulation of the problem ($Lu = \mathbb{M}_0 f$) instead of the strong one ($\mathbb{L}u = f$) since the weakly defined cousins of strongly defined Laplacians satisfy properties (SYM) and (PSD), which allows for efficient linear solvers for this problem.

We now turn to the question of which properties are desirable for discrete Laplacians on top of (SYM) and (PSD) and if it is furthermore possible to recover *all* properties of smooth Laplacians in the discrete case. We follow [410], on which this section is largely based.

Smooth Laplacians are differential operators that act *locally*. Locality can be represented in the discrete case by working with (weakly defined) discrete Laplacians based on edge weights:

(LOC) vertices i and j do not share an edge $\Rightarrow$ $\omega_{ij} = 0$.

This property reflects locality of action by ensuring that if vertices i and j are not connected by an edge, then changing the function value u_j at a vertex j does not alter the value $(Lu)_i$ at vertex i. Property (LOC) results in sparse matrices, which can be treated efficiently in computations.

Keeping an eye on the relation between discrete Laplacians and barycentric coordinates, it is desirable (or even mandatory) to require *linear reproduction*; see requirement (1.9) in Chapter 1. In the smooth setting, linear reproduction corresponds to the fact that linear functions on $\mathbb{R}^n$ are in the kernel of the standard Laplacian on Euclidean domains. For discrete Laplacians on a graph Γ, linear reproduction means that $(Lu)_i = 0$ at each interior vertex whenever Γ is embedded into the plane with straight edges and u is a *linear* function on the plane, point-sampled at the vertices of Γ,

(LIN) $\Gamma \subset \mathbb{R}^2$ embedded and $u\colon \mathbb{R}^2 \to \mathbb{R}$ linear $\Rightarrow$ $(Lu)_i = 0$ at interior vertices.

In applications this property is desirable for denoising of surface meshes [120] (where one expects to remove normal noise only but not to introduce tangential vertex drift), mesh parameterization [148] (where one expects planar regions to remain invariant under parameterization), and plate bending energies [341] (which must vanish for flat configurations).

Furthermore, it is often natural and desirable to require positive edge weights:

$$\text{(Pos)} \qquad \text{vertices } i \text{ and } j \text{ share an edge} \quad \Rightarrow \quad \omega_{ij} > 0.$$

This requirement implies (PSD) and is a sufficient (but by no means necessary) condition for a *discrete maximum principle* (MAX). In diffusion problems corresponding to $u_t = -\Delta u$, (POS) ensures that flow travels from regions of higher potential to regions of lower potential. Additionally, as discussed above, positive edge weights establish a connection to barycentric coordinates.

The combination (LOC)+(SYM)+(POS) is related to Tutte's embedding theorem for planar graphs [172, 400]: Positive weights associated to edges yield a straight-line embedding of an abstract planar graph for a fixed convex boundary polygon. Tutte's embedding is unique for a given set of positive edge weights, and it satisfies (LIN) by construction since each interior vertex (and therefore its x- and y-coordinate) is a convex combination of its adjacent vertices with respect to the given edge weights.

It is natural to ask which of the properties (LOC), (SYM), (POS), and (LIN) are satisfied by the discrete Laplacians considered so far.

As the simplest example, consider again the graph Laplacian ($\omega_{ij} = 1$). This Laplacian clearly satisfies (LOC)+(SYM)+(POS), but in general fails to satisfy (LIN). Next, consider the cotan Laplacian. The edge weights of the cotan Laplacian turn out to be a special case of weights arising from *orthogonal duals*. To define those, consider a graph embedded into the plane with straight edges that do not cross. An orthogonal dual is a realization of the dual graph in the plane, with straight edges orthogonal to primal edges (viewed as vectors in the plane). Edge weights can then be constructed as the ratio between the *signed* lengths of dual edges and the unsigned lengths of primal edges,

$$\omega_e = \frac{|\star e|}{|e|}.$$

Here, $|e|$ denotes the usual Euclidean length, whereas $|\star e|$ denotes the *signed* Euclidean length of the dual edge. The sign is obtained as follows. The dual edge $\star e$ connects two dual vertices $\star f_1$ and $\star f_2$, corresponding to the primal faces f_1 and f_2, respectively. The sign of $|\star e|$ is positive if along the direction of the ray from $\star f_1$ to $\star f_2$, the primal face f_1 lies before f_2. The sign is negative otherwise.

For the cotan Laplacian, the dual graph is obtained by connecting circumcenters of (primal) triangles by straight edges; see Figure 5.1 (left). More generally, discrete Laplacians derived from orthogonal duals on arbitrary (including non-planar) triangular surfaces were considered in [166]. These Laplacians always satisfy (LOC)+(SYM)+(LIN) for the case of planar primal graphs, but fail to satisfy (POS) in general. Indeed, for the cotan Laplacian one has $(\cot \alpha_{ij} + \cot \beta_{ij}) > 0$ if and only if $(\alpha_{ij} + \beta_{ij}) < \pi$. This is the case for all edges (i, j) if and only if the triangulation is Delaunay. One may restore (POS) by successive edge flips (thereby changing the

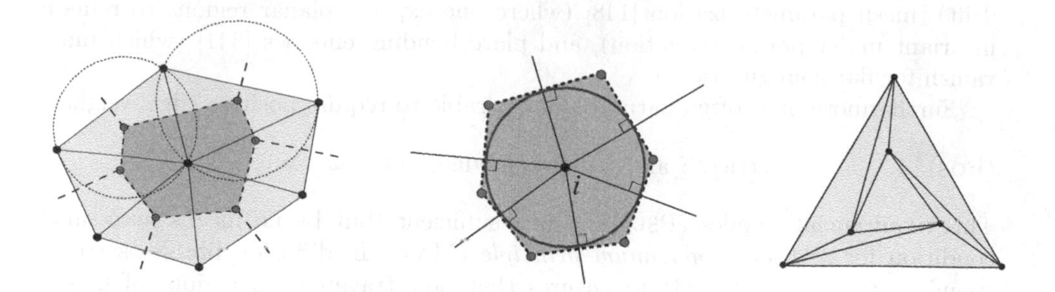

Figure 5.1 Left: Primal graph (solid lines) and orthogonal circumcentric dual graph (dashed lincs) defining the cotan Laplacian. Middle: Mean value weights correspond to dual edges tangent to the unit circle around a primal vertex. Right: The projection of the Schönhardt polytope does not allow for a discrete Laplacian satisfying (SYM)+(LOC)+(LIN)+(POS).

combinatorics) until one arrives at a Delaunay triangulation [57]. Unfortunately, the number of required edge flips to obtain a Delaunay triangulation from an arbitrary given triangulation cannot be bounded a priori. Therefore, this approach fails to satisfy locality in general.

More generally, so-called *weighted Delaunay triangulations* turn out to be the only triangulations that possess positive orthogonal duals and thus admit discrete Laplacians that satisfy (LOC)+(SYM)+(LIN)+(POS). Like Rippa's theorem (see above) this fact provides an instance of the intricate connection between properties of discrete differential operators (the Laplacian) and purely geometric properties (weighted Delaunay triangulations).

Finally, if one drops the requirement of symmetric edge weights, then one enters the realm of barycentric coordinates. In this case one may still obtain an orthogonal *dual face per primal vertex*, but these dual faces no longer fit into a consistent dual graph; see Figure 5.1 (middle). Hence, for the case of dual edges with positive lengths, one obtains edge weights satisfying (LOC)+(LIN)+(POS) but not (SYM). Floater et al. [149] explored a subspace of this case: a one-parameter family of linear precision barycentric coordinates, including mean value and Wachspress coordinates (see Section 1.2.4). Langer et al. [232] showed that each member of this family corresponds to a specific choice of orthogonal dual face per primal vertex.

The fact that no discrete Laplace operator satisfies all of the desired properties simultaneously is not a coincidence. It can be shown that general simplicial meshes do not allow for discrete Laplacians that satisfy (LOC)+(SYM)+(LIN)+(POS); see [410] for more on this topic and Figure 5.1 (right) for a simple example. This limitation provides a taxonomy on existing literature and explains the plethora of existing discrete Laplacians: Since not all desired properties can be fulfilled simultaneously, it depends on the application at hand to design discrete Laplacians that arc tailored towards the specific needs of a concrete problem.

Another important desideratum is *convergence*: In the limit of refinement of simplicial manifolds that approximate a smooth manifold, one seeks to approximate the smooth Laplacian by a sequence of discrete ones. For applications this is important in terms of obtaining discrete operators that are as mesh independent as possible—re-meshing a given shape should not result in a drastically different Laplacian.

A closely related concept to convergence is *consistency*. A sequence of discrete Laplacians $(\Delta_n)_{n\in\mathbb{N}}$ is called consistent if $\Delta_n u \to \Delta u$ for all appropriately chosen functions u. For example, it can be shown that Laplacians on point clouds, such as those considered in [35] are consistent; see [122].

Convergence is more difficult to show than consistency since it additionally requires that the solutions u_n to the Poisson problems $\Delta_n u_n = f$ converge (in an appropriate norm) to the solution u of $\Delta u = f$. Discussing convergence in detail is beyond the scope of this short survey. Roughly speaking, Laplacians on simplicial manifolds converge to their smooth counterparts (in an appropriate operator norm) if the inner products on discrete k-forms used for defining simplicial Laplacians converge to the inner products on smooth k-forms. In this case, one obtains convergence of solutions to the Poisson problem, convergence of the components of the Hodge decomposition, convergence of eigenvalues and eigenfunctions [124, 134, 191, 409, 427], and (using different techniques) convergence of Cheeger cuts [398].

II

Applications in Computer Graphics

CHAPTER 6

Mesh Parameterization

Bruno Lévy

Inria Nancy Grand Est, Villers-lès-Nancy, France

CONTENTS

THIS CHAPTER introduces mesh parameterization, an application of generalized barycentric coordinates. By "unfolding" a surface onto a 2D space, mesh parameterization has many possible applications, such as texture mapping. In this chapter we present some basic notions of topology that characterize the class of surfaces that admit a parameterization. Then we focus on the specific case of a topological disk with its boundary mapped to a convex polygon. In this setting, Tutte's barycentric mapping theorem not only gives sufficient conditions, but also a practical algorithm to compute a parameterization. We outline the main argument of the simple and elegant proof of Tutte's theorem by Gortler, Gotsman, and Thurston [172]. It is remarkable that their proof solely uses basic topological notions together with a counting argument. Finally, we mention the importance of the weights in the quality of the result, and demonstrate how mean value coordinates can be used to reduce the distortions.

6.1 INTRODUCTION

Before diving into the formal definition, we first give an intuitive explanation of what a mesh parameterization is. Intuitively, given a triangulated mesh (see Figure 6.1, left), a parameterization "unfolds" the mesh in 2D (see Figure 6.1, right). Such an "unfolding" is an interesting representation that can be considered as a "map" of the 3D object, and that can be exploited by several applications.

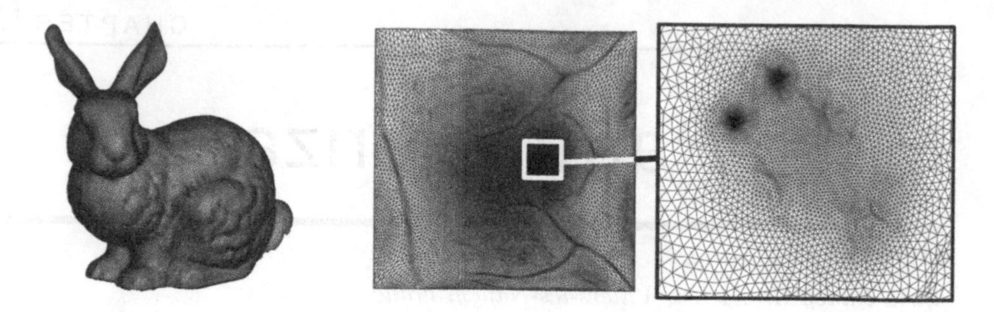

Figure 6.1 Left: a 3D mesh. (Model courtesy of Stanford University.) Right: a 2D parameterization of the 3D mesh and a zoom into the region of the head.

More formally, if we denote by $\boldsymbol{p}_i = (x_i, y_i, z_i)$ the 3D position of the mesh vertex $\mathbf{v}_i$, $i = 1, \ldots, n_{\mathrm{V}}$, where n_{V} denotes the number of vertices, then a parameterization of the mesh is a set of two additional coordinates $\boldsymbol{u}_i = (u_i, v_i)$ associated with each mesh vertex.

In addition, these coordinates need to be such that there is no intersection between the triangles in 2D. This chapter presents a method to generate such (u_i, v_i) coordinates. Before entering the heart of the matter, to motivate the subject, we present several applications of mesh parameterization. We suppose for now that the (u_i, v_i) coordinates are already given; how to compute them will be explained later.

6.2 APPLICATIONS OF MESH PARAMETERIZATION

Texture mapping, that is, "wrapping" images around 3D objects, is one of the main applications of mesh parameterization (see Figure 6.2). Texture mapping uses a one-to-one correspondence between the 3D mesh and a 2D domain (referred to as *parametric space*). It is possible to define such a correspondence by linearly interpolating the (u_i, v_i) coordinates associated with each vertex of the triangulation. Suppose you want to know the 3D point that corresponds to a given point $\boldsymbol{u} = (u, v)$ in parameter space. If the point $\boldsymbol{u}$ corresponds to one of the vertices $\boldsymbol{u}_i = (u_i, v_i)$, then it is mapped to $\boldsymbol{p}_i = (x_i, y_i, v_i)$. If the point $\boldsymbol{u}$ is arbitrary, then it is mapped as follows:

1. Determine the triangle with 2D vertices $\boldsymbol{u}_i = (u_i, v_i), \boldsymbol{u}_j = (u_j, v_j), \boldsymbol{u}_k = (u_k, v_k)$ that contains the point $\boldsymbol{u} = (u, v)$;
2. determine the *barycentric coordinates* (see Section 1.1.1) $\phi_1(\boldsymbol{u}), \phi_2(\boldsymbol{u}), \phi_3(\boldsymbol{u})$ of the point $\boldsymbol{u}$ with respect to the triangle $[\boldsymbol{u}_i, \boldsymbol{u}_j, \boldsymbol{u}_k]$;
3. then $\boldsymbol{u}$ is mapped to $\phi_1(\boldsymbol{u})\boldsymbol{p}_i + \phi_2(\boldsymbol{u})\boldsymbol{p}_j + \phi_3(\boldsymbol{u})\boldsymbol{p}_k$.

Conversely, for an arbitrary 3D point $\boldsymbol{p} = (x, y, z)$ on the mesh, contained in the triangle $t = [\boldsymbol{p}_i, \boldsymbol{p}_j, \boldsymbol{p}_k]$, one can retrieve its barycentric coordinates $\phi_1(\boldsymbol{p})$, $\phi_2(\boldsymbol{p})$, $\phi_3(\boldsymbol{p})$ with respect to t by working in the supporting plane of t, and map it to the

Figure 6.2 A mesh parameterization can be used to "wrap" a 3D object with an image. The used image is a regular grid (left) and a "fur" texture (right). (Model courtesy of Stanford University and *fur* texture courtesy of `www.myfreetextures.com`, used under the Creative Commons license.)

point $\boldsymbol{u} = \phi_1(\boldsymbol{p})\boldsymbol{u}_i + \phi_2(\boldsymbol{p})\boldsymbol{u}_j + \phi_3(\boldsymbol{p})\boldsymbol{u}_k$. One can check that this procedure defines the inverse of the $(u, v) \rightarrow (x, y, z)$ mapping defined above.

Equipped with this extended, linearly-interpolated mapping, we now have a one-to-one correspondence between all the 3D points of the surface and all the 2D points of the parametric domain. This correspondence can be used for many different applications. For instance, *texture mapping* consists of drawing an image in parametric space, and transporting it onto the surface through the linearly-interpolated mapping. An example is shown in Figure 6.2. This technique is widely used both in the movie industry and in interactive 3D graphics (visualization, video games). In this latter context, graphics processing units (GPUs) have very efficient optimized silicon for evaluating barycentric coordinates and linearly interpolating (u, v) coordinates. Texture mapping can also be used to associate a high-resolution attribute to a coarse mesh, as shown in Figure 6.3. By encoding the fine-scale

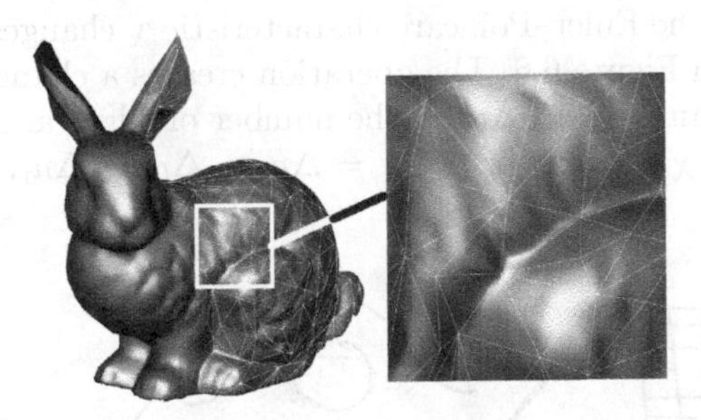
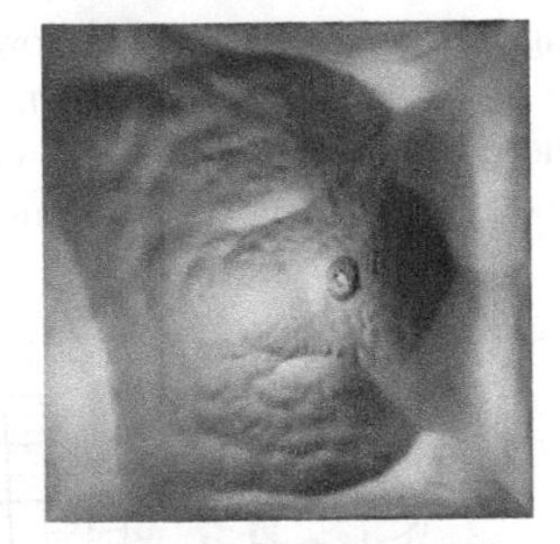

Figure 6.3 Application of mesh parameterization: normal mapping. A coarse mesh (left, compare with Figure 6.1) is textured-mapped with an image that encodes the fine geometric details of the initial high-resolution mesh (right).

geometric details in an image (here the normal vectors), it is possible to "replace triangles with pixels," and substantially lower rendering cost and time.

6.3 NOTIONS OF TOPOLOGY

Clearly, not all surfaces admit a parameterization. For instance, a sphere cannot be mapped to the plane without cutting it or poking a hole (see Figure 6.4, left). Therefore, it is a necessary condition that the surface has a boundary. However, it is not a sufficient condition. Consider the surface in Figure 6.4 (right): due to the handle, it is impossible to unfold it to 2D without intersections. Therefore, it is necessary to be able to determine whether a given mesh has a handle or not. The number of handles of a surface is referred to as its *genus*. A sphere (see Figure 6.5 A, B), has genus 0, a torus (see Figure 6.5 C), has genus 1, a double torus (see Figure 6.5 E), has genus 2.

The genus (denoted by g in what follows) is a global property of surfaces that is at first sight non-trivial to determine. Fortunately, for a closed connected surface, it can be deduced from the Euler–Poincaré characteristic χ, another quantity that is very easy to compute, given by $\chi = n_\mathrm{v} - n_\mathrm{e} + n_\mathrm{f}$, where n_v denotes the number of vertices, n_e the number of edges, and n_f the number of facets. Both the Euler–Poincaré characteristic and the genus are *topological invariants* that depend neither on the shape of the surface nor on the way it is decomposed into polygons. For instance, consider the sphere in Figure 6.5 (A), decomposed into 8 vertices, 12 edges, and 6 facets in a cube-like manner. Its Euler–Poincaré characteristic is $\chi = 8 - 12 + 6 = 2$. Considering now the tetrahedron-like decomposition of the sphere in Figure 6.5 (B), one gets $\chi = 4 - 6 + 4 = 2$ again. Whatever the considered decomposition of the sphere, one always gets $\chi = 2$. Let us take a look now at the torus (see Figure 6.5 C), decomposed as shown in Figure 6.5 (D). It has 16 vertices, 32 edges, and 16 faces, thus $\chi = 16 - 32 + 16 = 0$.

Now we would like to know the Euler–Poincaré characteristic of a double torus (see Figure 6.5 E), that is, a surface of genus $g = 2$. One possibility would be to "mesh" the double torus and count the number of elements. Another more interesting method is to analyze how the Euler–Poincaré characteristic χ changes when one creates a handle, as shown in Figure 6.6. The operation creates a change Δn_v of the number of vertices and changes $\Delta n_\mathrm{e}, \Delta n_\mathrm{f}$ of the number of edges and faces, respectively. In turn, χ becomes $\chi + \Delta\chi$, where $\Delta\chi = \Delta n_\mathrm{v} - \Delta n_\mathrm{e} + \Delta n_\mathrm{f}$. In the

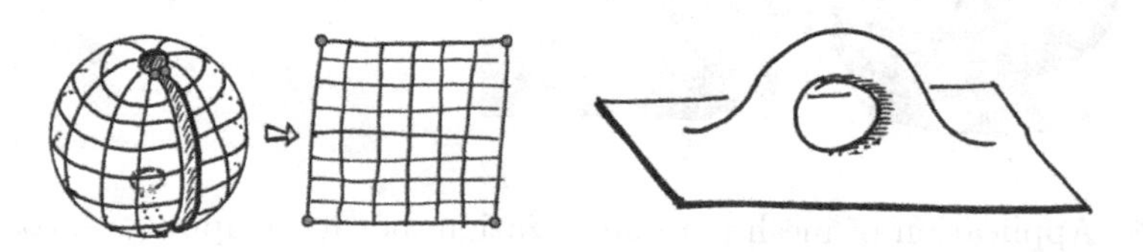

Figure 6.4 Some surfaces cannot be unfolded to the plane without cutting.

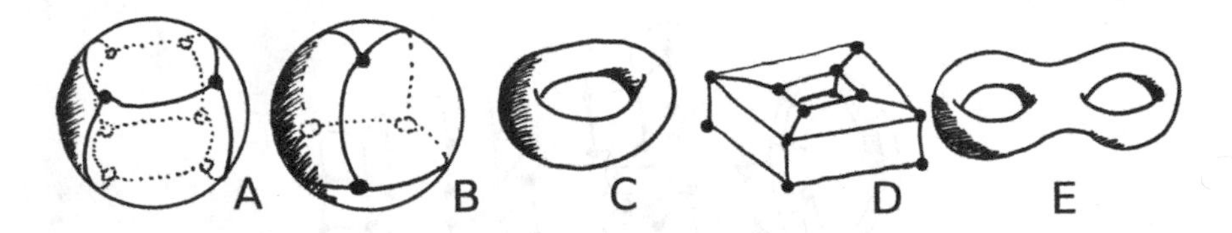

Figure 6.5 Determining the genus (number of handles) using the Euler–Poincaré characteristic.

first step (A), two faces are deleted, therefore n_f is decreased twice ($\Delta n_f = -2$), as well as $\chi = n_v - n_e + n_f$. In the second step (B), the two boundary components created by deleting the two faces are merged. This merges the vertices and the edges on the boundary components, therefore both n_v and n_e are decreased by the same number (there are the same number of vertices and edges on both boundary components; that is, four in the present example). The changes of n_v and n_e cancel out when taken into account in $\Delta\chi$, therefore the second step does not change χ. To summarize, this operation decreases χ by two. Note that this is completely independent from the number of vertices in the two facets that are removed; they just need to match. As a consequence, starting from a sphere that has genus zero and Euler–Poincaré characteristic $\chi = 2$ (see Figure 6.5), each time one creates a handle, g is increased by one (by the definition of g) and χ is decreased by two. To summarize, for a connected surface without any boundary, we have

$$\chi = n_v - n_e + n_f = 2 - 2g.$$

For example, for the double torus (see Figure 6.5 E), we have $g = 2$, thus $\chi = 2 - 2g = -4$.

Let us now turn to surfaces with multiple connected components and boundaries. It is easy to see what the formula becomes: since for a topological sphere, $\chi = 2$, each connected component adds two to the Euler–Poincaré characteristic. Consider now the "topological surgery" operation that creates a boundary by deleting a facet. It decreases n_f, thus it also decreases χ. To summarize, for a surface with multiple connected components and multiple boundary components, we have

$$\chi = n_v - n_e + n_f = 2n_{cc} - 2g - n_{bc},$$

where n_{cc} denotes the number of connected components and n_{bc} the number of boundary components.

In what follows, we consider topological disks, that is, surfaces with genus zero, a single connected component, and a single boundary component. They are characterized by

$$g = 0, \qquad n_{cc} = 1, \qquad n_{bc} = 1.$$

Using the previous identity, for a topological disk, it is easy to check that $\chi = 1$. To test whether a given surface mesh is a topological disk, we first test the value of $\chi = n_v - n_e + n_f$. Then we need to determine whether the surface is connected. One

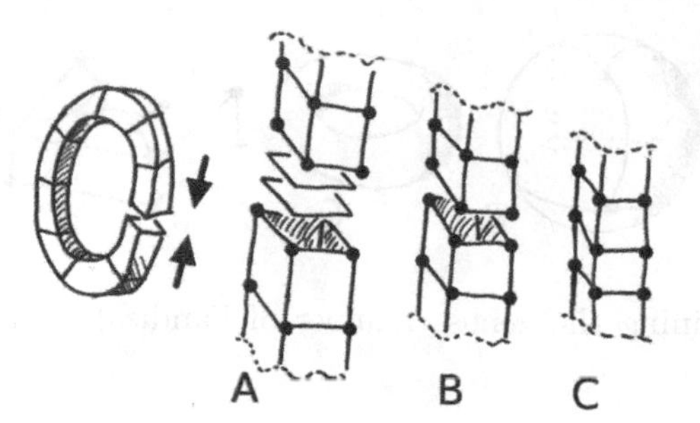

Figure 6.6 Increasing the genus by "topological surgery." A: delete two facets ($\Delta n_\mathrm{f} = -2$); B: glue the two holes ($\Delta n_\mathrm{v} = -4$; $\Delta n_\mathrm{e} = -4$); C: the result ($\Delta\chi = \Delta n_\mathrm{v} - \Delta n_\mathrm{e} + \Delta n_\mathrm{f} = -2$).

can use a recursive traversal algorithm from an arbitrary facet and test whether it visits all the facets. Finally, to determine the number of boundary components, one can iterate along the edges on the boundary component incident to an arbitrary boundary edge and test whether there exists another boundary edge that was not visited.

6.4 TUTTE'S BARYCENTRIC MAPPING THEOREM

We now focus on Tutte's barycentric mapping theorem [400], which not only characterizes a family of mesh parameterizations, but can also be used as a computational algorithm to create a parameterization [143]. In simple form, this theorem may be stated as follows.

Theorem 6.1. *Consider a 3-connected mesh that is a topological disk and with vertices in 2D. If the boundary vertices are positioned on a (not necessarily strictly) convex polygon, and if each interior vertex is a convex combination of its neighbors, then there is no intersection between the edges and each face is strictly convex.*

A mesh is *3-connected* if it remains connected after removing any two vertices and their incident edges. This condition is required to avoid some degeneracies (collapsed triangles) that occur, for instance, if a long interior edge connects two vertices on the boundary.

More formally, each interior vertex is a linear combination of its neighbors, if

$$\begin{aligned}
\sum_{j\in N_i} w_{ij}u_j &= u_i, \quad i = 1,\ldots,(n_\mathrm{v} - n_\mathrm{bv}),\\
\sum_{j\in N_i} w_{ij}v_j &= v_i, \quad i = 1,\ldots,(n_\mathrm{v} - n_\mathrm{bv}),\\
u_i &= b_i^u, \quad i = (n_\mathrm{v} - n_\mathrm{bv} + 1),\ldots,n_\mathrm{v},\\
v_i &= b_i^v, \quad i = (n_\mathrm{v} - n_\mathrm{bv} + 1),\ldots,n_\mathrm{v},
\end{aligned} \tag{6.1}$$

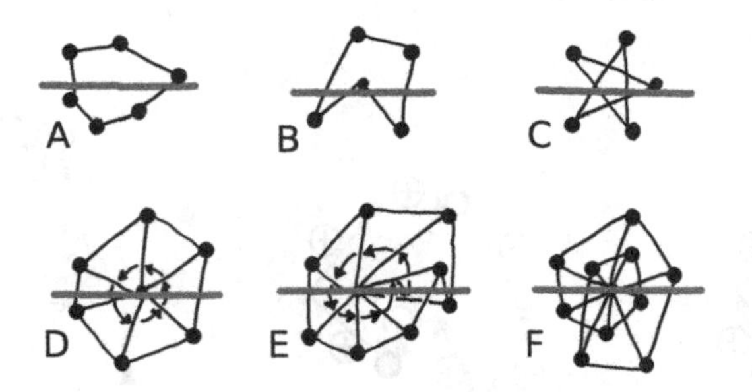

Figure 6.7 Convex face (A) and non-convex faces (B, C); wheel vertex (D) and non-wheel vertices (E,F).

where N_i denotes the neighbors of vertex i (that is, all vertices connected to i by an edge), n_{bv} is the number of boundary vertices, and where vertices are ordered in such a way that interior vertices appear first. The given coefficients w_{ij} associated with the half-edges (i, j) are positive and such that $\sum_j w_{ij} = 1$. The coefficients w_{ij} are not required to be symmetric (in general, $w_{ij} \neq w_{ji}$). In other words, Condition (6.1) means that each vertex is inside the convex hull of its neighbors (hence the term *convex combination*).

From the given positions of the boundary vertices (u_i, v_i), $i = 1, \ldots, n_{\text{v}} - n_{\text{bv}}$ and the weights w_{ij}, one can construct a parameterization by solving the two linear systems in (6.1), as suggested in [143] and explained in the next section. Before explaining how, the rest of this section deals with the proof of Tutte's theorem.

The original proof by Tutte [400] requires a substantial background from graph theory. Later, another proof was proposed by Gortler et al. [172]. This latter proof is much simpler and only requires elementary concepts. In order to better grasp an intuition of the structure of the proof, we adopt here a top-down approach (starting from the conclusion) and skip the details for keeping only the main arguments.

The main argument of the proof, exposed later, is that if (6.1) is satisfied, then each face is strictly convex and each interior vertex is a wheel. Then it is quite easy to deduce that no two faces can have an intersection. A *wheel vertex* $\mathbf{v}$ is such that the oriented angles between each pair of consecutive oriented edges emanating from $\mathbf{v}$ are all positive and sum to 2π. Examples of non-wheel vertices are shown in Figure 6.7 (E) (one of the angles is negative) and Figure 6.7 (F) (the vertex is a double-wheel, with an angle sum of 4π instead of 2π). Any line in generic position (thick gray) intersects a convex polygon in zero or two edges (A). Any line that passes through the center of a wheel intersects two wedges (D). For non-convex faces (non-wheel vertices, respectively), there exist lines with a larger number of intersections.

We now show that if (6.1) is satisfied, then all polygons are convex and all vertices are wheels, by studying the relation between a straight line with a direction vector (α, β) and the polygons' and vertices' neighborhoods, and showing that configurations such as B, C, E, F cannot be encountered.

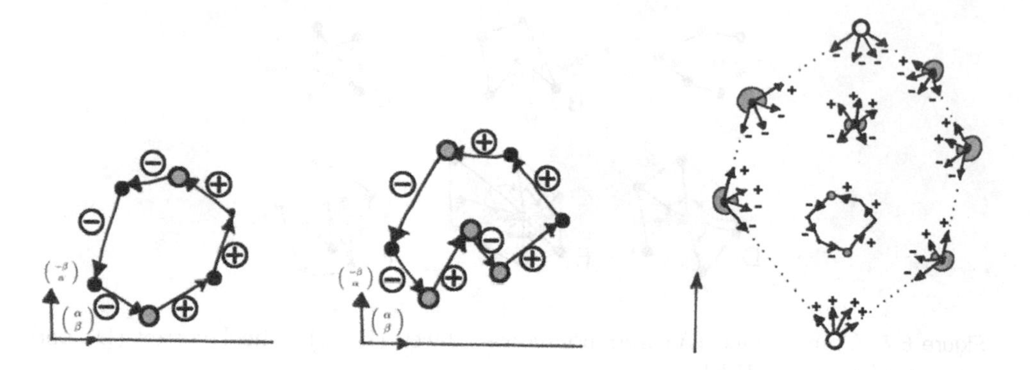

Figure 6.8 Left and center: the maximum number of intersections between a polygon and a line is bounded by the number of sign changes. Right: global structure of sign changes in the vertices and facets of a mesh with (u, v) coordinates that satisfy the conditions of Tutte's theorem.

Consider the configurations depicted in Figure 6.8, with a line of direction vector (α, β). Note that the figure is rotated to make the oriented line direction (α, β) horizontal. We classify each oriented edge (i, j) as ascending $(+)$ or descending $(-)$, depending on the sign of the dot product $(\boldsymbol{u}_j - \boldsymbol{u}_i) \cdot \boldsymbol{n}$, where $\boldsymbol{n} = (-\beta, \alpha)$ denotes the normal vector of the line.

The edges intersected by a line appear in $(+, -)$ pairs, therefore their number is bounded by the number of sign changes (the vertices that correspond to sign changes are highlighted in gray in the figure). A convex polygon has exactly two sign changes for any (α, β) in generic position (not collinear to an edge), whereas a non-convex polygon admits (α, β) directions with a larger number of sign changes. The same reasoning can be applied to the vertices' neighborhoods.

We now consider a topological disk and transform it into a topological sphere by adding a face that closes its boundary. The main argument of the proof shows that if the (u_i, v_i) coordinates satisfy the conditions of Tutte's theorem (boundary vertices on a convex polygon and interior vertices satisfying (6.1)), then for any line direction (α, β), the structure of the sign changes in the overall mesh is as depicted in Figure 6.8 (right): all the interior vertices and facets have two sign changes, two boundary vertices have no sign change (topmost and bottommost), and all the other $n_{\mathsf{bv}} - 2$ vertices have two sign changes. In the figure, we took a convention that is different from [172]: the arrows correspond to the oriented edges emanating from each vertex and to the oriented edges that form the boundary of the facets. Sign changes are highlighted in gray. Note that there is an infinite outer facet (the one that was added to turn the topological disk into a topological sphere), thus some sign-changes wedges are on the exterior of the boundary.

To prove that the structure induced by the solution of (6.1) matches this configuration, Gortler et al. [172] make the observation that the set of values $\boldsymbol{n} \cdot (\boldsymbol{u}_j - \boldsymbol{u}_i)$ attached to the oriented half-edges of the mesh can be considered as a discrete vector field (or 1-form) and that it obeys a discrete equivalent of the Poincaré–Hopf theorem [181], which relates the topology of a vector field (here represented by the sign changes) to the topology of the underlying surface (here a topological sphere). In a nutshell, this theorem states that the number of singularities of the vector field is equal to the Euler–Poincaré characteristic of the surface. In the present case, we have a sphere with $\chi = 2$ and two singularities (topmost and bottommost vertices). Several discrete versions of this theorem were proposed [323, 324]. The discrete Poincaré–Hopf theorem in [172] can be stated as follows.

Theorem 6.2. *For any set of non-vanishing scalar values z_{ij} attached to the half-edges of an oriented mesh with genus g and no boundary and such that $z_{ji} = -z_{ij}$, we have*

$$\sum_{i=1}^{n_\mathrm{v}} \mathrm{ind}(\mathbf{v}_i) + \sum_{j=1}^{n_\mathrm{f}} \mathrm{ind}(\mathbf{f}_j) = \chi = 2 - 2g.$$

The integer quantities $\mathrm{ind}(\mathbf{v}) = (2 - \mathrm{sc}(\mathbf{v}))/2$ and $\mathrm{ind}(\mathbf{f}) = (2 - \mathrm{sc}(\mathbf{f}))/2$ are called the *index* of vertex $\mathbf{v}$ and the index of facet $\mathbf{f}$ in z. The integer quantities $\mathrm{sc}(\mathbf{v})$ and $\mathrm{sc}(\mathbf{f})$ denote the number of *sign changes* of z when traversing the half-edges emanating from $\mathbf{v}$ and the half-edges turning around $\mathbf{f}$, respectively. The index can take the following values:

- ind = 1: z_{ij} has the same sign for all half-edges of the vertex/face. A vertex with index one is called a *source* (+) or a *sink* (−). A face with index one is called a *vortex*;
- ind = 0: a vertex or facet with zero index is said to be *non-singular*;
- ind < 0: a vertex or facet with negative index is called a *saddle*.

Proof. The last statement follows, because

$$\begin{aligned}\sum_{i=1}^{n_\mathrm{v}} \mathrm{ind}(\mathbf{v}_i) + \sum_{j=1}^{n_\mathrm{f}} \mathrm{ind}(\mathbf{f}_j) &= \frac{1}{2}\sum_{i=1}^{n_\mathrm{v}}(2 - \mathrm{sc}(\mathbf{v}_i)) + \frac{1}{2}\sum_{j=1}^{n_\mathrm{f}}(2 - \mathrm{sc}(\mathbf{f}_j)) \\ &= n_\mathrm{v} + n_\mathrm{f} - \frac{1}{2}\left(\sum_{i=1}^{n_\mathrm{v}} \mathrm{sc}(\mathbf{v}_i) + \sum_{j=1}^{n_\mathrm{f}} \mathrm{sc}(\mathbf{f}_j)\right) \\ &= n_\mathrm{v} + n_\mathrm{f} - n_\mathrm{e}.\end{aligned}$$

The last step of the derivation above uses the observation that each half-edge introduces exactly one sign change in a vertex and one sign change in a facet: consider the half-edge h_1 in Figure 6.9 and its opposite half-edge h_2 (with opposite sign). There are two cases: either h_1 and h_2 have the same sign, then the wedge h_2, h_3 is a sign change in vertex $\mathbf{v}$, or h_1 and h_2 have opposite signs, then the corner h_1, h_2 is a sign change in facet $\mathbf{f}$. □

Equipped with this discrete Poincaré–Hopf theorem, we can examine the structure of sign changes induced by a set of values $\boldsymbol{n}\cdot(\boldsymbol{u}_j - \boldsymbol{u}_i)$ attached to the half-edges

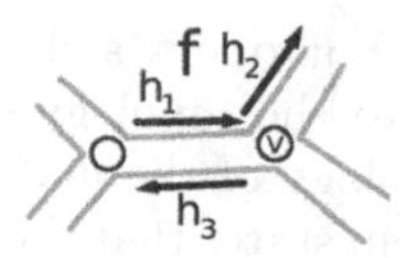

Figure 6.9 There is exactly one sign change in a vertex and one sign change in a facet per half-edge.

(i, j) of a mesh, where the $\boldsymbol{u}_i = (u_i, v_i)$ are given by the solution of Tutte's equations in (6.1). Note that the faces cannot be vortices (that is, with index one), since this would mean that one would not end up at its original position when turning around a facet.

Since the $\boldsymbol{u}_i$'s satisfy Tutte's equation, it is easy to verify that the values z_{ij} associated with the half-edges emanating from an interior vertex $\mathbf{v}_i$ satisfy the condition

$$\sum_{j \in N_i} w_{ij} z_{ij} = 0.$$

Since all the w_{ij} coefficients are positive and z_{ij} is non-vanishing, the sign of z_{ij} needs to change, thus $\mathbf{v}_i$ can be neither a source nor a sink. To summarize,

$$\begin{aligned} \operatorname{ind}(v_i) &\le 0, & i &= 1, \ldots, n_{\mathrm{v}} - n_{\mathrm{bv}}, \\ \operatorname{ind}(f_j) &\le 0, & j &= 1, \ldots, n_{\mathrm{f}}. \end{aligned}$$

Now, remembering that the maximum possible value for an index is one, the only possibility we have for reaching the overall value of two dictated by the Poincaré–Hopf theorem (for $\chi = 2$),

$$\sum_{i=1}^{n_{\mathrm{v}}} \operatorname{ind}(\mathbf{v}_i) + \sum_{j=1}^{n_{\mathrm{f}}} \operatorname{ind}(\mathbf{f}_j) = 2,$$

is to have index zero for all facets and all interior vertices, index one for two boundary vertices, and index zero for all the remaining boundary vertices.

Note that throughout the proof we assume that all the considered geometrical objects are in generic position. The initial article includes perturbation lemmas that handle the cases where an edge is collinear with the direction (α, β) of the intersected line. The original article also includes results for multiple possibly non-convex boundaries and higher-genus objects that we do not describe here. In the frame of this chapter, we restrict ourselves to the case of a single topological disk with a convex boundary. For more general cases, the reader is referred to the original article [172] and to [61, 66, 201].

6.5 SOLVING THE LINEAR SYSTEMS

Given a topological disk, the 2D positions of its boundary vertices, and the weights w_{ij}, a parameterization can be obtained by solving (6.1). We explain here several methods for solving such linear systems. Tutte's equation in its original form corresponds to the linear systems

$$u_i = \sum_{j \in N_i} w_{ij} u_j \qquad i = 1, \ldots, n_{\mathrm{v}} - n_{\mathrm{bv}},$$
$$v_i = \sum_{j \in N_i} w_{ij} v_j \qquad i = 1, \ldots, n_{\mathrm{v}} - n_{\mathrm{bv}}.$$

Note that the right-hand side contains a mixture of unknowns (the u_j's and v_j's associated to interior vertices) and fixed values (the u_j's and v_j's associated to boundary vertices). To re-organize the equation in a form that gathers the fixed values on the right-hand side, one can extend the definition of w_{ij} with the following conventions: $w_{ii} = -1$ and $w_{ij} = 0$ if $j \notin N_i$. We then have

$$\sum_{j=1}^{n_{\mathrm{v}}-n_{\mathrm{bv}}} w_{ij} u_j = -\sum_{j=n_{\mathrm{bv}}+1}^{n_{\mathrm{v}}} w_{ij} u_j \qquad i = 1, \ldots, n_{\mathrm{v}} - n_{\mathrm{bv}},$$
$$\sum_{j=1}^{n_{\mathrm{v}}-n_{\mathrm{bv}}} w_{ij} v_j = -\sum_{j=n_{\mathrm{bv}}+1}^{n_{\mathrm{v}}} w_{ij} v_j \qquad i = 1, \ldots, n_{\mathrm{v}} - n_{\mathrm{bv}}.$$

In matrix form, the two linear systems can be written as

$$W\boldsymbol{X}^u = \boldsymbol{Y}^u,$$
$$W\boldsymbol{X}^v = \boldsymbol{Y}^v,$$

where W is an $(n_{\mathrm{v}} - n_{\mathrm{bv}}) \times (n_{\mathrm{v}} - n_{\mathrm{bv}})$ matrix, and $\boldsymbol{X}^u, \boldsymbol{Y}^u, \boldsymbol{X}^v, \boldsymbol{Y}^v$ are $(n_{\mathrm{v}} - n_{\mathrm{bv}})$-dimensional vectors.

We now outline several methods to solve such linear systems. Considering a system in the form $W\boldsymbol{X} = \boldsymbol{Y}$, with $n = n_{\mathrm{v}} - n_{\mathrm{bv}}$, Gauss–Seidel relaxation is a method that is very simple to implement. Intuitively, it can be understood by considering the equation in the form of the linear system

$$\begin{aligned}
w_{1,1}x_1 + w_{1,2}x_2 + \quad & \cdots & + w_{1,n}w_n &= y_1, \\
& & &\vdots \\
w_{i,1}x_1 + w_{i,2}x_2 + \cdots + & w_{i,i}x_i + \cdots & + w_{i,n}w_n &= y_i, \\
& & &\vdots \\
w_{n,1}x_1 + w_{n,2}x_2 + \quad & \cdots & + w_{n,n}w_n &= y_n.
\end{aligned}$$

The algorithm iterates over and over through the n equations. For each equation i, it "pretends" that all the variables x_j for $j \neq i$ are known and updates x_i accordingly as

$$x_i \leftarrow \frac{1}{w_{ii}} \left(y_i - \sum_{j=1}^{n_{\mathrm{v}}-n_{\mathrm{bv}}} w_{ij} x_j \right).$$

Replacing the right-hand side component y_i with the expression $-\sum_{j=n_{\rm bv}+1}^{n_{\rm v}} w_{ij}x_j$ and w_{ii} with -1, we get the update

$$x_i \leftarrow -\left(-\sum_{j=n_{\rm bv}+1}^{n_{\rm v}} w_{ij}x_j - \sum_{j=1}^{n_{\rm v}-n_{\rm bv}} w_{ij}x_j\right)$$

or simply

$$x_i \leftarrow \sum_{j\in N_i} w_{ij}x_j.$$

To summarize, applying Gauss–Seidel relaxation to Tutte's equation simply means iteratively moving each interior vertex to the weighted barycenter of its neighbors. This method is very simple to implement, and gives results in a reasonable time for small meshes (with a few thousand elements). If larger meshes need to be processed, then one can consider using more efficient iterative methods such as the bi-conjugate gradient method, as suggested in [143]. If memory consumption is not an issue, sparse direct solvers such as SuperLU [246] can be an even more efficient alternative. See also the chapter on numerical aspects in [66].

6.6 CHOOSING THE WEIGHTS

Depending on the application, it is often desired to create a parameterization that minimizes deformations between the 2D parametric space and the 3D surface. Using uniform weights $w_{ij} = 1/|N_i|$, where $|N_i|$ denotes the number of neighbors of vertex $\mathbf{v}_i$, the obtained parameterization is valid, but can be highly distorted. Figure 6.10 (left) displays the deformations by painting a regular grid in the 2D parametric space and transforming it onto the 3D surface.

One can use instead as weights w_{ij} the mean value coordinates [144, 200], which take into account the geometry of the input surface and are positive as required by Tutte's theorem (see Section 1.2.3). Note that to satisfy the requirements of Tutte's theorem, they need to be normalized as in (1.12). The resulting parameterization is displayed in Figure 6.10 (right).

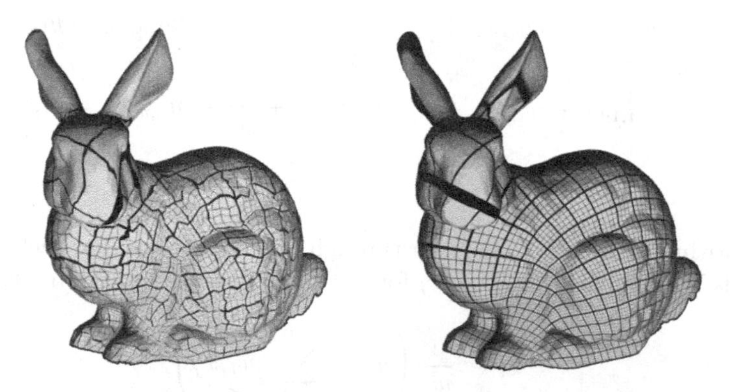

Figure 6.10 Uniform weights (left) and mean value coordinates (right).

CHAPTER 7

Planar Shape Deformation

Ofir Weber
Bar-Ilan University, Ramat Gan, Israel

CONTENTS

BARYCENTRIC COORDINATES are extremely useful for deforming shapes. They lead to smooth deformations that are simple and highly efficient to compute. In this chapter, we focus on the special case of *planar* shape deformation, where the barycentric map can be interpreted as a complex-valued function. We generalize barycentric coordinates from real to complex-valued functions and introduce the notion of complex barycentric coordinates. Complex coordinates can lead to planar barycentric maps with unique shape preserving properties. We provide a general construction for complex barycentric coordinates (Section 7.2.2) and show how to design custom-made complex coordinates (Section 7.2.3). We derive holomorphic coordinates (Section 7.2.1), relying on the rich theory of complex analysis, and explore their intimate relation to conformal maps (Section 7.4).

7.1 INTRODUCTION

Interactive shape deformation is a fundamental problem in computer graphics and geometry processing. It is essential for designing and modeling variations of shapes and for generating believable animation sequences. In this chapter, we focus on the special case of *planar* shape deformation, for which the domain Ω to be deformed is an open and connected subset of the plane. For simplicity, we assume that the domain is bounded by a simple polygon P. The codomain is again a subset of the plane, where in some scenarios (as in Chapter 4), the user explicitly dictates its shape (e.g., to be a polygon as well), while in other scenarios, more freedom is allowed, and the shape freely evolves. More formally, our goal is to compute a map (a vector function) $\boldsymbol{f}(\boldsymbol{x})\colon \Omega \to \mathbb{R}^2$ that maps points in the domain to points in the codomain, such that $\boldsymbol{f}$ satisfies some boundary constraints while having certain desirable properties. Some properties, such as smoothness of $\boldsymbol{f}$, are easier to attain, whereas other properties, such as local injectivity, are more challenging to guarantee. The complete list of desired properties will be revealed as the discussion of this chapter evolves. However, we briefly mention and motivate below the most important properties and requirements.

The domain Ω is equipped with a colored texture and we are interested in visualizing the map $\boldsymbol{f}$ by rendering a high quality image of the codomain. The algorithms that we consider for computing the underlying map $\boldsymbol{f}$ are indifferent to the choice of such texture. These algorithms are solely based on geometric considerations; however, for the result to be valuable from the graphics application standpoint, some requirements have to be added. A high quality result should preserve the fine details of the underlying texture. Hence, smoothness of the map is often required. Luckily, this is relatively easy to obtain when working with barycentric coordinates as it is implied by the smoothness of the coordinates themselves. In addition to smoothness, we often desire that the amount of metric distortion induced by the map is as low as possible. We consider both the conformal and the isometric distortion, where conformal distortion refers to the amount by which angles between arbitrary intersecting curves are altered under the action of the map, and isometric distortion refers to the amount of change in the length of arbitrary curves. A map is said to be conformal if it has zero conformal distortion everywhere. The Jacobian matrix $J_{\boldsymbol{f}}$ of a conformal map is a similarity transformation, allowing only rotations with isotropic scale, but it can have arbitrarily high isometric distortion. Ideally, a map should be isometric, with a Jacobian that is restricted to be a rotation (without scale). Unfortunately, unlike the space of conformal maps, the space of isometries in $\mathbb{R}^2$ is fairly restricted and contains only rigid motions (a *global* rotation and translation) and reflections. Hence, we can only hope for maps that are close to isometry, having a low amount of isometric and conformal distortion.

In order to control the end result, a user interface is provided, where some boundary conditions for the map $\boldsymbol{f}$ can be prescribed. For example, the user can prescribe the shape of the entire target boundary. Since the target boundary is a curve with an infinite number of degrees of freedom, a common choice in practice is to assume that the image of the map is bounded by a polygon. The user only determines the vertex positions, and the map is implicitly assumed to be linear along the polygon edges. Notwithstanding, fixing the entire boundary can be overly restrictive. It requires more than a bare minimum input from the user and can lead to less visually pleasing deformations, where the underlying map $\boldsymbol{f}$ has excessive metric distortion. One way to alleviate this is to use less restricting constraints and to allow the algorithm to deduce the missing degrees of freedom automatically by solving a variational problem (Section 7.3). For example, the user can prescribe the behavior of the map $\boldsymbol{f}$ on a sparse set of interior points rather than on the entire boundary (Section 7.3.2).

Another requirement is that the underlying algorithms need to be relatively efficient to compute. This allows the users to explore the space of valuable deformations interactively and to reach their envisioned artistic design by trial and error. Interactivity and simplicity of the user interface also allow novice users to use the application. As we will see, many of the involved computations can be done in a preprocessing step, leaving a marginal amount of computation to be done in run-time during interaction with the system. Moreover, the deformation can often be computed in parallel, utilizing the graphics hardware, where the computation time is of the same order of magnitude as the rendering time itself (Section 7.5).

Working with the special case of maps from $\mathbb{R}^2$ to $\mathbb{R}^2$ allows us to use an alternative representation by identifying $\mathbb{R}^2$ with the complex plane $\mathbb{C}$. Points $\boldsymbol{x} = (x, y) \in \mathbb{R}^2$ are identified with complex numbers $z \in \mathbb{C}$ by $z = x + iy$. The map $\boldsymbol{f}$ becomes a complex-valued function of a single complex variable $f \colon \Omega \to \mathbb{C}$. Utilizing complex notation often leads to simplified expressions and allows us to rely on powerful mathematical results from the rich theory of complex analysis. This is particularly apparent when the underlying complex functions are holomorphic (complex analytic) and/or harmonic, which are of particular value in shape deformation due to their smoothness and other mathematical properties.

7.2 COMPLEX BARYCENTRIC COORDINATES

Let $z_1, \ldots, z_n \in \mathbb{C}$ be the vertices of a simple polygon P (not necessarily convex) in counter-clockwise orientation (see Figure 7.1). We sometimes refer to P as the *cage*, as it encapsulates the part of the plane that we would like to deform. The interior of P is treated as an open domain denoted by Ω. A set of complex-valued functions $\phi_i \colon \Omega \to \mathbb{C}$, $i = 1, \ldots, n$ are called *complex barycentric coordinates* [415] if they satisfy the complex analogue of the Properties (1.8) and (1.9) in Definition 1.1,

$$\sum_{i=1}^{n} \phi_i(z) = 1, \tag{7.1}$$

$$\sum_{i=1}^{n} \phi_i(z) z_i = z \tag{7.2}$$

for all points $z \in \Omega$.

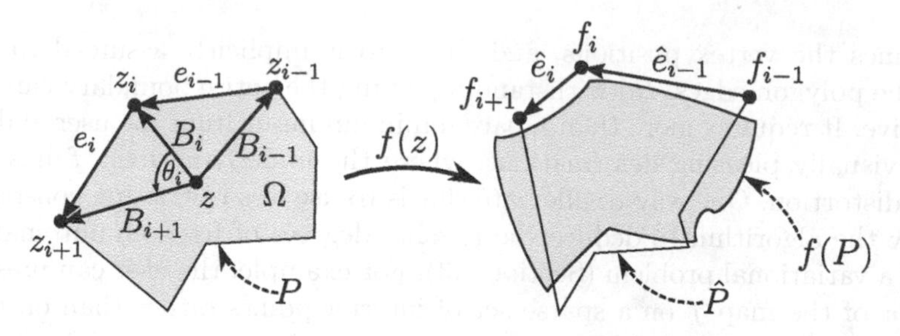

Figure 7.1 Notation used for complex barycentric maps. Left: the source polygon P. Note that $B_i(z) = z_i - z$ are functions of z, while the edge vectors $e_i = z_{i+1} - z_i$ are independent of z. Right: the target polygon $\hat{P}$ as well as the image of a complex holomorphic barycentric map $f\colon \Omega \to \mathbb{C}$. Note that $f(z)$ does not interpolate the target polygon.

A *complex barycentric map* is a planar map expressed as a complex-valued function $f\colon \Omega \to \mathbb{C}$ with

$$f(z) = \sum_{i=1}^{n} \phi_i(z) f_i,$$

where $f_i \in \mathbb{C}$ are some complex coefficients. Similarly to the real case, Property (7.1) is called the *constant precision* property (or sometimes denoted as the partition of unity property). It requires that the real parts of the coordinates sum to 1, and in addition, that their imaginary parts sum to 0. Algebraically, it implies that if g is a *complex* constant function over Ω, and $f_i = g(z_i)$, then $f(z) = g(z)$. In other words, f reproduces complex constant functions from their vertex values. Geometrically, (7.1) means that the barycentric map is invariant to translations applied to the cage vertices. Property (7.2) is called the *linear precision* property. Algebraically, it implies that f reproduces *complex* linear functions over Ω from their vertex values. In particular, it reproduces the (linear) identity function $g(z) = z$, and hence it is sometimes referred to as the identity reproduction property. Together, Properties (7.1) and (7.2) imply that f reproduces *complex* affine functions, that is, functions $\psi\colon \mathbb{C} \to \mathbb{C}$ of the form $\psi(z) = \alpha z + \beta$, where $\alpha, \beta \in \mathbb{C}$ are complex constants. However, it is important to note that unlike real barycentric maps, complex barycentric maps do not, in general, reproduce planar affine maps, that is, maps $\boldsymbol{g}\colon \mathbb{R}^2 \to \mathbb{R}^2$ of the form $\boldsymbol{g}(\boldsymbol{x}) = \boldsymbol{A}\boldsymbol{x} + \boldsymbol{b}$, where $\boldsymbol{A}$ is a 2×2 real matrix and $\boldsymbol{x}, \boldsymbol{b} \in \mathbb{R}^2$ are real vectors (see Figure 7.2). While this may be initially perceived as a limitation, for the particular application of planar shape deformation, it is mostly an advantage. The matrix $\boldsymbol{A}$ encodes rotation, isotropic scale, and shear, whereas in the complex case, the multiplication αz is equivalent to a multiplication of $\boldsymbol{x}$ by a similarity matrix $\left(\begin{smallmatrix} \alpha_x & -\alpha_y \\ \alpha_y & \alpha_x \end{smallmatrix}\right)$, where $\alpha = \alpha_x + \mathrm{i}\alpha_y$. In other words, complex barycentric maps reproduce similarity transformations (translation, rotation, and isotropic scale) but do not reproduce, in general, anisotropic scale and shears, which lead to distortion of angles.

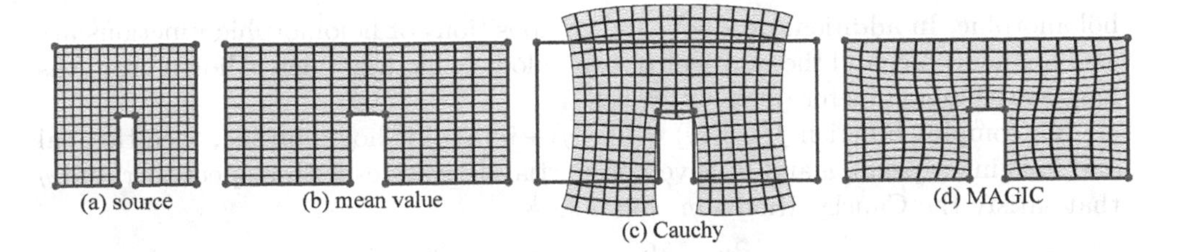

Figure 7.2 Comparison of real and complex barycentric coordinates. The cage of a pants shape in (a) is transformed in (b–d) by the affine transformation $\boldsymbol{g}\colon (x, y) \to (2x, y)$. In (b) the mean value barycentric map reproduces $\boldsymbol{g}$. In (c) and (d), the Cauchy and magic complex barycentric coordinates produce maps that do not reproduce $\boldsymbol{g}$. The map in (c) is conformal. See color insert.

In addition to constant and linear precision, smoothness is also required, that is,

$$\phi_i \in C^k(\Omega), \qquad i = 1, \ldots, n,$$

where k is at least 1 (continuously differentiable), but typically we work with C^∞ (infinitely differentiable) functions.

Note that unlike their real counterpart, complex barycentric coordinates do not have a sign, hence the positivity property is omitted. The Lagrange property (interpolation of the boundary data) is possible to attain but we will see that it can interfere with other (later to be defined) properties such as holomorphicity, hence it is not considered a mandatory property.

The first question that comes to mind is how can we come up with a set of functions ϕ_i that satisfy (7.1) and (7.2)? A thorough answer to this question is provided in Section 7.2.2. However, we start by deriving a special type of complex barycentric coordinates, which are denoted as Cauchy coordinates. Unlike all real barycentric coordinates, these coordinates have the special property of being holomorphic functions.

7.2.1 Holomorphic functions

Holomorphic functions (also known as complex analytic functions) have remarkable mathematical properties and they are the central objects of study in the mathematical field of complex analysis. We provide below a short introduction to holomorphic functions and some of their fascinating properties, and refer the reader to [5] for an in-depth review. A holomorphic function $f\colon \Omega \to \mathbb{C}$ is a complex-valued function that is complex differentiable in a neighborhood of every point $z \in \Omega$. Complex differentiability is a strong condition implying that f is infinitely differentiable. Furthermore, holomorphic functions are integrable (on a simply connected domain) an infinite number of times, and these derivatives and anti-derivatives are all holomorphic as well. Holomorphic functions form a linear space, hence sums of holomorphic functions and multiplications of a holomorphic function with a complex scalar are holomorphic. In addition, products and compositions of holomorphic functions are also holomorphic, and the quotient of two holomorphic functions is holomorphic as long as the denominator does not vanish.

If a complex function $f(x+iy) = u(x,y) + iv(x,y)$ is holomorphic, then the real and imaginary parts, u and v, have first partial derivatives with respect to x and y that satisfy the Cauchy–Riemann equations

$$\frac{\partial u}{\partial x} = \frac{\partial v}{\partial y}, \qquad \frac{\partial u}{\partial y} = -\frac{\partial v}{\partial x}.$$

The converse is also true under very mild assumptions; that is, if the partial derivatives of f are continuous and satisfy the Cauchy–Riemann equations, then f is holomorphic. Treating f as a planar map, we can express the Jacobian of f as

$$J_f = \begin{pmatrix} \frac{\partial u}{\partial x} & \frac{\partial u}{\partial y} \\ \frac{\partial v}{\partial x} & \frac{\partial v}{\partial y} \end{pmatrix} = \begin{pmatrix} \frac{\partial u}{\partial x} & -\frac{\partial v}{\partial x} \\ \frac{\partial v}{\partial x} & \frac{\partial u}{\partial x} \end{pmatrix} = \begin{pmatrix} a & -b \\ b & a \end{pmatrix},$$

where the second equality is due to the Cauchy–Riemann equations. The first complex derivative of a holomorphic function with respect to z is often denoted by f' and satisfies $f' = a + ib$. As can be seen above, the Jacobian matrix of a holomorphic function (when treated as a planar map) has the structure of a similarity matrix. Its determinant is given by $\det(J_f) = a^2 + b^2$ and is always nonnegative; hence, holomorphic maps cannot reverse orientation. If a holomorphic map f has nonvanishing derivative $f'(z) \neq 0$ (which is equivalent to $\det(J_f) > 0$) for every $z \in \Omega$, then the map f is called a *conformal map*. For conformal maps, the Jacobian above can be expressed as

$$J_f = r \begin{pmatrix} \cos\theta & -\sin\theta \\ \sin\theta & \cos\theta \end{pmatrix}, \tag{7.3}$$

where $r = |f'| = \sqrt{a^2 + b^2} = \sqrt{\det(J_f)}$ is the modulus (length) of the complex derivative and $\theta = \operatorname{Arg} f'$ is its complex argument, that is, the angle between the complex number viewed as a vector and the positive real axis. Equation (7.3) provides a geometric interpretation for conformal maps as those maps that locally scale (isotropically), rotate, and translate. As such, conformal maps preserve angles between intersecting curves in the sense that if the angle between the tangents of two intersecting curves at a point $z_0 \in \Omega$ is α, then the angle between the images of these curves at $f(z_0)$ is also α. Conformality also means that small (infinitesimal) circles are mapped to small circles (rather than ellipses). Conformal maps, in general, are not isometric as they distort the length of curves. The parameter r in (7.3) characterizes the amount of change that the conformal map induces on the original metric. The quantity r^2 is called the *pullback metric*, and the quantity $\ln r$ is called the *conformal scale factor*. The isometric distortion $\tau(z)$ of a conformal map at a point $z \in \Omega$ measures the amount of shrinkage and expansion in a symmetric manner. Popular definitions are: $|\ln r|$, $\max(r, \frac{1}{r})$, or $r + \frac{1}{r}$.

A real-valued function $u(x,y)$ is harmonic if it satisfies the Laplace equation

$$\Delta u = \frac{\partial^2 u}{\partial x^2} + \frac{\partial^2 u}{\partial y^2} = 0.$$

A complex-valued function $f(x + iy) = u(x, y) + iv(x, y)$ is harmonic if both u and v are harmonic. It is easy to verify (by differentiation and using the Cauchy–Riemann equations) that if f is holomorphic then both u and v are harmonic. Hence, a holomorphic map is also a harmonic map. The converse is not true in general, with many harmonic planar maps not being holomorphic.

Transfinite complex barycentric coordinates

The derivation of holomorphic complex barycentric coordinates can be obtained through the generalization of the concept of complex barycentric coordinates to the transfinite case. The polygonal region is substituted with an arbitrary domain Ω bounded by a simple contour $\partial\Omega$ with arbitrary shape. Real transfinite barycentric coordinates are discussed in detail in Chapter 3. Here we focus on the complex counterpart.

To this end, we use the notion of a kernel function $\phi\colon \partial\Omega \times \Omega \to \mathbb{C}$. Loosely speaking, the continuous contour can be viewed as a polygon with an infinite number of vertices, and the complex kernel $\phi(w, z)$ is viewed as an analogue for the i-th coordinate where w, a boundary point, substitutes the i-th vertex. Analogously to (7.1) and (7.2) in the discrete polygonal case, we say that $\phi(w, z)$ is a *complex barycentric kernel*, if the properties

$$\int_{\partial\Omega} \phi(w, z)\, dw = 1, \tag{7.4}$$

$$\int_{\partial\Omega} \phi(w, z) w\, dw = z \tag{7.5}$$

hold for every point $z \in \Omega$. Then, a *transfinite complex barycentric map* is a complex-valued function $f\colon \Omega \to \mathbb{C}$ of the form

$$f(z) = \int_{\partial\Omega} \phi(w, z) f(w)\, dw, \tag{7.6}$$

where the differential dw in the contour integrals in (7.4), (7.5), and (7.6) is complex.

Somewhat surprisingly, there is a fairly simple choice for a kernel $\phi(w, z)$ that satisfies (7.4) and (7.5), which is called the Cauchy kernel [36]

$$\phi_{\mathrm{C}}(w, z) = \frac{1}{2\pi i} \frac{1}{w - z}.$$

To see why this is true, let us recall Cauchy's integral formula, which is one of the most fundamental statements in complex analysis. The statement asserts that the value of a holomorphic function $h\colon \Omega \to \mathbb{C}$ at a point $z \in \Omega$, is fully determined by its values on the boundary of the domain, and can be expressed explicitly by evaluating the contour integral,

$$h(z) = \frac{1}{2\pi i} \int_{\partial\Omega} \frac{h(w)}{w - z}\, dw. \tag{7.7}$$

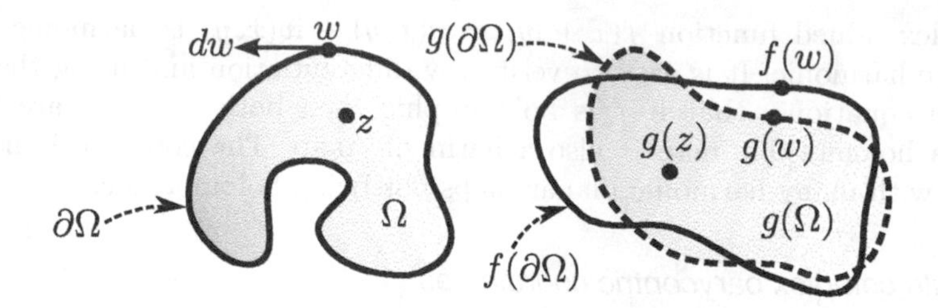

Figure 7.3 The Cauchy transform. The domain on the left is mapped holomorphically to the domain on the right using the Cauchy transform (7.8). However, note that $g(w) \neq f(w)$.

Applying Cauchy's integral formula to the holomorphic functions $h(z) = 1$ and $h(z) = z$ results in (7.4) and (7.5), respectively. The underlying assumption for Cauchy's formula to hold is that the function h in (7.7) is holomorphic. At this point, an interesting question arises. What if we substitute $h(w)$ on the right-hand side of (7.7) with an arbitrary complex-valued function $f\colon \Omega \to \mathbb{C}$ that is not holomorphic? In other words, let us find out what we can say about the function $g\colon \Omega \to \mathbb{C}$ with

$$g(z) = \frac{1}{2\pi \mathrm{i}} \int_{\partial\Omega} \frac{f(w)}{w - z}\, dw. \tag{7.8}$$

Apparently, under very mild assumptions on the smoothness of f in the vicinity of the boundary, the function g is holomorphic on Ω, hence (7.8) can be seen as a projection operator that acts on an arbitrary function f and returns a holomorphic function g. This operator is sometimes referred to as the *Cauchy transform* [36]. It is important to note that if f in (7.8) is not holomorphic, then the holomorphic function g will not, in general, interpolate f on the boundary, that is, $g(w) \neq f(w)$, $w \in \partial\Omega$ (see Figure 7.3). In other words, one cannot expect to prescribe the values of a function f on $\partial\Omega$, and obtain through (7.8) a holomorphic function f in Ω having these boundary values. To conclude, if the boundary values of f are such that there exists a holomorphic function that realizes them, then (7.8) precisely reproduces them. Otherwise, if a holomorphic function with such boundary values does not exist, (7.8) obviously does not reproduce it, but still results in a holomorphic function. It is important to note that a limitation of holomorphic functions is that a holomorphic function with arbitrary prescribed boundary values does not exist.

Derivation of Cauchy coordinates

The next step toward derivation of complex barycentric coordinates is to discretize the Cauchy transform (7.8) by assuming that f is a function that belongs to a finite-dimensional function space. In the language of boundary element methods (BEM), such f is called a *trial* function. While many choices are possible, a simple choice is to assume that f is a complex piecewise affine continuous function, that is, f is a (different) similarity transformation on each polygon edge. The space of piecewise affine continuous functions on the boundary of P can be parameterized by a set of complex coefficients f_i, $i = 1, \ldots, n$. Geometrically, one can think of f_i as the vertices of a target polygon $\hat{P}$ that is in one-to-one correspondence with the source polygon P. Each edge of the source polygon (z_i, z_{i+1}) is then mapped to a corresponding edge (f_i, f_{i+1}) in the target polygon by the similarity transformation (see Figure 7.1)

$$S_i(z) = \frac{B_{i+1}(z) f_i - B_i(z) f_{i+1}}{e_i}, \tag{7.9}$$

where $B_i(z) = z_i - z$, and $e_i = z_{i+1} - z_i$. The contour integral in (7.8) is then split into a sum of integrals over the polygon edges,

$$g(z) = \frac{1}{2\pi \mathrm{i}} \sum_{i=1}^{n} \int_{z_i}^{z_{i+1}} \frac{S_i(w)}{w - z}\, dw, \tag{7.10}$$

where the integral over each edge has a simple closed-form expression,

$$\int_{z_i}^{z_{i+1}} \frac{S_i(w)}{w - z}\, dw = S_i(z) \operatorname{Log}\left(\frac{B_{i+1}(z)}{B_i(z)} \right) + \hat{e}_i, \tag{7.11}$$

where $\hat{e}_i = f_{i+1} - f_i$, and Log is the principal branch of the complex logarithm. When summing over all edges, the term $\hat{e}_i$ in (7.11) vanishes since the target polygon is closed. Plugging the expression $S_i(z) \operatorname{Log}(\frac{B_{i+1}(z)}{B_i(z)})$ into (7.10) and rearranging the terms yields the *discrete Cauchy transform* of f,

$$g(z) = \sum_{i=1}^{n} C_i(z) f_i, \tag{7.12}$$

where the functions $C_i(z)$ are called the *Cauchy barycentric coordinates* and have the closed-form expression

$$C_i(z) = \frac{1}{2\pi \mathrm{i}} \left(\frac{B_{i+1}(z)}{e_i} \operatorname{Log}\left(\frac{B_{i+1}(z)}{B_i(z)} \right) - \frac{B_{i-1}(z)}{e_{i-1}} \operatorname{Log}\left(\frac{B_i(z)}{B_{i-1}(z)} \right) \right). \tag{7.13}$$

A similar derivation with other types of trial functions f in (7.8) can be obtained. For example, a higher-order polynomial per edge can be used, as well as discontinuous piecewise smooth functions. Such choices typically provide better approximation power and lead to more accurate results [414, 417] but are excluded here to keep the discussion simple.

Cauchy coordinates are shown in [415, Theorem 3] to be equivalent to Lipman's Green coordinates [252]. Green coordinates are a generalization of real barycentric coordinates where the barycentric map is expressed as a sum of two terms, the first blends the target vertex positions, and the second blends the normals to the target edges. The derivation of Green coordinates is done through the application of

Green's third integral identity to harmonic functions, which is equivalent in 2D to Cauchy's integral formula [5]. The (real-valued) expressions for Green coordinates are more complicated, compared to (7.13), and working with complex numbers is a great advantage; however, in contrast to Cauchy coordinates, Green coordinates have a natural generalization to 3D where they lead to volumetric harmonic maps [42, 252].

7.2.2 General construction of complex barycentric coordinates

In the previous section we derived Cauchy complex barycentric coordinates. Our construction was based on Cauchy's kernel, which reproduces holomorphic functions in the transfinite case. Holomorphic functions are extremely valuable in the context of planar shape deformation due to their relation to conformal maps. However, holomorphic maps form a rather small subspace out of the wider space of complex-valued maps. In particular, they do not allow for the Lagrange (interpolation) property to hold. Non-holomorphic complex barycentric coordinates clearly exist, since any real barycentric coordinates are also complex barycentric coordinates. As we will see next, there exist complex barycentric coordinates that are neither real nor holomorphic, and in contrast to Cauchy's coordinates, some of them possess the Lagrange property.

It is convenient to think about complex barycentric coordinates as a special type of complex basis functions that are associated with the cage's vertices. However, unlike arbitrary basis functions, the additional requirements (7.1) and (7.2) make it challenging to design such coordinates. To overcome this problem, we follow the same approach as in Section 1.1.2.

Definition 7.1. A set of complex-valued functions $w_i \colon \Omega \to \mathbb{C}$, $i = 1, \ldots, n$ that satisfy

$$\sum_{i=1}^{n} w_i(z)(z_i - z) = 0 \tag{7.14}$$

at every point $z \in \Omega$ are called *complex homogeneous coordinates.*

Normalizing these weight functions then gives complex barycentric coordinates.

Proposition 7.2. *Let $w_i(z)$ be complex homogeneous coordinates, such that their sum at every point $z \in \Omega$ is nonvanishing. Then the normalized functions*

$$\phi_i(z) = \frac{w_i(z)}{\sum\limits_{j=1}^{n} w_j(z)} \tag{7.15}$$

are complex barycentric coordinates.

Proof. The partition of unity property (7.1) follows immediately from the normalization in (7.15). The linear precision property (7.2) is due to

$$\sum_{i=1}^{n} \phi_i(z) z_i = \sum_{i=1}^{n} \frac{w_i(z) z_i}{\sum\limits_{j=1}^{n} w_j(z)} = \frac{1}{\sum\limits_{j=1}^{n} w_j(z)} \sum_{i=1}^{n} w_i(z) z_i = \frac{1}{\sum\limits_{j=1}^{n} w_j(z)} \sum_{i=1}^{n} w_i(z) z = z,$$

where the third inequality follows from (7.14). □

Proposition 7.2 is useful because finding complex homogeneous coordinates is simpler than finding complex barycentric coordinates. Rather than vertex-based weight functions, let us consider weight functions γ_i that are associated conceptually with the i-th edges.

Proposition 7.3. *Let $\gamma_i\colon \Omega \to \mathbb{C}$, $i = 1, \ldots, n$ be arbitrary weight functions, then the functions*

$$w_i(z) = \gamma_i(z)\frac{B_{i+1}(z)}{e_i} - \gamma_{i-1}(z)\frac{B_{i-1}(z)}{e_{i-1}} \tag{7.16}$$

are complex homogeneous coordinates.

Proof. The proof follows directly by plugging (7.16) into (7.14),

$$\begin{aligned}\sum_{i=1}^{n} w_i(z)(z_i - z) = \sum_{i=1}^{n} w_i(z)B_i(z) &= \sum_{i=1}^{n} \gamma_i(z)\frac{B_{i+1}(z)B_i(z)}{e_i} - \gamma_{i-1}(z)\frac{B_{i-1}(z)B_i(z)}{e_{i-1}}\\ &= \sum_{i=1}^{n} \gamma_i(z)\frac{B_{i+1}(z)B_i(z)}{e_i} - \sum_{i=1}^{n} \gamma_{i-1}(z)\frac{B_{i-1}(z)B_i(z)}{e_{i-1}}\\ &= \sum_{i=1}^{n} \gamma_i(z)\frac{B_{i+1}(z)B_i(z)}{e_i} - \sum_{i=1}^{n} \gamma_i(z)\frac{B_i(z)B_{i+1}(z)}{e_i} = 0,\end{aligned}$$

where in the third row, we shifted the summation index on the right-hand side by one. □

Proposition 7.4. *The sum of the weight functions γ_i at every point in the domain is identical to the sum of the homogeneous coordinates w_i in (7.16).*

Proof.

$$\begin{aligned}\sum_{i=1}^{n} w_i(z) &= \sum_{i=1}^{n} \gamma_i(z)\frac{B_{i+1}(z)}{e_i} - \gamma_{i-1}(z)\frac{B_{i-1}(z)}{e_{i-1}}\\ &= \sum_{i=1}^{n} \gamma_i(z)\frac{B_{i+1}(z)}{e_i} - \sum_{i=1}^{n} \gamma_{i-1}(z)\frac{B_{i-1}(z)}{e_{i-1}}\\ &= \sum_{i=1}^{n} \gamma_i(z)\frac{B_{i+1}(z)}{e_i} - \sum_{i=1}^{n} \gamma_i(z)\frac{B_i(z)}{e_i}\\ &= \sum_{i=1}^{n} \gamma_i(z)\frac{B_{i+1}(z) - B_i(z)}{e_i} = \sum_{i=1}^{n} \gamma_i(z)\frac{z_{i+1} - z - (z_i - z)}{e_i}\\ &= \sum_{i=1}^{n} \gamma_i(z)\frac{z_{i+1} - z_i}{e_i} = \sum_{i=1}^{n} \gamma_i(z).\end{aligned}$$

□

As an immediate corollary of Proposition 7.4, we have that if the γ_i have nonvanishing sum, then the homogeneous coordinates w_i of (7.16) also have nonvanishing sum.

We have basically shown that for any set of arbitrary complex weight functions γ_i with nonvanishing sum, there is a corresponding set of complex barycentric coordinates. Constructing barycentric coordinates from the weight functions is straightforward. Simply plug the γ_i weight functions into (7.16) to obtain the homogeneous coordinates w_i (Proposition 7.3) and normalize (Proposition 7.2) by either $\sum_{i=1}^n w_i$ or $\sum_{i=1}^n \gamma_i$ to obtain complex barycentric coordinates ϕ_i. Hence, we simplified the task of constructing complex barycentric coordinates to the task of finding arbitrary weight functions with nonvanishing sum. Moreover, the above construction covers the entire space of complex barycentric coordinates in the sense that for any complex barycentric coordinates ϕ_i there exist (not necessarily unique) weights γ_i that result in ϕ_i if the above construction is applied. The proof is done by choosing γ_1 arbitrarily and recursively computing all other γ_i using (7.16) [415].

We now provide an alternative interpretation for the weight functions γ_i. Any set of complex-valued weight functions γ_i with a nonvanishing sum everywhere inside the domain corresponds to the complex barycentric map [416]

$$f(z) = \sum_{i=1}^{n} S_i(z)\tilde{\gamma}_i(z), \tag{7.17}$$

where S_i are the unique similarity transformations that map the edges of the source cage to the edges of the target cage (recall (7.9) for their expression) and $\tilde{\gamma}_i = \gamma_i / \sum_{j=1}^n \gamma_j$ are the normalized weights. The converse is also true. That is, any complex barycentric map can be expressed as a (normalized) blending of edge-to-edge similarity transformations. In particular, it is easy to verify that the choice of $\gamma_i = \mathrm{Log}(\frac{B_{i+1}}{B_i})$ leads to Cauchy's coordinates. This alternative viewpoint is somewhat more intuitive geometrically than (7.16). It emphasizes that the weight functions can be associated with the edges and is useful for deriving custom-made barycentric coordinates (Section 7.2.3). Moreover, it allows us to extend the concept of barycentric maps in two possible ways. First, we can use (7.17) for an arbitrary collection of segments that do not necessarily form a closed polygon, allowing us to define "partial" cages or to augment a given cage with additional interior segments for extra user control. The second possible extension is to substitute the similarity transformations S_i in (7.17) with more general edge-to-edge affine transformations A_i,

$$f(z) = \sum_{i=1}^{n} A_i(z)\tilde{\gamma}_i(z).$$

Note that such planar maps cannot in general be expressed as real or complex barycentric maps. However, their evaluation is straightforward and as computationally efficient as the evaluation of barycentric maps, since the weights $\tilde{\gamma}_i$ can be computed and stored in a preprocessing step.

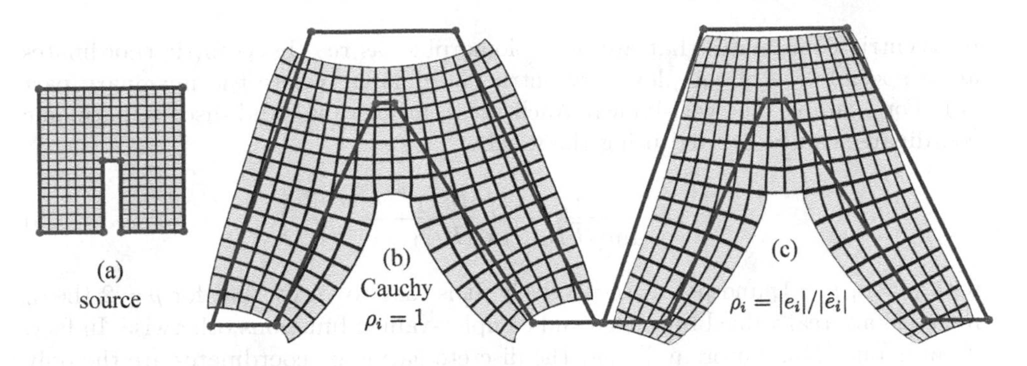

Figure 7.4 Blending edge-to-edge affine maps. A source cage of a pants shape in (a) is deformed identically in (b) and (c) to create an effect of legs stretching. In (b) the edge-to-edge affine maps are chosen to be similarities. The map from (a) to (b) is the holomorphic Cauchy barycentric map. In (c) the edge-to-edge affine maps stretch only in the direction of the edges but not in the orthogonal direction, resulting in a non-holomorphic map with a more natural look that better approximates the target cage. See color insert.

Planar affine maps $A(z) = az + b\overline{z} + c$ have three complex degrees of freedom $a, b, c \in \mathbb{C}$. Two are nailed down by the requirement to map the source edge endpoints to the target ones, $A_i(z_i) = f_i$ and $A_i(z_{i+1}) = f_{i+1}$. The third degree of freedom controls the behavior of the map in the direction orthogonal to the edge. Let $e_i = z_{i+1} - z_i$ and $\hat{e}_i = f_{i+1} - f_i$, then the (unnormalized) normals to the source and target edges are $n_i = \mathrm{i}e_i$ and $\hat{n}_i = \mathrm{i}\hat{e}_i$. We require that A_i transforms the vector n_i to the vector $\rho_i \hat{n}_i$, where $\rho_i \in \mathbb{C}$, and it turns out that the edge-to-edge affine maps can be expressed as

$$A_i(z) = f_i - \frac{(\rho_i + 1)\hat{e}_i B_i(z)}{2e_i} + \frac{(\rho_i - 1)\hat{e}_i \overline{B_i}(z)}{2\overline{e_i}}.$$

If $\rho_i = 1$, then the anti-holomorphic term on the right-hand side above vanishes and $A_i(z) = S_i(z)$. An alternative choice is to use $\rho_i = |e_i|/|\hat{e}_i|$, which maps the source normal to a vector that is orthogonal to the target edge while maintaining its length. This encourages the map to allow stretch only in the direction tangential to the boundary. Figure 7.4 shows a comparison between two types of maps having $\gamma_i = \mathrm{Log}\left(\frac{B_{i+1}}{B_i}\right)$. The first map uses $\rho_i = 1$, and is hence identical to Cauchy's coordinates, while the second map uses $\rho_i = |e_i|/|\hat{e}_i|$.

7.2.3 Magic coordinates

In the previous section we showed how to construct complex barycentric coordinates based on arbitrary weight functions γ_i. Since the similarity transformations S_i are holomorphic functions, it is clear that if γ_i are holomorphic, then the complex barycentric map (7.17) is also holomorphic. Obviously, there exist complex barycentric coordinates that are not holomorphic, as real barycentric coordinates are a special case of complex barycentric coordinates, where the imaginary part is 0. For example, the well-known Wachspress, mean value, and discrete harmonic coordinates can be derived using the weights

$$\gamma_i(z) = \frac{e_i}{\operatorname{Im}(\overline{B_i}B_{i+1})}\left(\frac{|B_{i+1}|^p}{B_{i+1}} - \frac{|B_i|^p}{B_i}\right), \tag{7.18}$$

with $p = 0$, $p = 1$, and $p = 2$, respectively. It is easy to verify that for $p = 2$ the γ_i in (7.18) are real-valued functions and complex-valued functions otherwise. In fact, it turns out [416, Theorem 3] that the discrete harmonic coordinates are the only real-valued barycentric coordinates with a real-valued γ_i function.

Holomorphicity is a strong condition that conflicts with the need to interpolate the boundary data. While all holomorphic barycentric maps are non-interpolating, almost all real barycentric maps are interpolating. It is interesting then to explore new types of complex barycentric coordinates that are interpolating, yet not purely real. These are obviously not holomorphic.

Sufficient conditions for interpolation of complex barycentric coordinates are discussed in [416]. Loosely speaking, each weight function γ_i should approach infinity on its corresponding edge (including its endpoints) while being finite elsewhere. This ensures that the limit of the normalized weight $\tilde{\gamma}_i$ is 1 when approaching a point on the open (i.e., excluding the endpoints) i-th edge from within the domain, while all other weights $\tilde{\gamma}_j$, $j \neq i$ approach zero. Hence, the barycentric map $f = \sum_{i=1}^{n} S_i \tilde{\gamma}_i$ maps the open i-th edge as S_i. For the exact sufficient conditions, as well as the conditions to interpolate the vertices, see [416, Section 3.1].

Let us demonstrate how to design "custom-made" coordinates with desirable properties, in particular the Lagrange property. To interpolate the vertices, we include in the weight function γ_i the term $1/|B_i(z)B_{i+1}(z)|$, which grows to infinity at the i-th edge endpoints and decays as the distance to the edge increases. This term is finite on the edge itself, hence we include the term $1/(\pi - \theta_i(z))$, where $\theta_i(z) = \operatorname{Im}(\operatorname{Log}(B_{i+1}(z)/B_i(z)))$ is the signed angle of the triangle $[z, z_i, z_{i+1}]$ at z (see Figure 7.1). The angle $\theta_i(z) \in (-\pi, \pi]$ approaches π as z approaches the open edge (z_i, z_{i+1}) from within the cage, causing the weight function to explode, providing interpolation. Finally, we add a third term that weighs the contribution of each edge according to its length $|e_i|$. Combining the three terms results in the weight functions of the *Made-to-order Angle Guided Interpolating Coordinates* (MAGIC)

$$\gamma_i(z) = \left|\frac{e_i}{B_i(z)B_{i+1}(z)}\right| \frac{1}{\pi - \theta_i(z)}.$$

Since these weight functions are strictly positive everywhere, it is clear that their sum is non-vanishing, hence they correspond to barycentric coordinates. Moreover, γ_i are C^∞ except at the edges where they have a jump discontinuity, hence the magic coordinates are smooth inside the cage but do not extend smoothly to the exterior of the cage. Figure 7.5 provides a comparison between magic and (real) harmonic coordinates [215].

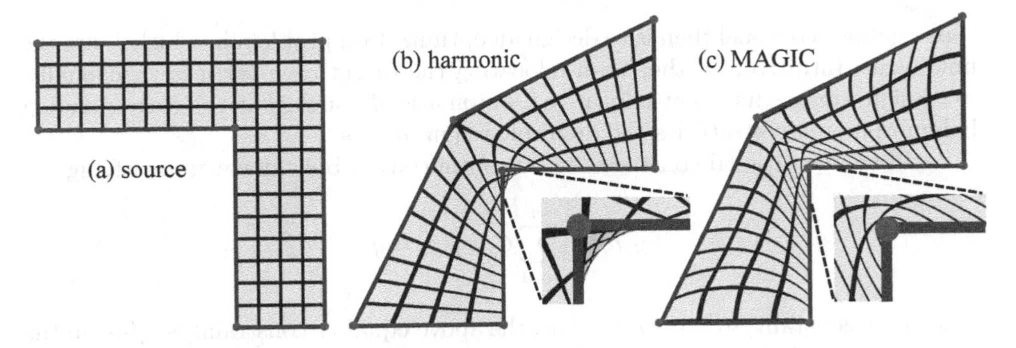

Figure 7.5 Comparison of magic and harmonic coordinates. The upper part of the cage in (a) is bent to the right. (b) The result of harmonic coordinates. Note the spilling of the image outside the cage in the vicinity of the corner where the map is not bijective (see the zoom-in images). (c) The result with the magic coordinates. See color insert.

7.3 VARIATIONAL BARYCENTRIC COORDINATES

So far we have constructed complex barycentric coordinates with a closed-form expression. Such coordinates are efficient to compute, simple to implement, and have high numerical accuracy. We now introduce other types of coordinates that possess additional properties that are highly desirable in the context of planar shape deformation. These coordinates, unfortunately, do not possess a closed-form expression and are approximated numerically. While their derivation is more involved, the additional computational effort can be performed in a preprocessing step. Once the computation of the coordinates is completed, in runtime, the evaluation of the barycentric map is as efficient as any other barycentric map. This new type of coordinate obeys some variational principles and is considered optimal with respect to such principles. Such coordinates are obtained through the solution of a numerical optimization problem.

7.3.1 Point-based barycentric maps

A limitation of holomorphic maps is that they cannot, in general, satisfy positional boundary conditions at *every* boundary point, therefore holomorphic barycentric maps do not possess the Lagrange property. Nonetheless, we can always find a holomorphic map with some desirable values at a *finite* set of points. Prescribing the exact behavior of the map at such points allows the users to better control the deformation and fulfill their design goals. Let $r_j \in \Omega$, $j = 1, \ldots, p$ be a set of user-defined points inside the domain. We would like to find a holomorphic function $g\colon \Omega \to \mathbb{C}$ that maps r_j to prescribed target positions q_j. In order to guarantee that the map is holomorphic, we use the discrete Cauchy transform $g(z) = \sum_{i=1}^{n} C_i(z) f_i$ in (7.12), where for any choice of the coefficients f_i, the map g is holomorphic by construction. Our goal then is to design an optimization problem in which f_i are the unknowns. Intuitively, rather than relocating the target cage vertices f_i manually, we design a procedure that relocates f_i automatically such that the corresponding holomorphic map g satisfies the desirable requirements.

Adding a positional constraint to our optimization boils down to satisfying

$$g(r_j) = \sum_{i=1}^{n} C_i(r_j) f_i = q_j.$$

Since r_j is constant, so is $C_i(r_j)$. Thus, the above equality constraint is affine in the variables f_i and easy to enforce. Setting $p = n$, such constraints correspond to a nonsingular linear system of equations with a unique solution. One possible choice is to use $r_j = z_j$, that is, to design holomorphic complex barycentric coordinates that interpolate the vertices of the cage. The edges themselves will not be interpolated, though.

Trying to add more than n constraints leads to an overdetermined system that has no solution. A possible remedy is to obtain a least-squares solution that minimizes an energy that measures the Euclidean distances between the image of the points r_j and their desired target positions q_j,

$$\min_{f_1,\ldots,f_n} \sum_{j=1}^{p} |q_j - g(r_j)|^2 = \min_{f_1,\ldots,f_n} \sum_{j=1}^{p} \left| q_j - \sum_{i=1}^{n} C_i(r_j) f_i \right|^2 . \tag{7.19}$$

In [415], the points r_j are taken to be dense uniform samples on the source cage, and q_j are the corresponding samples on the target cage. The solution to the least-squares problem (7.19) is obtained by solving a linear system, leading to a new type of barycentric coordinates that are called the Szegő coordinates. Since the energy is not zero, in general, the Szegő barycentric map does not interpolate the target cage. However, it provides better approximation compared to the Cauchy barycentric map, which has a higher energy (7.19).

7.3.2 Point-to-point barycentric coordinates

Till now, the user interacted with the deformation system by manipulating the vertices of the cage. In many cases, the user is not interested in prescribing the exact shape of the boundary of the deformed shape. Especially for shapes with a complicated boundary that consists of hundreds of vertices, manipulating each vertex separately is time-consuming. Instead, it makes more sense to prescribe the position of a *small* set of $p < n$ points (also called P2P handles) that are not necessarily part of the boundary, while allowing the boundary to deform freely. In this scenario, the corresponding linear system is underdetermined and has an infinite number of solutions. This provides an opportunity to pick a particular solution with additional desirable geometric properties. A simple, yet powerful choice is to regularize the solution, allowing the barycentric map to be as smooth as possible. This can be achieved by minimizing some aggregated differential quantity. More specifically, we wish to solve the optimization problem

$$\min_{f_1,\dots,f_n} \quad \frac{1}{2}\int_{\partial\Omega} |g''(w)|^2 ds, \qquad (7.20)$$
$$\text{s.t.} \quad g(r_j) = q_j, \qquad j = 1,\dots,p,$$

where $f_i \in \mathbb{C}$ are the variables, $g(z)$ is the discrete Cauchy transform, and $g''(z) = \sum_{i=1}^{n} C_i''(z) f_i$ is its second complex derivative. The choice of g'' rather than g' in (7.20) is due to the fact that the second complex derivative of a similarity transformation $S(z) = \alpha z + \beta$ is zero everywhere. Therefore, if the user transforms the P2P handles using a similarity $S(z)$, the solution to the optimization is $f_i = S(z_i)$ with the corresponding map $g(z) = S(z)$ having zero energy. In particular, the map g reproduces the identity map if the P2P handles are fixed, that is, if $g(r_j) = r_j, j = 1,\dots,p$, then $g(z) = z$.

Luckily, the derivatives of Cauchy coordinates possess simple closed-form expressions [414, Appendix C–E]. In particular the second derivative is given by

$$C_i''(z) = \frac{1}{2\pi\mathrm{i}}\left(\frac{1}{B_{i-1}(z)B_i(z)} - \frac{1}{B_i(z)B_{i+1}(z)}\right). \qquad (7.21)$$

The boundary integral in (7.20), however, is approximated numerically. The simplest way to approximate the integral is to substitute it with a summation over a dense set of $k \gg n$ boundary samples, where the sampling should be as uniform as possible. This allows us to assume that the differential ds is approximately constant. While Cauchy coordinates have well-defined limits at the vertices [414, Appendix B], the second derivative in (7.21) is singular at the vertices. There are two possible ways to avoid these singularities. One is to exclude the cage vertices from the set of samples (note that (7.21) is well-defined on the open edges). A second choice is to offset the source cage in the outward normal direction by a very small amount, and to use the outward offset cage as our new cage. The original cage (which is strictly inside the offset cage) is then sampled. This choice also makes the evaluation of the Cauchy coordinates themselves simpler.

With the integral being approximated as a sum, we can express (7.20) in matrix form

$$\min_{\boldsymbol{f}} \quad \frac{1}{2}\|\boldsymbol{E}\boldsymbol{f}\|^2, \qquad (7.22)$$
$$\text{s.t.} \quad \boldsymbol{A}\boldsymbol{f} = \boldsymbol{q},$$

where $\boldsymbol{f} \in \mathbb{C}^{n\times 1}$ is the complex column vector with f_i as entries, $\boldsymbol{E} \in \mathbb{C}^{k\times n}$ is the matrix of second complex derivatives of the Cauchy coordinates evaluated at the samples, $\boldsymbol{A} \in \mathbb{C}^{p\times n}$ is the matrix of Cauchy coordinates evaluated at the P2P handles, and $\boldsymbol{q} \in \mathbb{C}^{p\times 1}$ is the vector with the user-specified target P2P positions.

Equation (7.22) is a linearly constrained least-squares problem. Its solution can be obtained using the method of Lagrange multipliers. Let $\boldsymbol{\lambda} \in \mathbb{C}^{p\times 1}$ be additional variables (the Lagrange multipliers), then the solution to (7.22) can be obtained by solving the linear system

$$\begin{pmatrix} M & A^H \\ A & 0 \end{pmatrix} \begin{pmatrix} f \\ \lambda \end{pmatrix} = \begin{pmatrix} 0 \\ q \end{pmatrix}, \tag{7.23}$$

where $|\cdot|^H$ stands for the conjugate transpose matrix, and the matrix $M \in \mathbb{C}^{n \times n}$ is given by $E^H E$. Let $B \in \mathbb{C}^{n+p \times n+p}$ be the block matrix on the left-hand side of (7.23). Let the block matrix $B^{-1} = \begin{pmatrix} \cdot & T \\ \cdot & \cdot \end{pmatrix}$ be the inverse of B, with $T \in \mathbb{C}^{n \times p}$, then the solution to (7.22) is given by $f = Tq$. The matrix T represents a linear transformation that maps the user-prescribed target positions of the P2P handles q_j to the (unknown) target cage vertices f_i. Computing T boils down to evaluating B and inverting it. Evaluating B is embarrassingly parallel and can be computed efficiently using modern GPUs. Since B is typically small, the time for inverting it is often short and the typical runtime for computing T for a complicated shape is less than a second. More importantly, this computation is done once in preprocessing.

In a typical deformation application, the barycentric coordinates are also evaluated in a preprocessing step on a set of m points inside Ω on which we want to evaluate the barycentric map. Let $C \in \mathbb{C}^{m \times n}$ be a matrix of Cauchy coordinates evaluated at these points, then during interaction with the system, the only computation that is done is that of the vector $d \in \mathbb{C}^{m \times 1}$ of deformed points, which is given by the matrix-vector multiplication $d = Cf$. Since $f = Tq$, we have $d = Pq$, where $P = CT$ is in $\mathbb{C}^{m \times p}$. The matrix P can be computed in preprocessing as well, hence during interaction we only need to compute the multiplication Pq, which is more efficient than Cf since the number of P2P handles p is typically much smaller than the number of vertices in the cage n.

Finally, the columns of P can be interpreted as a special type of complex barycentric coordinates, where the vertices of the cage in the barycentric map are substituted with the P2P handles. We call them the P2P-Cauchy coordinates. Calling them barycentric coordinates is justified by the fact that they have constant precision (7.1) and linear precision (7.2). Figure 7.6 shows two examples of image deformation made with the P2P-Cauchy coordinates. An alternative derivation of P2P coordinates can be done by substituting the equality constraints of (7.20) with soft constraints [415]. This results in a smoother deformation at the expense of only approximating the P2P handles.

It is important to note that the quality of variational coordinates like the Szegő and P2P-Cauchy coordinates can be dramatically improved at the expense of longer preprocessing times. Whereas the discrete Cauchy transform always produces holomorphic functions, in contrast to the smooth Cauchy transform (7.8), the discrete transform only spans a finite-dimensional subspace of holomorphic functions. As the number of cage vertices increases, the subspace becomes richer. A simple way to enrich this subspace is to insert "virtual" vertices along the cage edges without altering the geometry of the cage. Typically, the cage is up-sampled as uniformly as possible such that the total number of vertices is at least 200. A non-uniform sampling strategy is also possible, and the actual number of vertices depends on the complexity of the shape and the intended effect.

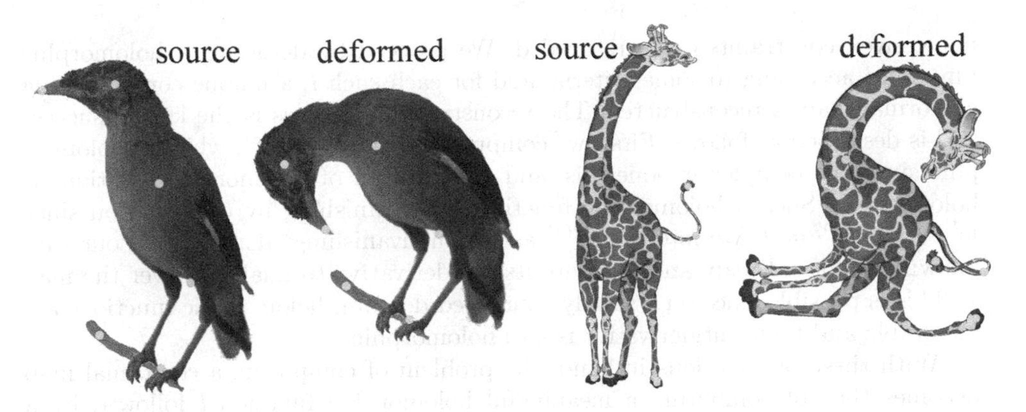

Figure 7.6 Deformation with P2P-Cauchy coordinates. The P2P handles are depicted by cyan disks. See color insert.

7.4 CONFORMAL MAPS

A common mistake is to identify holomorphic functions with planar conformal maps. While a planar conformal map can be expressed as a holomorphic function, the converse is not always true. A holomorphic function $f\colon \Omega \to \mathbb{C}$ is conformal only if its first complex derivative f' is nonvanishing everywhere in the domain. If f is holomorphic and $f'(z_0) = 0$ for a point $z_0 \in \Omega$, then angles between intersecting curves at z_0 are not preserved. Furthermore, the isometric distortion induced by the map is infinite at the singular point z_0, and high in its vicinity. The turning number of the image of the boundary curve increases, and the map is not locally injective at z_0. For planar shape deformation, we are mostly interested in locally injective maps with low distortion and artifacts like these should be avoided.

7.4.1 Log derivative construction

As we saw in the previous sections, generating holomorphic maps is quite simple, but unfortunately, conformal maps do not form a linear subspace and are harder to control. If f is Cauchy's barycentric map represented as $f(z) = \sum_{i=1}^{n} C_i(z) f_i$, the additional condition $|f'(z_0)| > 0$ that guarantees that f is conformal at z_0 is a non-convex constraint. Non-convex constraints pose a great challenge for numerical optimization and cannot be enforced efficiently in general. We can address this difficulty by parameterizing the space of conformal maps, by using alternative variables. On a simply connected domain $\Omega \in \mathbb{C}$, there is a one-to-one correspondence between the space of conformal maps and the space of holomorphic maps [241, 417]. If f is conformal, it is also holomorphic. The derivative of a holomorphic map is holomorphic, hence f' is holomorphic and nonvanishing. The complex logarithm of a nonvanishing holomorphic function is holomorphic, meaning that if f is conformal, then $l(z) = \log(f'(z))$ is holomorphic. With this observation in hand, we represent our conformal map based on the holomorphic function l. No special non-convex constraints on l are needed. We proceed by designing a holomorphic function l according to some criteria, and for each such l, a unique corresponding conformal map is reconstructed. The reconstruction process is the key to success and is described as follows. First, we compute the function $\mathrm{e}^{l(z)}$, which is holomorphic since the complex exponent is, and composition of holomorphic functions is holomorphic. Such a holomorphic function is nonvanishing by construction since $\left|\mathrm{e}^{l(z)}\right| = \mathrm{e}^{|l(z)|} \neq 0$. We interpret $\mathrm{e}^{l(z)}$ as the (nonvanishing) derivative of our (unknown) conformal map, and compute its antiderivative to finally recover the map f. This is possible since on a simply connected domain, holomorphic functions are integrable and their antiderivative is also holomorphic.

With these observations in hand, the problem of computing a conformal map becomes that of computing a meaningful holomorphic function l followed by a reconstruction step. Recalling the discussion on the derivative of holomorphic functions in Section 7.2.1, and the definition of the complex logarithm, we have that the function l encodes the amount of local scale and rotation in its real and imaginary parts, respectively,

$$l(z) = \log(f'(z)) = \ln(|f'(z)|) + \mathrm{i} \arg f'(z).$$

We discretize the function l as

$$l(z) = \sum_{i=1}^{n} D_i(z) l_i, \tag{7.24}$$

with the basis functions

$$D_i(z) = \frac{1}{2\pi \mathrm{i}} \gamma_i^{\text{Cauchy}}(z) = \frac{1}{2\pi \mathrm{i}} \operatorname{Log}\left(\frac{B_{i+1}(z)}{B_i(z)}\right).$$

Other choices for discretization are possible, including higher order approximations [417], but this simple choice suffices for the sake of discussion. The coefficients l_i in (7.24) can be interpreted as the piecewise constant values of $\log(f'(z))$ on the edges of the cage and our goal is to estimate them from the data provided by the user, that is, the shape of the target cage. Our simple choice is to aim at designing a conformal map f that acts as a (different) similarity transformation on each edge. Given the unique similarity S_i from (7.9) that maps the source edge to the target edge, the derivative is given by $S_i'(z) = \hat{e}_i / e_i$ and we use $l_i = \log(\hat{e}_i / e_i)$.

When computing l_i in practice, care should be taken when evaluating the complex logarithm, which is a multi-valued function. Standard C++/MATLAB software implementations only produce the single-valued principal branch, and using it independently on each edge may lead to wrong assignment for the imaginary part that encodes the rotation of the edge in radians. Instead, we suggest the following strategy to recover the correct angles. The rotation of the first edge is computed as $l_1 = \operatorname{Log}(\hat{e}_1 / e_1)$, where Log is the standard principal branch implementation with imaginary values in the range $(-\pi, \pi]$. To compute the subsequent l_i, we first compute the difference of corner vertex angles between the source and target cages,

$$d_i = \operatorname{Log}\left(\frac{\hat{e}_i}{\hat{e}_{i-1}}\right) - \operatorname{Log}\left(\frac{e_i}{e_{i-1}}\right),$$

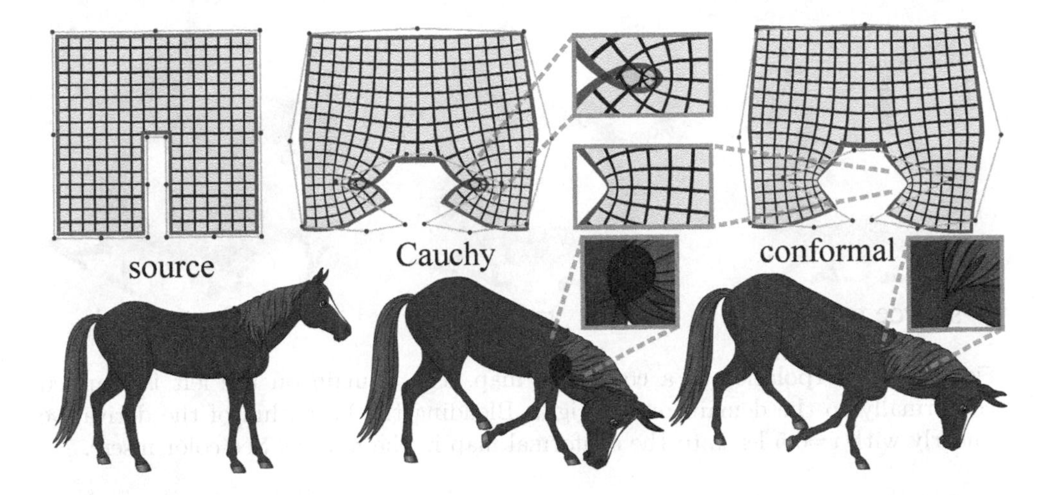

Figure 7.7 A comparison between holomorphic and conformal maps. Left: source domains. Middle: deformation with Cauchy's coordinates. Note the vanishing derivative which is emphasized in the zoom-in images. Right: conformal maps obtained from the same target cage. See color insert.

and then accumulate these values to obtain the correct rotation of the edges,

$$l_i = l_{i-1} + d_i, \qquad i = 2, \ldots, n.$$

We assume that the turning number of the target cage is 1, or equivalently that $\sum_{i=1}^{n} d_i = 0$. This makes sense since a necessary condition for local injectivity of any map (in particular, conformal) is that the image of the boundary curve under the map forms a self-overlapping curve (which implies that the turning number is one) [419].

Plugging l_i into (7.24) provides a closed-form expression for the logarithm of the derivative $l(z)$ of our conformal map $f(z)$ at any point $z \in \Omega$. Furthermore, the derivative is given by the single-valued function $\mathrm{e}^{l(z)}$, which also has a closed form. Yet in order to obtain the actual map f, we need the antiderivative $\int \mathrm{e}^{l(z)}$, which cannot be expressed in a straightforward manner. This antiderivative is holomorphic (hence precisely integrable) but it does not belong, in general, to the finite-dimensional space spanned by the discrete Cauchy transform. To alleviate this, we can either project $\int \mathrm{e}^{l(z)}$ to the Cauchy's finite-dimensional space [417] (which boils down to solving a small dense linear system), or approximate the integral numerically [101, 241]. Both alternatives are very efficient to compute.

The reconstruction of the conformal map from its derivative is only unique up to a complex constant that corresponds to a global translation. The simplest way to nail down this degree of freedom is to allow the user to pick an anchor point $z_0 \in \Omega$, and set $f(z_0) = \sum_{i=1}^{n} \phi_i(z_0) f_i$, where ϕ_i are barycentric coordinates (we used Cauchy's coordinates in Figure 7.7), and f_i are the target cage vertices. Since

Figure 7.8 Interpolation of a conformal map. The domain on the left is mapped conformally to the domain on the right. Blending the logarithm of the derivative linearly with t=0.5 leads to the conformal map in the middle. See color insert.

the reconstructed conformal map does not interpolate the target cage, a slightly more sophisticated alternative is to solve for the translation (and possibly even for a similarity) that will minimize the distance between the target cage and the image of the boundary. Figure 7.7 provides a comparison of deformations between holomorphic maps obtained with the cage-based Cauchy's coordinates and the simple cage-based conformal method that we described. Both methods are highly efficient to compute as they do not require solving a numerical optimization problem at runtime.

7.4.2 Shape interpolation

Planar shape interpolation is a classical problem in computer graphics. It is important for creating animation sequences and for exploring shape spaces. A naive and widely used interpolation approach is to linearly blend the input maps. If $f^1(z)$ and $f^2(z)$ are conformal, then $f^t(z) = (1-t)f^1(z) + tf^2(z)$, for $t \in [0,1]$ is holomorphic but is not, in general, conformal. However, generating conformal maps based on the logarithm of the derivative (see Section 7.4.1) also leads to a natural way for interpolating two or more conformal maps.

The idea is to linearly blend the log derivatives rather than the maps themselves, and then to reconstruct the map (which is conformal by construction). For brevity, let us consider the most common case where we want to blend a given conformal map f with the identity map. Since the logarithm of the derivative of the identity map is 0 we have $l^t(z) = t\,l(z)$, where $l(z)$ is the log derivative of $f(z)$. Blending l linearly leads to natural visually pleasing results, since rotations (the imaginary part of l) are blended linearly. Figure 7.8 shows such a result where we computed l using the simple technique described in Section 7.4.1. Then we reconstructed the conformal map as before and repeated the reconstruction process with $0.5\,l(z)$ as input.

7.4.3 Variational conformal maps

The simple strategy of constructing a conformal map from l can be extended by using a variational approach, designing a numerical optimization in which the real and imaginary parts of l_i in (7.24) are variables. We can add, for example, equality constraints on the real and/or imaginary parts of $l(r_j)$ at some user-defined points r_j at which the user can *precisely* control the amount of scale and/or rotation induced by the map. Such constraints are linear in the variables and can be easily enforced. In fact, in the continuous case, it is possible to prescribe the scale or the rotation (but not both) on the *entire* boundary (of a simply connected domain). In [417], the rotation is fully prescribed on the boundary and a linear system is solved to recover l. Furthermore, the method of [241] can be used to enforce additional *inequality* constraints that bound the isometric distortion (recall Section 7.2.1),

$$\tau(w) = \big|\ln|f'(w)|\big|.$$

The goal is to bound the distortion at a point y_j by some positive user-defined constant Γ. This constraint can be expressed as

$$\begin{aligned} \tau(y_j) \leq \Gamma \quad &\Leftrightarrow \\ -\Gamma \leq \ln|f'(y_j)| \leq \Gamma \quad &\Leftrightarrow \\ -\Gamma \leq \mathrm{Re}\,(\log(f'(y_j))) \leq \Gamma \quad &\Leftrightarrow \\ -\Gamma \leq \mathrm{Re}\,(l(y_j)) \leq \Gamma, \quad & \end{aligned}$$

where it is easy to see that the last row corresponds to two inequalities that are *linear* in our variables. Such inequalities are convex and can be enforced efficiently in practice using convex programming [67]. Furthermore, since $\mathrm{Re}\,(l(z))$ is the real part of a holomorphic function, it is harmonic, therefore its minimum and maximum are attained on the boundary. Hence, bounding the isometric distortion of a conformal map on the boundary implies a global bound throughout the entire domain, which guarantees that the conformal map in hand is of high quality. Bounding the distortion on the boundary is still challenging since there are still an infinite number of boundary points. However, due to the smoothness of the basis functions, this can be accomplished by forcing a bound on a finite dense set of boundary samples y_j [95].

The approach presented in this section for computing conformal maps is simple to implement and fairly efficient. Nevertheless, it does not allow for prescription of positional constraints (P2P handles) since the positions are highly nonlinear expressions in the l_i variables. A sophisticated approach based on non-convex programming was introduced in [95]. This method is order of magnitudes slower compared to the ones presented here, but it produces remarkable results and can still run at interactive rates.

7.5 IMPLEMENTATION DETAILS

In this section we provide some directions for the efficient implementation of a planar deformation system and point out some practical aspects that should be considered.

7.5.1 Visualizing planar maps

Accurate and efficient visualization of barycentric maps is an important part of our shape deformation system. To this end, we should make an effort to perform as much as possible computations in preprocessing for the sake of faster computations during interaction. In addition, it is advised to utilize parallelism and in particular to shift computations to the graphics processer when possible.

The simplest and most efficient way to visualize the map is to use the GPU's ability to efficiently texture map triangle meshes. The user provides the polygonal cage that defines the source domain and a dense triangulation (a mesh) of the domain is computed (once) in preprocessing. We have found the code of [355] useful, though other high quality and efficient triangulation algorithms exist. At runtime, the map is evaluated on the vertices of this mesh and the GPU renders the deformed mesh with the original vertex positions as (u, v) texture coordinates. Since the GPU's rasterizer assumes that the map is piecewise linear on the triangles, it is preferable to use a high density mesh in order to obtain smooth results.

The next step is to compute the barycentric coordinates and store them in a large dense matrix. Some care should be taken with certain types of coordinates such as Cauchy's, since their expression (7.13) is singular on the boundary. The limits when approaching an edge or a vertex from within the cage exist [414], but special classification of the vertices according to their position is required. This complicates the implementation and is less suitable for implementation on the GPU, where regularity is crucial. A simple "trick" to avoid this problem is to offset the cage in the outward normal direction and to use the offset cage for the derivation of the coordinates. The amount of offsetting should be small enough to avoid visual artifacts and large enough to avoid numerical errors when evaluating the vertices that lie on the boundary. We used a default value of 0.0001, where the cages we used have a diameter of 10 units.

Evaluating basis functions is an embarrassingly parallel process. For each internal mesh vertex v_j and each cage vertex z_i the coordinate function $C_i(v_j)$ can be computed independently. Moreover, the regular memory access pattern is suited perfectly for a GPU implementation. Algorithm 1 shows our MATLAB code for computing a matrix of Cauchy's coordinates. The running time on a machine with Nvidia's Quadro K6000 processor for a mesh with $125,000$ vertices of the pants shape of Figure 7.2 is 17 milliseconds.

The most critical computation is the evaluation of the barycentric map, since this is done during interaction. Here again, there is room for exploiting parallelism, since the evaluation of $f(v_j) = \sum_{i=1}^{n} C_i(v_j) f_i$ is independent for each internal vertex v_j. This computation can be expressed as a matrix-vector multiplication,

Algorithm 1 Parallel MATLAB code for computing Cauchy's coordinates. `z` is a complex vector of internal mesh vertices, and `z_i` is a complex vector of offset cage vertices. The output is the coordinates matrix `C`. This code runs in parallel on the CPU. Moreover, if a CUDA-enabled GPU is available, and `z` is of type gpuArray, the computation is automatically performed on the GPU.

```
e_i = z_i([2:end 1], :) - z_i;
e_i_m = e_i([end 1:end-1], :);
B_i = bsxfun(@minus, z_i.', z.');
B_i_p = B_i(:, [2:end 1]);
B_i_m = B_i(:, [end 1:end-1]);
C = (1/(2*pi*1i))*(B_i_p.*bsxfun(@rdivide, log(B_i_p./B_i), e_i.')
    - B_i_m.*bsxfun(@rdivide, log(B_i./B_i_m), e_i_m.'));
```

for which many efficient parallel implementations are widely available. If the GPU is being used, it is preferable to compute and store the coordinates matrix on the GPU (during the preprocess) without transferring it to the CPU, and to perform the matrix-vector multiplication directly on the GPU. The Matlab code for doing it is as simple as `d=C*f`, where `C` is the matrix computed in Algorithm 1, `f` is the target cage vertices, and `d` is the deformed vertex positions. For maximum efficiency, it is possible to instruct the GPU to perform the rendering of the mesh directly according to `d`, which resides on the GPU memory, avoiding redundant data transfer between the GPU and the CPU.

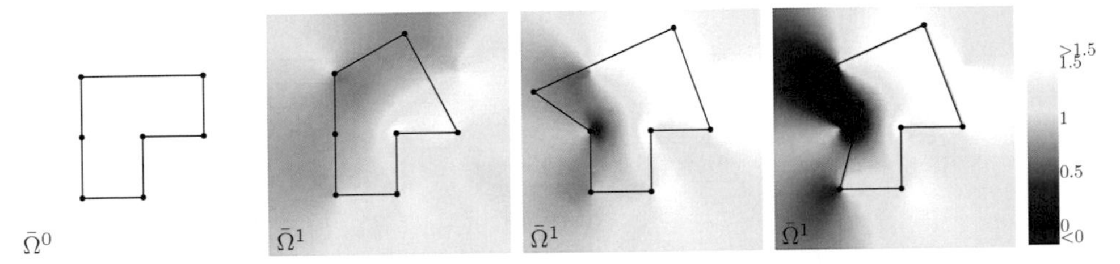

Figure 4.4 Color-coded plots of $J_{\boldsymbol{f}}$ for the mean value mapping $\boldsymbol{f}\colon \bar{\Omega}^0 \to \bar{\Omega}^1$ for different target polygons, which remains positive when the perturbation is small, but becomes negative for large deformations.

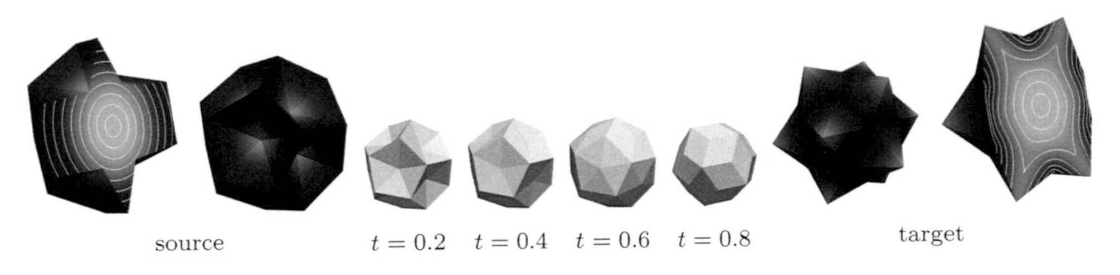

Figure 4.12 Example of a composite mean value mapping between two polyhedra. The color shows how the interior of the source polyhedron is morphed to the interior of the target polyhedron.

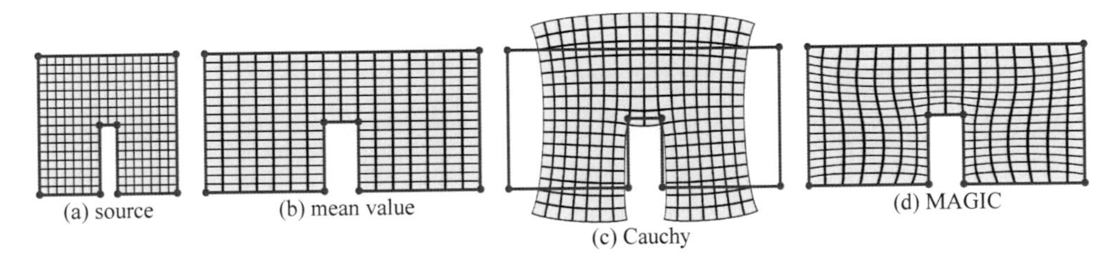

Figure 7.2 Comparison of real and complex barycentric coordinates. The cage of a pants shape in (a) is transformed in (b–d) by the affine transformation $\boldsymbol{g}\colon (x, y) \to (2x, y)$. In (b) the mean value barycentric map reproduces $\boldsymbol{g}$. In (c) and (d), the Cauchy and magic complex barycentric coordinates produce maps that do not reproduce $\boldsymbol{g}$. The map in (c) is conformal.

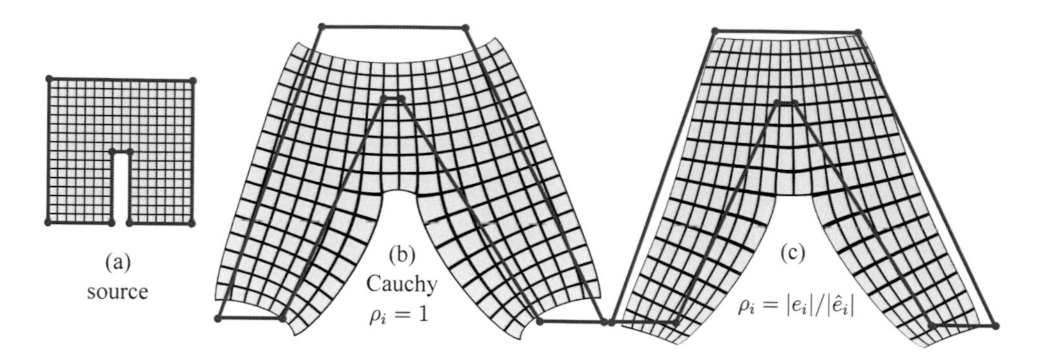

Figure 7.4 Blending edge-to-edge affine maps. A source cage of a pants shape in (a) is deformed identically in (b) and (c) to create an effect of legs stretching. In (b) the edge-to-edge affine maps are chosen to be similarities. The map from (a) to (b) is the holomorphic Cauchy barycentric map. In (c) the edge-to-edge affine maps stretch only in the direction of the edges but not in the orthogonal direction, resulting in a non-holomorphic map with a more natural look that better approximates the target cage.

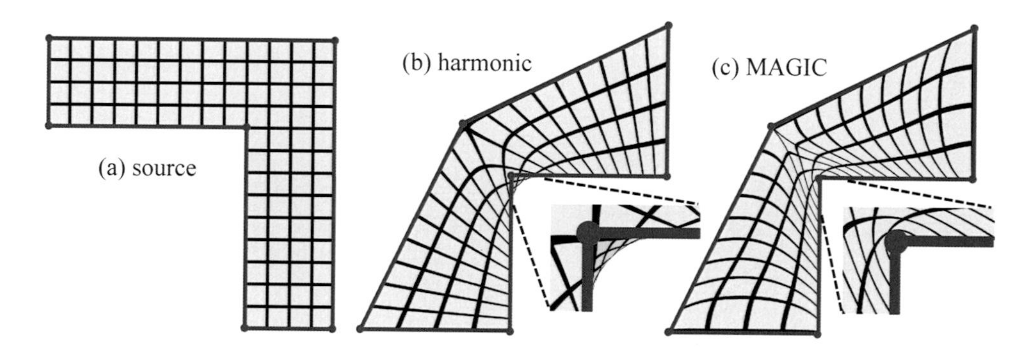

Figure 7.5 Comparison of magic and harmonic coordinates. The upper part of the cage in (a) is bent to the right. (b) The result of harmonic coordinates. Note the spilling of the image outside the cage in the vicinity of the corner where the map is not bijective (see the zoom-in images). (c) The result with the magic coordinates.

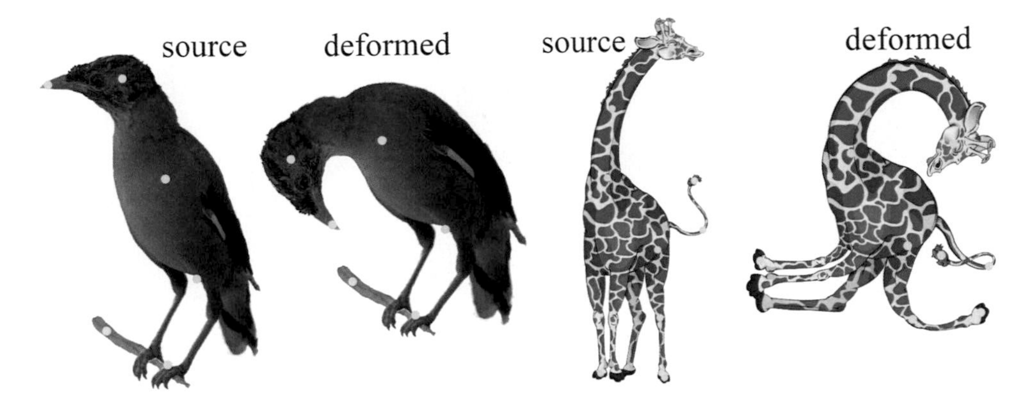

Figure 7.6 Deformation with P2P-Cauchy coordinates. The P2P handles are depicted by cyan disks.

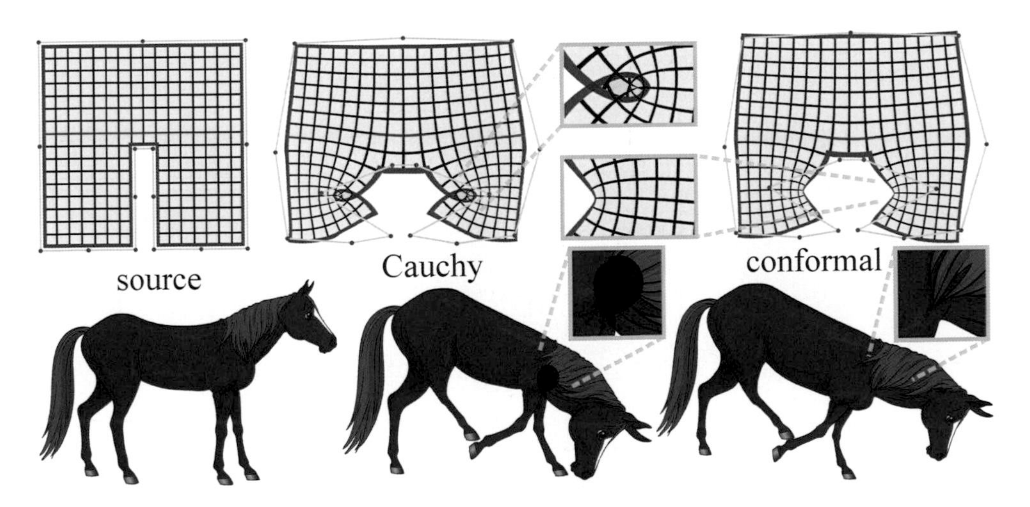

Figure 7.7 A comparison between holomorphic and conformal maps. Left: source domains. Middle: deformation with Cauchy's coordinates. Note the vanishing derivative which is emphasized in the zoom-in images. Right: conformal maps obtained from the same target cage.

Figure 7.8 Interpolation of a conformal map. The domain on the left is mapped conformally to the domain on the right. Blending the logarithm of the derivative linearly with t=0.5 leads to the conformal map in the middle.

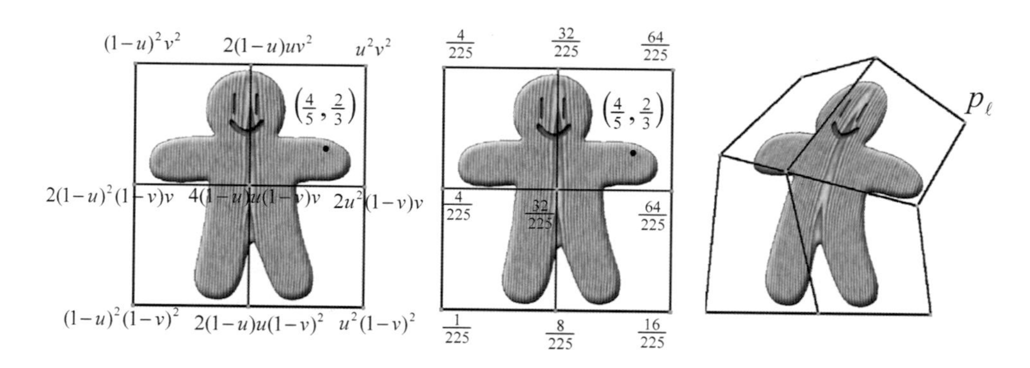

Figure 8.10 Image deformation using S-patches. The left image shows the image and a point to evaluate the deformation at which each control point lists its basis functions using a quadratic S-patch with Wachspress coordinates. The middle image shows the weights associated with the evaluation point after evaluating the basis functions. Moving the control points induces a deformation on the image shown on the right.

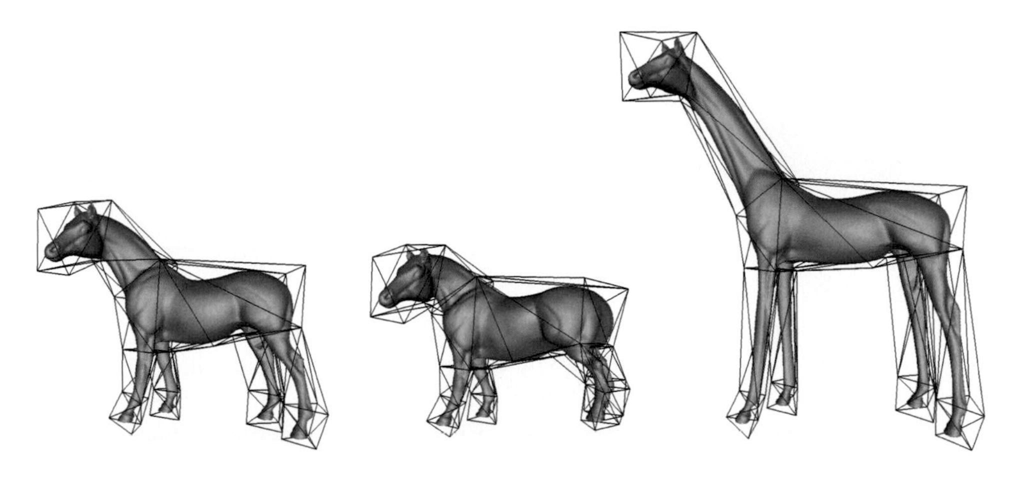

Figure 8.11 An example of surface deformation using linear S-patches. The initial surface of a horse is surrounded by a low resolution approximation called a cage (left). The right two images show different deformations where the head and torso are kept the same size but the neck and legs are compressed or stretched.

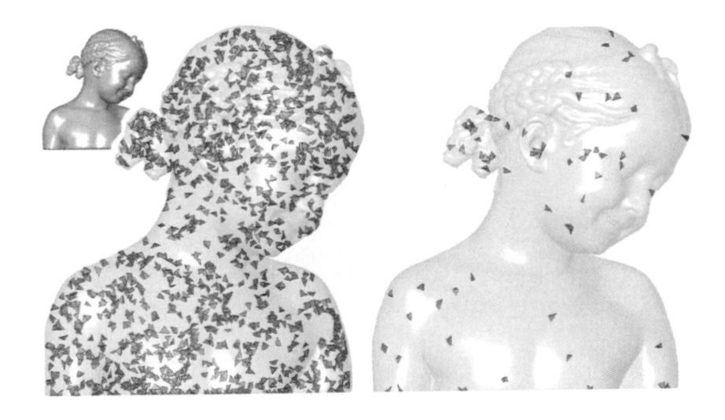

Figure 9.3 In a high-quality optimized Delaunay mesh of Bimba con Nastrino, we show the set of tetrahedra for which the distance between weighted circumcenter and barycenter is greater than a threshold, before (left) and after (right) the HOT optimization [289]. The primal triangulations are the same, and this significant optimization is done only on the dual, using the weights.

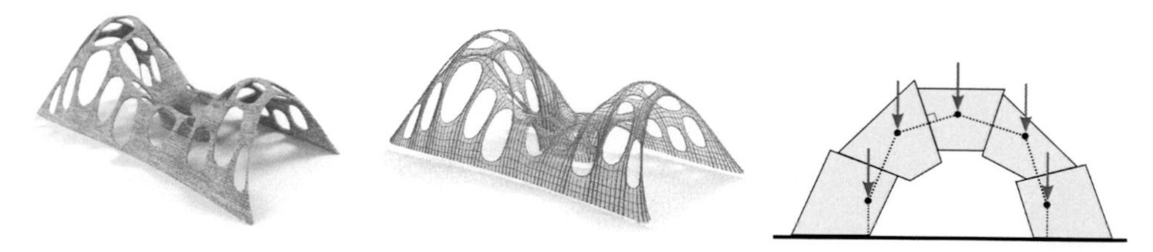

Figure 10.1 Left: A freeform masonry structure, composed of rigid, unreinforced blocks. (Image courtesy of Mathias Höbinger, used by permission.) Middle: The thrust network that certifies that this masonry structure is *stable* under its own weight. Each edge radius is proportional to the magnitude of compressive force carried by the edge. Right: An arch in 2D, with the self-load, and the thrust line certifying its stability, both illustrated.

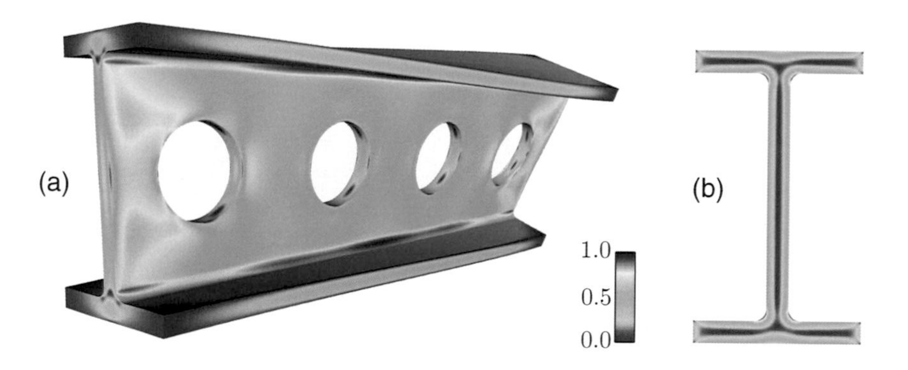

Figure 11.3 (a) Stress field (von Mises) resulting from a finite element analysis of the I-beam shown in Figure 11.1 with applied torsional tractions, (b) cross-section view. (Reproduced with permission from [48].)

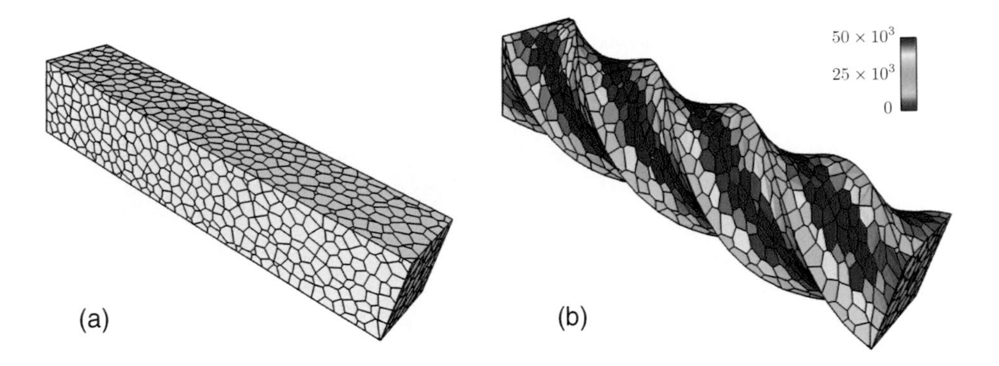

Figure 11.6 Large deformation example, a hyperelastic beam subjected to a torsion loading. One end is fixed while a full 360° rotation is applied to the opposite end. (a) original polyhedral mesh (Voronoi tessellation with maximal Poisson seeding), (b) deformed shape and von Mises stress field.

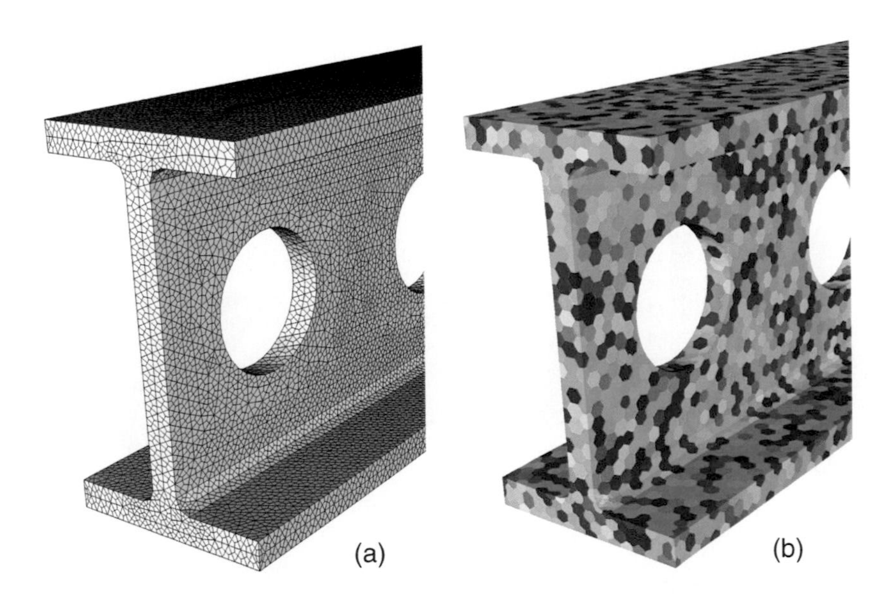

Figure 11.8 Example of polyhedral meshing using barycentric subdivision (see Figure 11.7(a)): (a) Barycentric dual of the tetrahedral mesh shown in Figure 11.1(c); (b) Each polyhedral cell is represented as a different color.

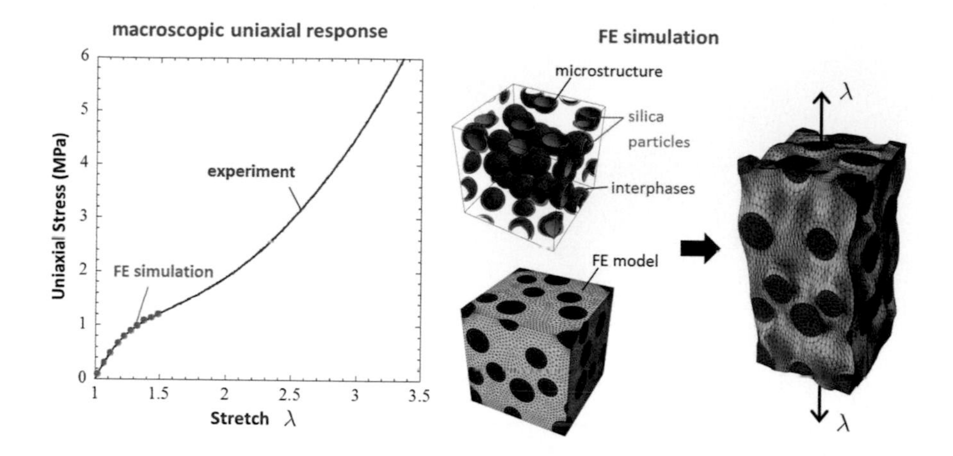

Figure 12.1 An illustration of the limited capability of standard tetrahedral elements to undergo large deformation. Experimentally, a synthetic rubber filled with 20% volume fraction of randomly distributed silica particles can be stretched more than four times its original length ($\lambda = 4$) without internal damage under uniaxial tension [315]. By contrast, a finite element model, based on standard 10-node hybrid elements with linear pressure, is only able to deform to a macroscopic stretch of $\lambda = 1.5$ [174].

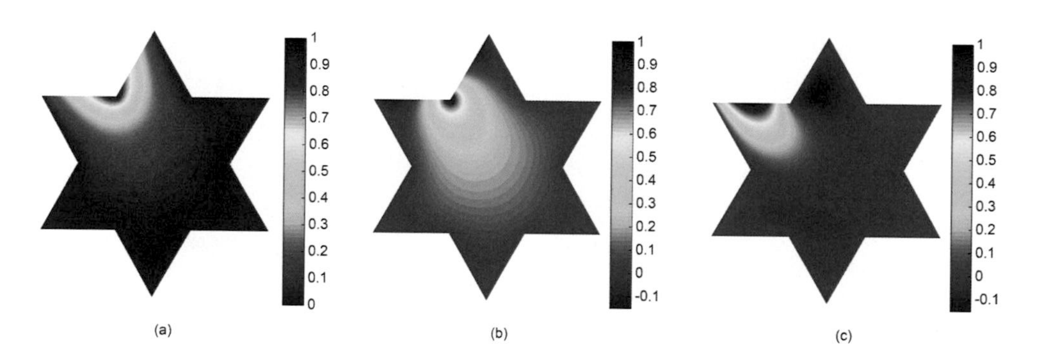

Figure 12.2 (a) Contour plot of a mean value basis $\phi_i^{(1)}$ over a concave polygon. (b) Contour plot of quadratic basis $\phi_i^{(2)}$ over a concave polygon associated with a vertex. (c) Contour plot of quadratic basis $\phi_i^{(2)}$ over a concave polygon associated with a mid-side node.

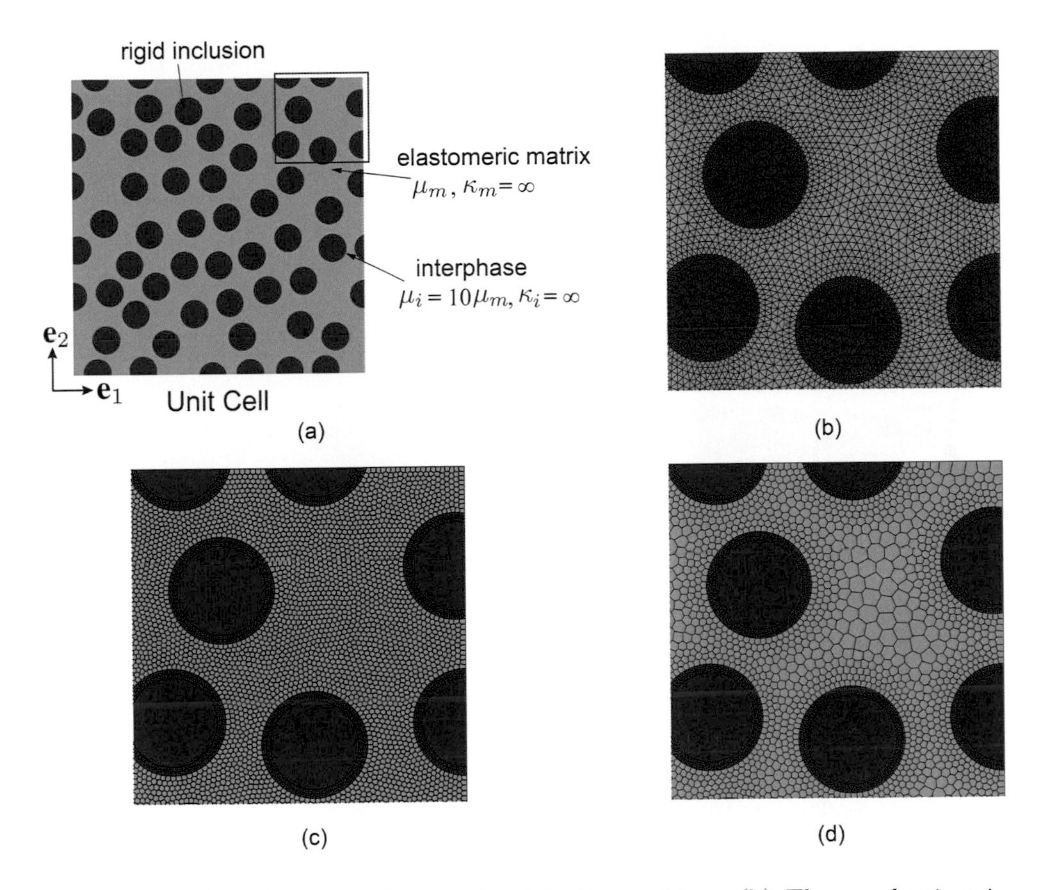

Figure 12.11 (a) The unit cell considered in the problem. (b) The quadratic triangular mesh comprised of 58,814 CPE6MH elements and 118,141 nodes in total, with 71,179 nodes in the matrix phase. (c) The linear polygonal mesh comprised of 39,738 elements and 76,020 nodes in total, with 66,095 nodes in the matrix phase. (d) The quadratic polygonal mesh comprised of 20,205 elements and 95,233 nodes in total, with 67,325 nodes in the matrix phase. This figure is adapted from [100].

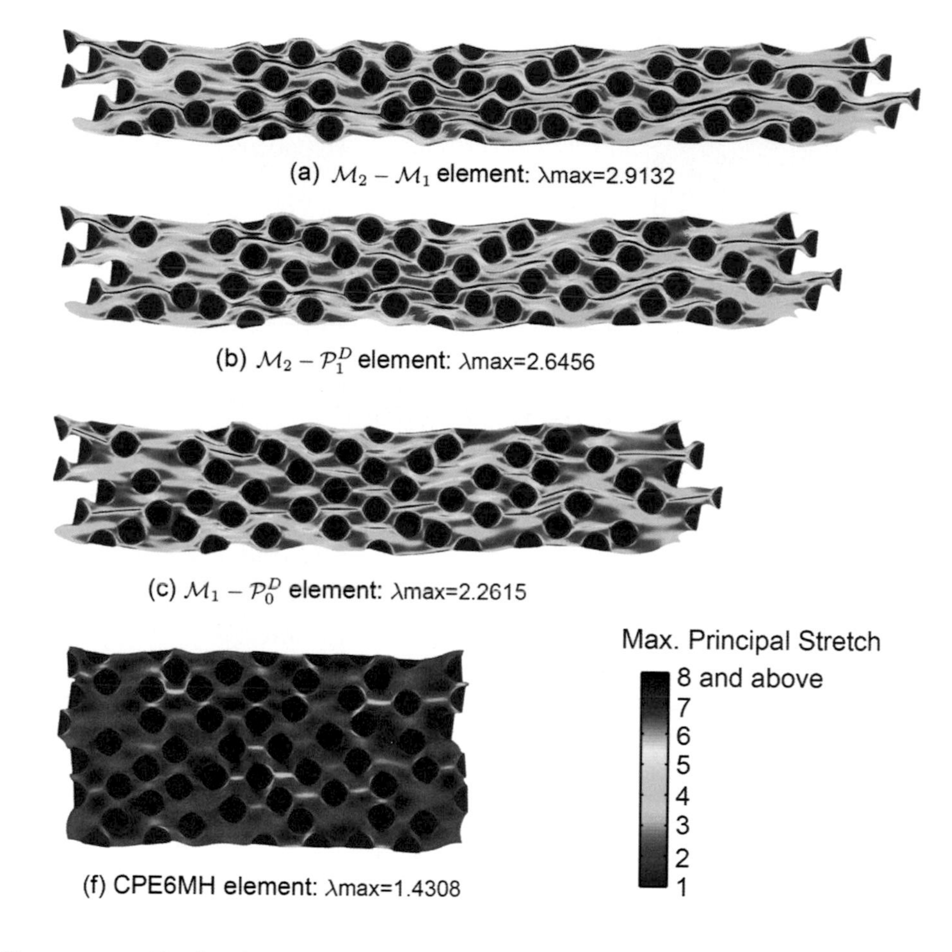

Figure 12.12 For loading condition (i), displaying pure shear with average deformation gradient $\langle \boldsymbol{F} \rangle = \lambda \boldsymbol{e}_1 \otimes \boldsymbol{e}_1 + \lambda^{-1} \boldsymbol{e}_2 \otimes \boldsymbol{e}_2$, $\lambda \geq 1$, the figure illustrates the final deformed configuration reached by (a) $\mathcal{M}_2 - \mathcal{M}_1$ elements, (b) $\mathcal{M}_2 - \mathcal{P}_1$ elements, (c) $\mathcal{M}_1 - \mathcal{P}_0$ elements, and (d) CPE6HM elements (solved in ABAQUS). This figure is adapted from [100].

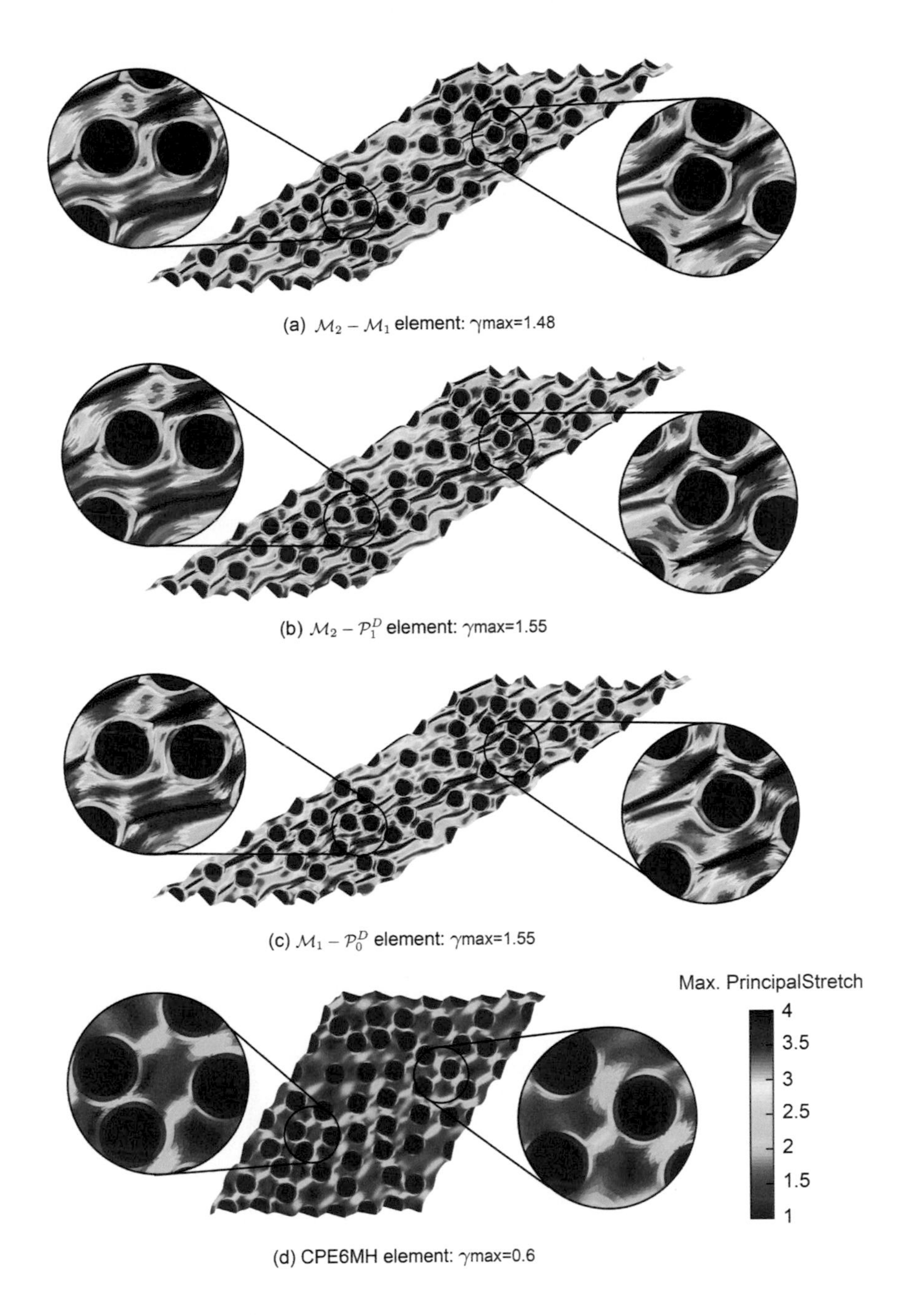

Figure 12.14 For loading condition (ii), displaying $\langle \boldsymbol{F} \rangle = \boldsymbol{I} + \gamma \boldsymbol{e}_1 \otimes \boldsymbol{e}_2$, $\gamma \geq 0$, the figure illustrates the final deformed configurations reached by (a) $\mathcal{M}_2 - \mathcal{M}_1$ elements, (b) $\mathcal{M}_2 - \mathcal{P}_1^D$ elements, (c) $\mathcal{M}_1 - \mathcal{P}_0^D$ elements, and (d) CPE6HM elements (solved in ABAQUS). Both matrix and interphase are considered to be typical silicone rubber.

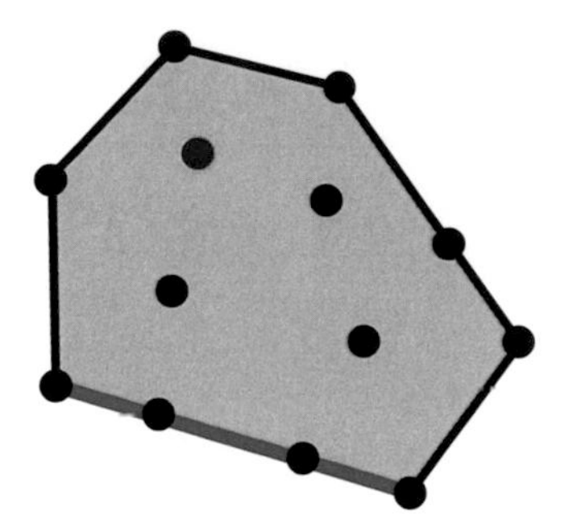

Figure 13.1 Set of vertices V in two dimensions, along with the convex hull P.

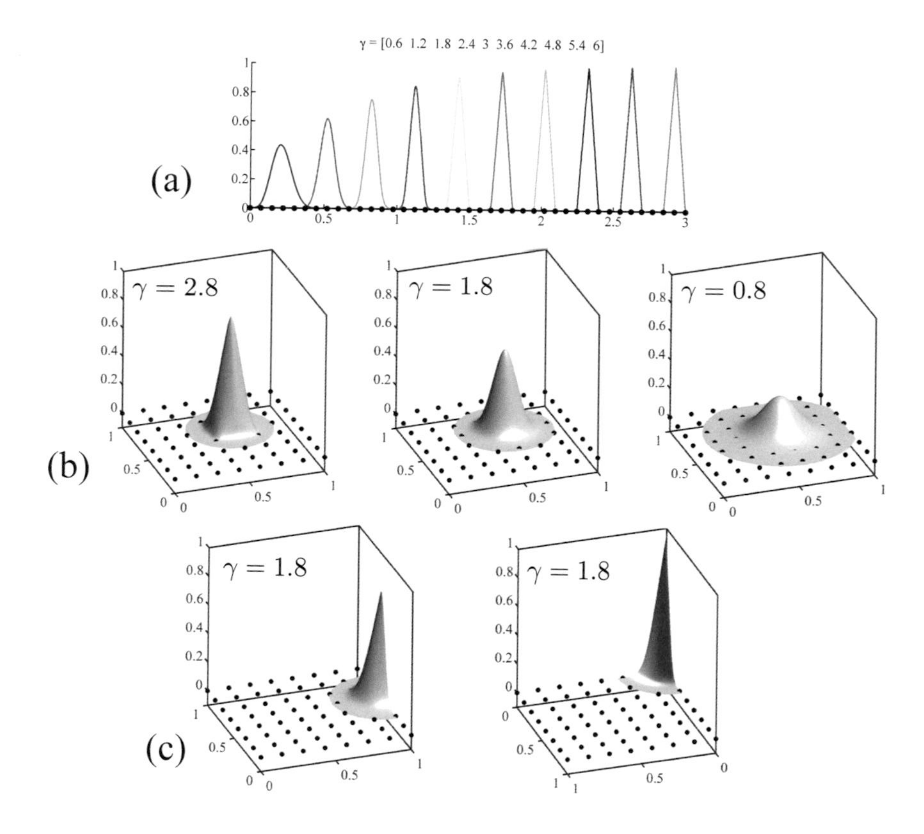

Figure 13.3 Local maximum-entropy (LME) basis functions. (a) Selected basis functions for a one-dimensional vertex set, with varying non-dimensional aspect ratio parameter $\gamma_i = \beta_i h^2$, see main text. (b) Basis functions for an interior node in a two-dimensional set of vertices and different aspect ratio parameters, and (c) representative basis functions of boundary nodes.

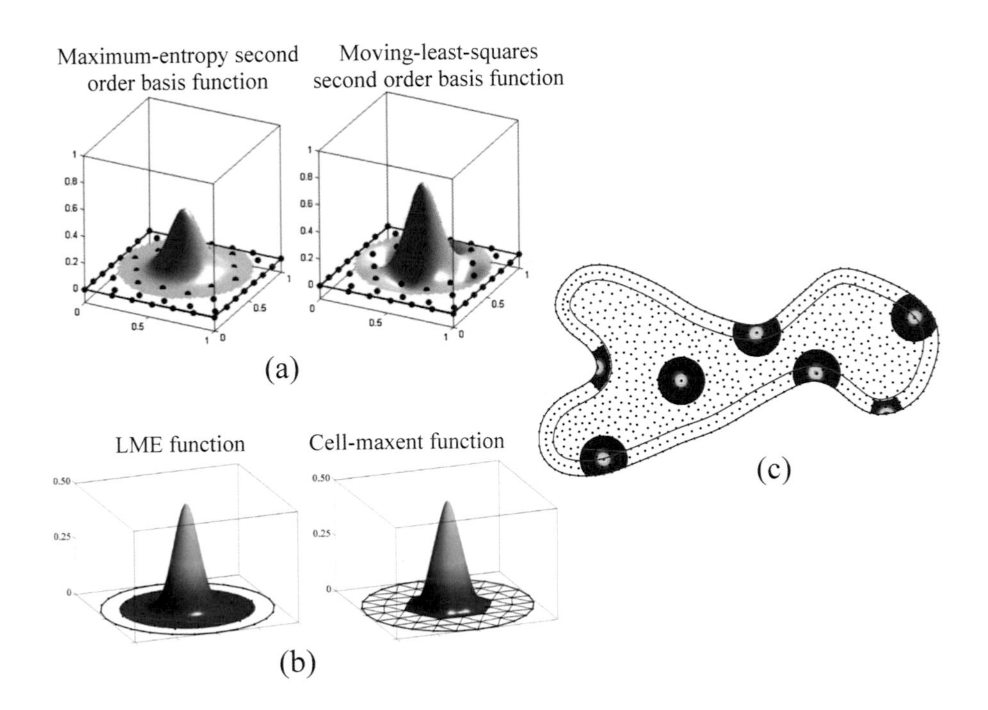

Figure 13.4 Extensions of maximum-entropy approximations. (a) Representative second-order maximum entropy basis function, compared to a moving-least-squares function, which is wiggly and negative at some points. (b) Comparison of a standard LME basis function and a cell-based maximum entropy basis function supported on a triangulation, which results in a structured adjacency relation and more efficient calculations to approximate PDEs for a comparable accuracy. (c) Combination of LME approximants with a boundary representation based on B-splines.

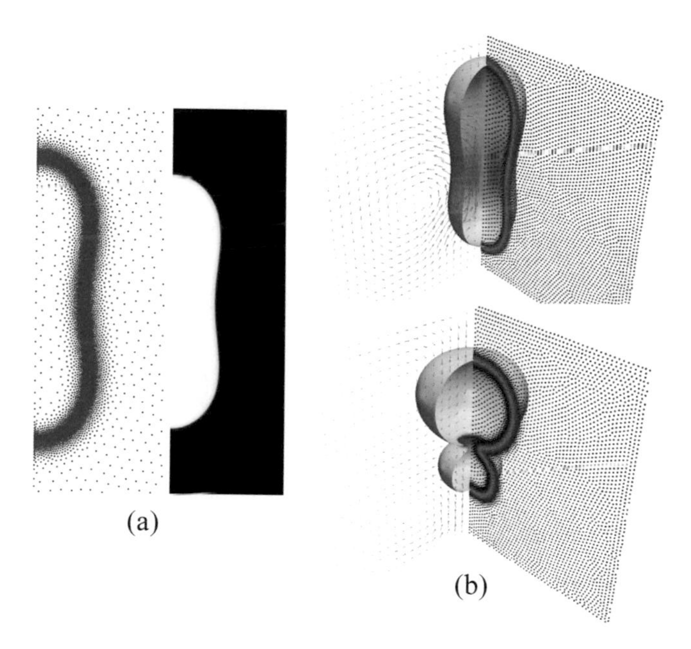

Figure 13.5 Application of the LME approximants to high-order PDEs. (a) In phase-field models of biomembranes, the unknown field defining the shape of a moving interface is governed by a fourth-order PDE. Therefore, a Galerkin method requires smooth basis functions (with square integrable second derivatives). The phase field develops sharp gradients, calling for an adaptive solution and an approximation scheme devoid of Gibbs phenomenon. All these conditions are met by LME approximations. (b) shows two snapshots in a dynamical simulation tracking the shape of a deforming vesicle made out of a fluid membrane with curvature elasticity, together with the hydrodynamics of the surrounding fluid. The resolution of the set of vertices follows the sharp variations of the phase-field.

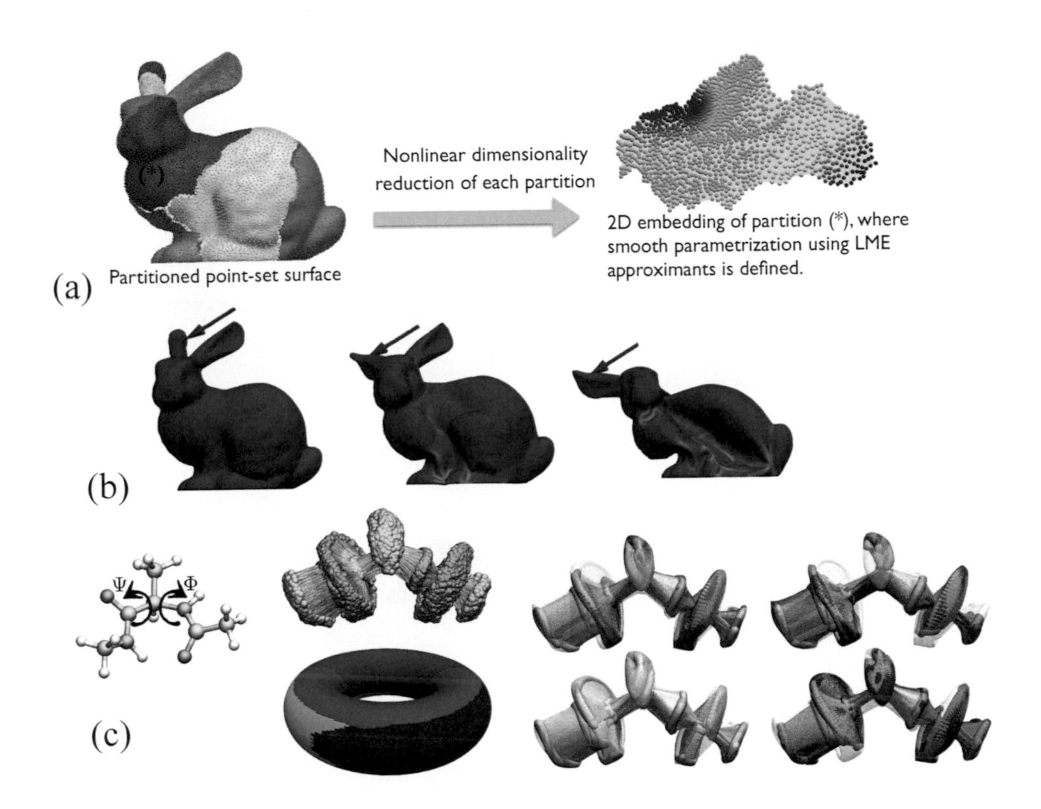

Figure 13.6 Approximation of manifolds using an atlas of smooth LME parametrizations. (a) Methodology to define local parametrizations mapping partitions of a surface defined by a set of points. (b) Application of this method to solve the high-order PDE for a nonlinear Kirchhoff–Love shell. (c) Application of the method to describe the molecular conformations of alanine dipeptide, a small molecule. An ensemble of conformations of this molecule samples an underlying configurational manifold homeomorphic to a torus. With the method presented in (a), this manifold is partitioned into four pieces, each of which is parametrized with LME approximants. This method allows us to define smooth collective variables for this system, required to enhance sampling in molecular dynamics simulations.

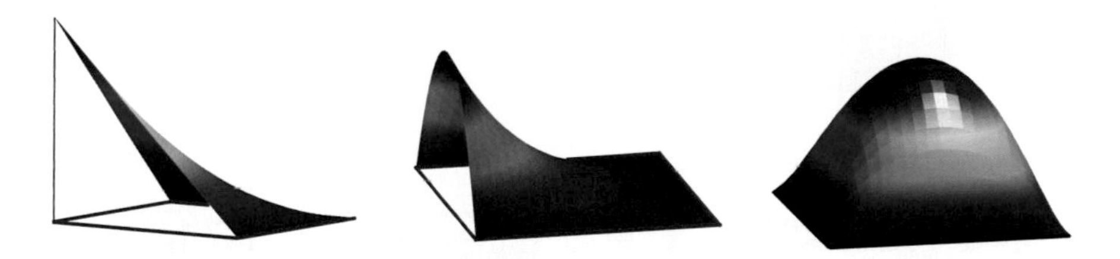

Figure 15.1 Vertex (left), edge (middle), and interior (right) polygonal basis functions for $k = 2$ on a square element.

CHAPTER 8

Multi-Sided Patches via Barycentric Coordinates

Scott Schaefer

Texas A&M University, College Station, USA

CONTENTS

CHAPTER 1 introduced several constructions of barycentric coordinates as well as their properties. In this chapter, we explore the deep connection between barycentric coordinates and higher order parametric representations of curves, surfaces, and volumes in arbitrary dimension. In the case of curves, these curves are known as Bézier curves, which are used in applications from font representations to controlling animations. The extension to convex surface patches, called S-patches [257], is more recent. As originally proposed, S-patches were parametric, multi-sided surface patches restricted to convex domains. However, these restrictions were more of a function of the limited set of generalized barycentric coordinates, namely Wachspress coordinates, available at that time. Today we have generalized barycentric coordinate functions that do not require convexity and extend to arbitrary dimension. Hence, we will investigate S-patches within their full generality, afforded by modern barycentric coordinates with generalized domains and in arbitrary dimension.

8.1 INTRODUCTION

8.1.1 Bézier form of curves

To begin, we consider one of the simplest instantiations of barycentric coordinates. Consider the domain defined by the interval $[0, 1]$. This interval yields barycentric coordinate functions

$$\begin{aligned} \phi_0(x) &= (1 - x), \\ \phi_1(x) &= x, \end{aligned} \tag{8.1}$$

which reproduce constant and linear functions,

$$\begin{aligned} \phi_0(x) + \phi_1(x) &= 1, \\ 0 \cdot \phi_0(x) + 1 \cdot \phi_1(x) &= x. \end{aligned}$$

If we examine the terms of the binomial expansion of $(\phi_0(x) + \phi_1(x))^m$, we obtain functions of the form

$$B_j^m(x) = \binom{m}{j}(1 - x)^{m-j}x^j, \tag{8.2}$$

where m is the degree of the functions $B_j^m(x)$. These functions are special functions called Bernstein basis functions. Associating these functions with control points f_j yields a Bézier curve

$$F(x) = \sum_{j=0}^{m} B_j^m(x) f_j.$$

Note that $F(x)$ trivially reproduces constant functions due to the fact that ϕ_0 and ϕ_1 form a partition of unity. For example, if $f_j = c$, then

$$c = \sum_{j=0}^{m} B_j^m(x)c = c\left((1 - x) + x\right)^m.$$

In addition, $F(x)$ can also reproduce all polynomial functions up to degree m. This last property follows directly from the linear reproduction property of barycentric coordinates. In particular, to reproduce a function x^k where $k \leq m$, there exists coefficients f_j such that

$$F(x) = (\phi_0(x) + \phi_1(x))^{m-k}(0 \cdot \phi_0(x) + 1 \cdot \phi_1(x))^k = 1^{m-k}x^k = x^k,$$

where the coefficients f_j are given by collecting the coefficients of $B_j^m(x)$ in the polynomial expansion above. For example, to reproduce the function x, the coefficients are given by $f_j = j/m$. Such properties also follow directly from the theory of blossoming and polar forms [316], which is beyond the scope of this chapter.

In addition, the barycentric coordinate functions ϕ_i also impart a number of useful geometric properties on the resulting Bézier curves. For example, Bézier curves interpolate their end points due to the Lagrange property of barycentric coordinates. Bézier curves also fall within the convex hull of their control points over the interval $[0, 1]$ since $0 \leq \phi_i(x) \leq 1$ for all $x \in [0, 1]$. The curves are also affinely

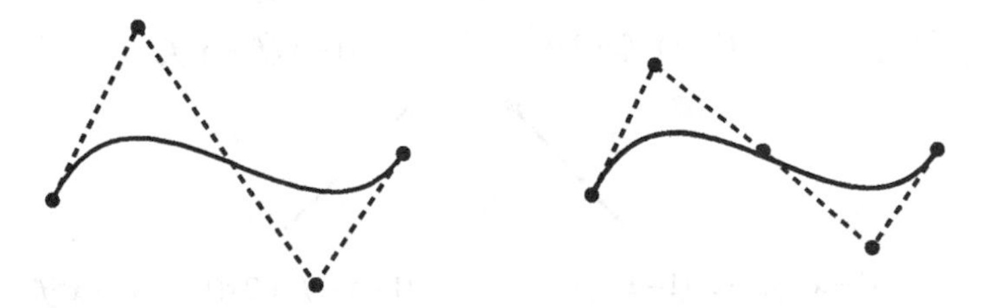

Figure 8.1 An example cubic Bézier curve (left) and the curve degree elevated to a quartic (right).

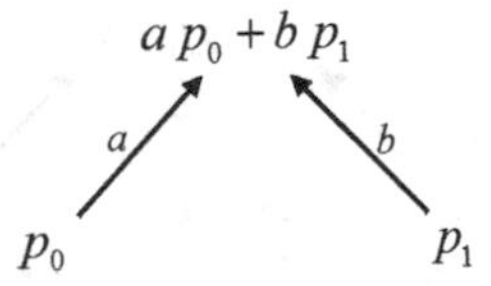

Figure 8.2 An example pyramid diagram. The values at the base are multiplied by the constants on the arrows and summed to produce the result at the apex of the pyramid.

invariant; that is, transforming each control point by an affine transformation T transforms the Bézier curve by T. Figure 8.1 shows an example of a Bézier curve where the f_j are points in $\mathbb{R}^2$.

8.1.2 Evaluation

While we can evaluate Bézier curves by evaluating the polynomial expressions in the Bernstein basis functions, a more elegant solution exists via de Casteljau's algorithm. To do so, we introduce a graphical notation for a linear combination of two control points. Figure 8.2 shows a simple pyramid diagram. The arrows denote taking the product of the value at the base of the arrow with the scalar value listed along the arrow. The result of this product is then added to the sum at the end of the arrow. Hence, this figure denotes taking f_0, f_1 and multiplying these values by a, b, respectively, to form the result $af_0 + bf_1$. Using this notation, de Casteljau's algorithm can be written in a very elegant fashion as a pyramid diagram shown in Figure 8.3 [167]. Note that each level of the pyramid produces lower order Bézier functions with the value of the curve appearing at the apex of the pyramid.

8.1.3 Degree elevation

Degree elevation for Bézier curves is simply finding a Bézier curve of degree $m+1$ to represent a Bézier curve of degree m. One benefit of performing degree elevation is that the process introduces an additional control point that can be used to control the shape of the curve. Furthermore, the addition of this new control point does

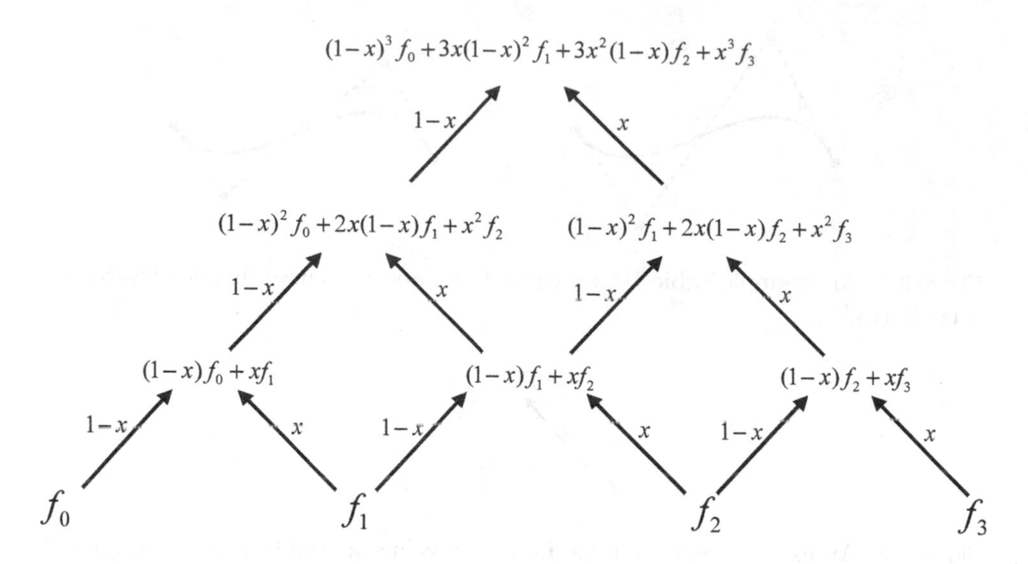

Figure 8.3 The de Casteljau algorithm for cubic Bézier curves.

not change the shape of the curve. Such degree elevation is always possible since every Bézier curve of degree m is also a Bézier curve of degree $m+1$.

To derive degree elevation, it is useful to notice a connection between the Bernstein basis functions of different degrees. Using (8.2), we can show that

$$B_j^{m+1}(x) = \frac{m+1}{m+1-j}\phi_0(x)B_j^m(x) = \frac{m+1}{j}\phi_1(x)B_{j-1}^m(x).$$

Therefore, we can elevate the degree of a Bézier curve by simply multiplying a degree m Bézier curve by $(\phi_0(x)+\phi_1(x))$, so that

$$\begin{aligned} F(x) &= \left(\sum_{j=0}^{m} B_j^m(x) f_j\right)(\phi_0(x)+\phi_1(x)) \\ &= \sum_{j=0}^{m+1} B_j^{m+1}(x)\left(\frac{j}{m+1}f_{j-1} + \frac{m+1-j}{m+1}f_j\right). \end{aligned}$$

Figure 8.1 (right) shows an example of elevating the degree of a Bézier curve.

8.2 MULTISIDED BÉZIER PATCHES IN HIGHER DIMENSIONS

Section 8.1.1 provides some hint as to how we might extend the univariate curve construction to domains such as polygons or even polytopes in higher dimensions through the use of generalized barycentric coordinates. In particular, we use the extension of (8.1) to more generalized domains. Given a polygon P with vertices

$\boldsymbol{v}_i \in \mathbb{R}^d$, the generalized barycentric coordinate functions ϕ_i satisfy

$$\begin{aligned} \sum_{i=1}^{n} \phi_i(\boldsymbol{x}) &= 1, \\ \sum_{i=1}^{n} \phi_i(\boldsymbol{x})\boldsymbol{v}_i &= \boldsymbol{x} \end{aligned} \tag{8.3}$$

for all points $\boldsymbol{x} \in \mathbb{R}^d$ (although $\boldsymbol{x}$ may be restricted to P for some barycentric coordinate constructions).

To build S-patches, we examine the functions that arise from the multinomial expansion of the barycentric basis functions

$$\left(\sum_{i=1}^{n} \phi_i(\boldsymbol{x})\right)^m = \sum_{|\boldsymbol{\ell}|=m} \binom{m}{\boldsymbol{\ell}} \prod_{i=1}^{n} \phi_i(\boldsymbol{x})^{\ell_i}$$

where the index $\boldsymbol{\ell} = (\ell_1, \ldots, \ell_n)$ is a vector of n non-negative integers, $|\boldsymbol{\ell}|$ is the sum of the entries of $\boldsymbol{\ell}$, and $\binom{m}{\boldsymbol{\ell}}$ is the multinomial coefficient $\binom{m}{\boldsymbol{\ell}} = \frac{m!}{\ell_1!\ell_2!\cdots\ell_n!}$. Setting $B^m_{\boldsymbol{\ell}}(\boldsymbol{x})$ to be the corresponding term in this expansion yields

$$B^m_{\boldsymbol{\ell}}(\boldsymbol{x}) = \binom{m}{\boldsymbol{\ell}} \prod_{i=1}^{n} \phi_i(\boldsymbol{x})^{\ell_i}. \tag{8.4}$$

These basis functions are the generalization of the Bézier basis functions from Section 8.1.1. In fact, if we use the barycentric basis functions from (8.1), we obtain the exact same curves albeit with a different indexing scheme. If we associate values $f_{\boldsymbol{\ell}}$ with each of these basis functions, we obtain a multi-sided Bézier function

$$f(x) = \sum_{|\boldsymbol{\ell}|=m} B^m_{\boldsymbol{\ell}}(x) f_{\boldsymbol{\ell}},$$

which is also known as an S-patch. Like univariate Bézier curves, S-patches can reproduce all polynomials up to total degree m, which follows directly from (8.3). For example, to reproduce $\boldsymbol{x}$, the control points are given by

$$f_{\boldsymbol{\ell}} = \sum_{i=1}^{n} \frac{\ell_i}{m} \boldsymbol{v}_i. \tag{8.5}$$

Unlike univariate curves, the barycentric coordinate functions in (8.3) are not necessarily polynomials. Nevertheless, we will refer to the S-patch basis functions $B^m_{\boldsymbol{\ell}}(\boldsymbol{x})$ as a function of degree m indicating the total degree of the individual barycentric coordinate functions ϕ_i.

8.2.1 Indexing for S-patches

For curves, indexing control points is trivial as each basis function is simply given an index $j = 0, \ldots, m$ based on an ordering of the Bernstein basis functions. However, this simple indexing scheme does not extend to higher dimensions. As already

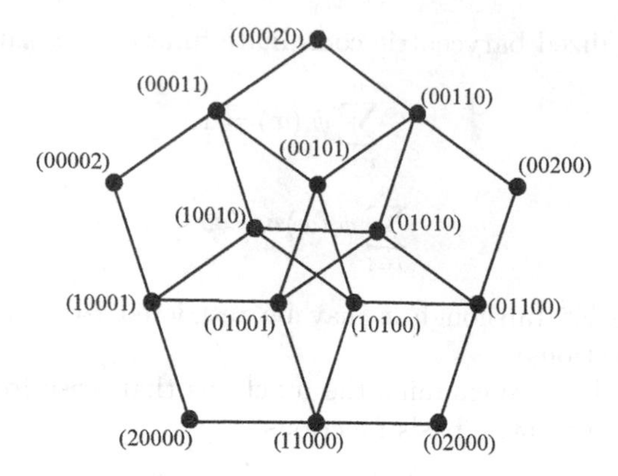

Figure 8.4 Indexing of a quadratic pentagonal S-patch using the specified connectivity rule.

alluded to in (8.4), S-patches use a multi-index to refer to control points. Each control point $f_{\boldsymbol{\ell}}$ is associated with an index vector of length n non-negative integers whose sum is the degree m of the patch. Given a polygon P, we typically draw the control points in canonical position such that $f_{\boldsymbol{\ell}}$ are placed so that $F(\boldsymbol{x}) = \boldsymbol{x}$. Luckily, this combination is solely a function of P, independent of what barycentric basis we choose, and is given by (8.5).

In addition, there is a simple rule for drawing connectivity of a simple 2D S-patch. We connect two control points $f_{\boldsymbol{\ell}}$ and $f_{\boldsymbol{h}}$ if there exists an index $1 \leq j \leq n$ such that

$$\begin{aligned} \ell_i &= h_i, \qquad i \neq j, j+1 \\ |\ell_j - h_j| &= 1 \\ |\ell_{j+1} - h_{j+1}| &= 1 \end{aligned}$$

where the arithmetic on j is performed modulo n. Figure 8.4 shows an example of this indexing for a quadratic pentagonal S-patch. While drawing this connectivity makes sense for 2D domains, S-patches extend beyond 2D to polytopes in higher dimensions. There, the cyclic ordering of the indices used in 2D to identify connectivity does not generalize to higher dimensions. While we can generalize the connectivity rules to higher dimensions, the information imparted by the connectivity in higher dimensions becomes more difficult to discern. This connectivity is not essential to evaluation algorithms or other properties of S-patches in higher dimensions, so we simply omit it.

Note that something curious happens to the S-patch control points for certain polygons. Figure 8.5 shows an example of two quadratic, quadrilateral patches. One example has 10 control points while the other appears to have 9 control points. The issue is that two of the control points corresponding to the (1010) and (0101) index overlap according to (8.5) in the left example and not the right example. To reproduce polynomial functions, these control points must have the same value, which leads to a loss of degrees of freedom. However, in the more general case of

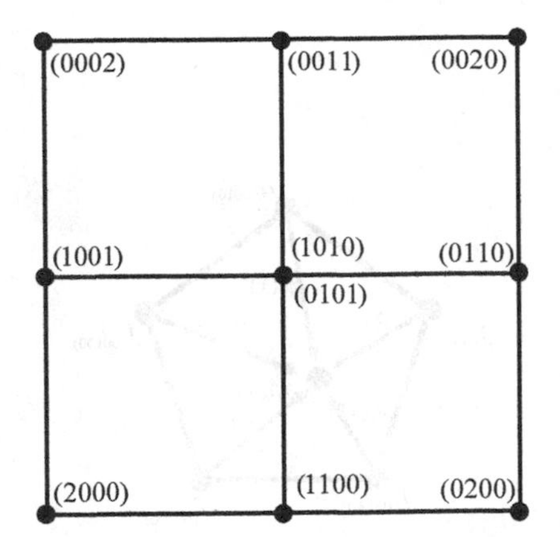

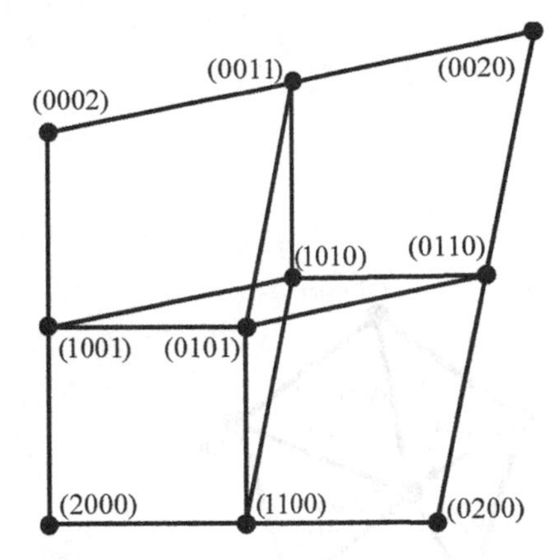

Figure 8.5 An example of ambiguous indexing for S-patches. The left shows an example where two control points overlap in the parametric domain. The right shows the same indexing for a different base polygon where the overlapping control points separate.

modeling shapes with S-patches, these two control points may have different values despite being drawn in the same location. It is useful to note that, in the situation on the left of the figure, if we require the overlapping control points to have the same function value, which effectively reduces the number of control points to 9, the S-patch is identical to a tensor-product Bézier patch when using Wachspress coordinates as the barycentric coordinate functions. Therefore, tensor product Bézier patches as well as triangular Bézier patches (S-patches with a simplicial domain) are all special cases of S-patches [257]. Indeed, the basis functions for Wachspress coordinates are identical for the (1010) and (0101) control points in this example. However, the basis functions for two overlapping control points do not have to be identical and, in fact, rarely are in the general case.

8.2.2 Evaluation

Similar to Bézier curves, one possible method for evaluating an S-patch is simply to multiply the control points by the corresponding basis functions from (8.4) evaluated at the point in question. However, there also exists a de Casteljau-like algorithm that utilizes the hierarchical relationship between the control points. Given an evaluation point $\boldsymbol{x}$, we compute the value of the barycentric basis functions $\phi_i(\boldsymbol{x})$ at the evaluation point using the polytope P. Now, we proceed in a recursive fashion. Let $f_{\boldsymbol{j}}$ be the control points at level $k-1$ where $|\boldsymbol{j}| = k-1$ and let $f_{\boldsymbol{\ell}}$ with $|\boldsymbol{\ell}| = m$ be the initial control points. Then, we compute the $f_{\boldsymbol{j}}$ for all $|\boldsymbol{j}| = k-1$

$$f_{\boldsymbol{j}} = \sum_{i=1}^{n} \phi_i(\boldsymbol{x}) f_{\boldsymbol{j}+I_i^n}$$

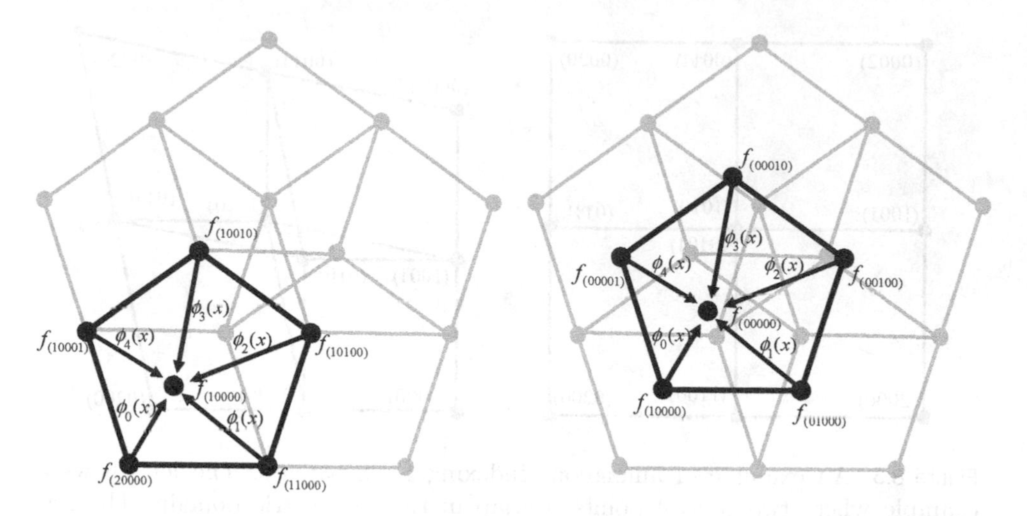

Figure 8.6 The de Casteljau algorithm applied to a quadratic, pentagonal S-patch. The left image depicts the calculation for one point of the first level of the algorithm. The right image shows the computation of the final evaluation point.

where I_i^n is the ith row of the $n \times n$ identity matrix. The value of the S-patch is then given by $f_{\mathbf{0}}$. Figure 8.6 depicts this hierarchical evaluation algorithm.

8.2.3 Degree elevation

Even though Loop and DeRose [257] mention degree elevation for S-patches, the authors provide no explicit formula. However, degree elevation is not difficult to derive. Let $f_{\boldsymbol{\ell}}$ be the values of the control points for an S-patch function of degree $|\boldsymbol{\ell}| = m$ and $\hat{f}_{\boldsymbol{j}}$ be control points of the same S-patch function of degree $|\boldsymbol{j}| = m+1$. Then,

$$\left(\sum_{|\boldsymbol{\ell}|=m} B_{\boldsymbol{\ell}}^m(\boldsymbol{x}) f_{\boldsymbol{\ell}} \right) \left(\sum_{i=1}^{n} \phi_i(\boldsymbol{x}) \right) = \sum_{|\boldsymbol{j}|=m+1} B_{\boldsymbol{j}}^{m+1}(\boldsymbol{x}) \hat{f}_{\boldsymbol{j}}. \tag{8.6}$$

Note that the basis functions of various degrees are related to each other via

$$B_{\boldsymbol{j}}^{m+1}(\boldsymbol{x}) = B_{\boldsymbol{j}-I_i^n}^m(\boldsymbol{x}) \frac{m+1}{j_i} \phi_i(\boldsymbol{x}) \tag{8.7}$$

for $j_i > 0$. Expanding the left-hand side of (8.6) and using (8.7) yields the formula for the $m+1$ degree control points $\hat{f}_{\boldsymbol{j}}$,

$$\hat{f}_{\boldsymbol{j}} = \sum_{j_i>0} \frac{j_i}{m+1} f_{\boldsymbol{j}-I_i^n}.$$

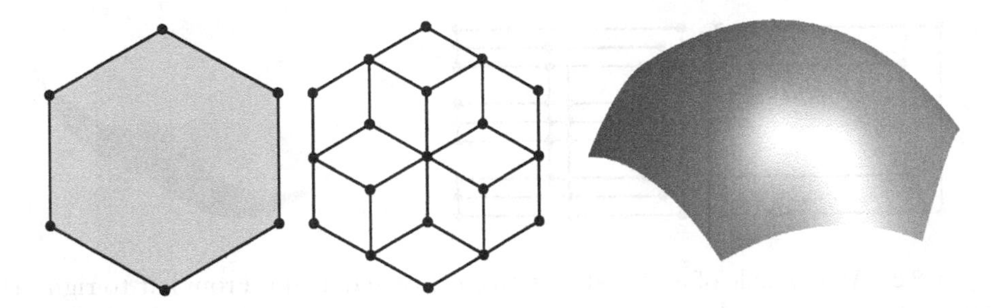

Figure 8.7 An example of an S-patch with a convex domain. From left to right: the base polygon P shaded gray, the control point structure for the quadratic S-patch, and an example S-patch in 3D.

8.3 APPLICATIONS

While the original definition of S-patches [257] used convex, multi-sided polygon domains, this restriction was more a function of the barycentric coordinates available at that time [407] than any limitation of the construction. Many different types of barycentric coordinates have been developed since then (see Chapter 1). The majority of these constructions are not limited to a particular dimension for the domain and do not require convex domains. While any of these constructions can be used to create S-patches, we utilize mean value coordinates [144, 150, 200, 217] here in these examples.

8.3.1 Surface patches

Perhaps the most commonly used application of S-patches is to fill multi-sided holes in surfaces. Figure 8.7 shows an example of such a multi-sided patch using a convex domain. Due to the interpolatory properties of barycentric coordinates, the curves along the boundary of the domain are solely a function of the control points along that boundary and form a Bézier curve. In this case, the six-sided patch is bounded by six quadratic Bézier curves.

Given that the original definition of S-patches [257] used Wachspress coordinates [407], the S-patch domain was required to be convex. Convexity is no longer a requirement with many barycentric coordinate constructions, though the misconception of this requirement for S-patches persists. Figure 8.8 (right) shows an example of a quadratic S-patch with a concave domain. Figure 8.8 (middle) shows a 2D image of the control point structure of this patch. Unlike convex domains, the control points of S-patches with concave domains may lie outside of the original polytope.

S-patches can be formed with domains that are even more general than these convex or concave examples. In fact, S-patches can support domains of nearly arbitrary shape that may even contain one or more holes. The restrictions on the domain follow directly from the barycentric coordinates used to construct the S-

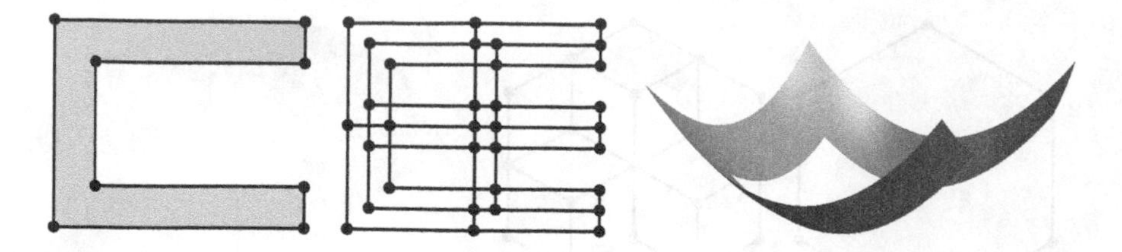

Figure 8.8 An example of an S-patch with a concave domain. From left to right: the base polygon P shaded gray, the control point structure for the quadratic S-patch, and an example S-patch in 3D. Note that, unlike the example in Figure 8.7, the control points do not all lie within the domain P.

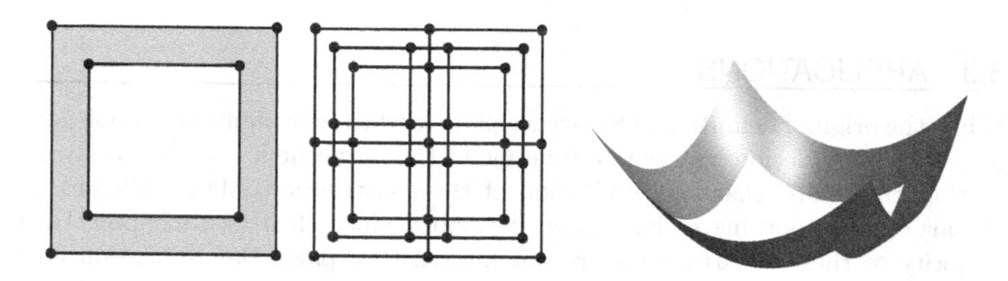

Figure 8.9 An example of an S-patch whose domain contains a hole. From left to right: the base polygon P shaded gray, the control point structure for the quadratic S-patch, and an example S-patch in 3D.

patch. For barycentric coordinate constructions like mean value coordinates, the domain may even contain holes like the example in Figure 8.9.

8.3.2 Spatial deformation

S-patches can be used for applications other than filling multi-sided holes on surfaces as well. Though rarely thought of in this way, S-patches can be used for image and surface deformation as well. In fact, free-form deformations [353] (FFDs) as well as cage-based deformations [215, 217] are all special cases of deformation using S-patches.

To perform such spatial deformation, we use S-patches to construct a map $\boldsymbol{F}\colon \mathbb{R}^d \to \mathbb{R}^d$ where $d = 2$ for images and $d = 3$ for surfaces and volumes. F is then given by

$$\boldsymbol{F}(\boldsymbol{x}) = \sum_{|\boldsymbol{\ell}|=m} B^m_{\boldsymbol{\ell}}(\boldsymbol{x})\boldsymbol{p}_{\boldsymbol{\ell}}.$$

Hence, we control the deformation by manipulating the control points $\boldsymbol{p}_{\boldsymbol{\ell}} \in \mathbb{R}^d$. The key to creating a map that is useful for deformation is to construct such a function that can produce the identity transformation; that is, $\boldsymbol{F}(\boldsymbol{x}) = \boldsymbol{x}$ for some

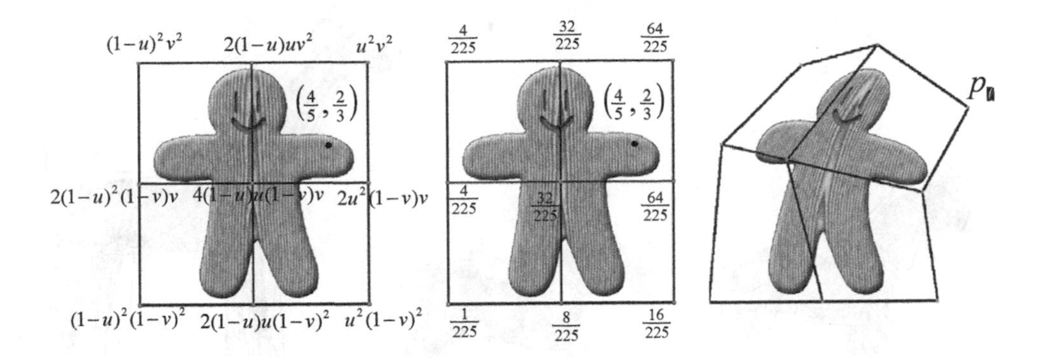

Figure 8.10 Image deformation using S-patches. The left image shows the image and a point to evaluate the deformation at which each control point lists its basis functions using a quadratic S-patch with Wachspress coordinates. The middle image shows the weights associated with the evaluation point after evaluating the basis functions. Moving the control points induces a deformation on the image shown on the right. See color insert.

configuration of control points. Luckily, (8.5) already provides the location of the control points to reproduce this function. We refer to this configuration of control points as the *bind pose*.

Now, given a shape, for each point $\boldsymbol{x}$ in the shape, we compute $\boldsymbol{x}$ as a weighted combination of the control points $\boldsymbol{p}_{\boldsymbol{\ell}}$ where the weights are simply given by the values of the basis functions $B^m_{\boldsymbol{\ell}}(\boldsymbol{x})$. Notice that the weights $B^m_{\boldsymbol{\ell}}(\boldsymbol{x})$ are constant for each point $\boldsymbol{x}$ and can be precomputed, which yields a fast deformation function. When $\boldsymbol{p}_{\boldsymbol{\ell}}$ satisfy (8.5), $F(\boldsymbol{x})$ is the identity map. However, as the user manipulates the control points away from the bind pose, the map produces a smooth deformation of the underlying shape.

Deforming a shape such as an image is simple with this function. Given an image such as the one shown in Figure 8.10, we surround the image with some base polygon P. In this example, we use a rectangle and compute the deformation using a quadratic S-patch. Then, we sample the deformation using a regular grid over the image. For each grid point, we compute the weights of the deformation function. As the user manipulates the control points, we simply apply the weights to the deformed control point positions to produce a deformed grid. Bilinearly interpolating the deformation within each grid cell generates the final deformation. Notice that the grid in this case may be as fine as the pixels in the input image, although such fine resolution is not typically necessary to create a smooth-looking deformation. Figure 8.10 shows the result of such a deformation. In this case, the deformation is equivalent to using a tensor-product Bézier patch when using Wachspress coordinates, which is called a free-form deformation [353] (a special case of an S-patch).

Surface deformation follows a similar route. In this case, we are typically given a triangulated surface that we wish to deform with vertices $\boldsymbol{x}_j$. Figure 8.11 (left)

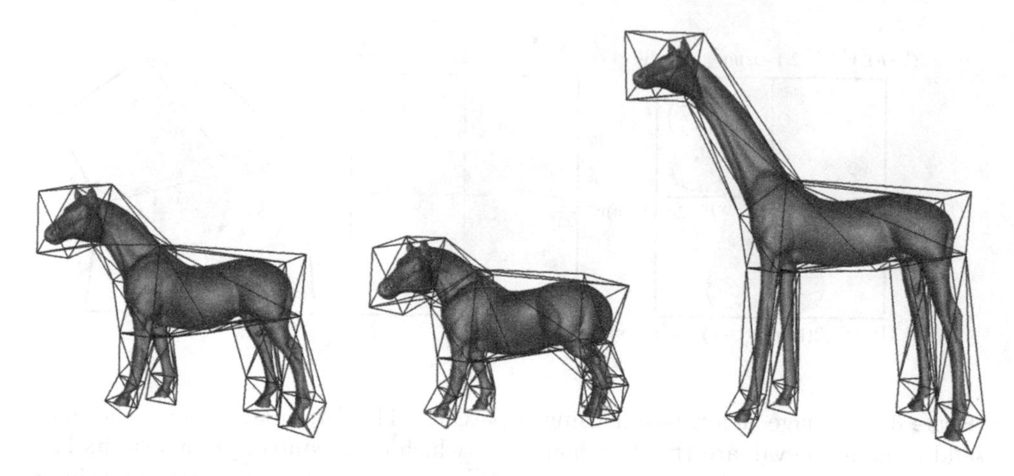

Figure 8.11 An example of surface deformation using linear S-patches. The initial surface of a horse is surrounded by a low resolution approximation called a cage (left). The right two images show different deformations where the head and torso are kept the same size but the neck and legs are compressed or stretched. See color insert.

shows an example of a 3D model of a horse containing 48,485 such vertices. First, we construct a polytope to surround the horse typically with fewer vertices $\boldsymbol{v}_{\boldsymbol{\ell}}$ than the actual horse. In this example, the cage, shown on the left of the figure, contains only 51 control points. Next, we compute each $\boldsymbol{x}_j$ as a weighted combination of the $\boldsymbol{v}_{\boldsymbol{\ell}}$. As the user manipulates the control points, we apply the constant weights for each vertex to the deformed control point locations to create the location of the deformed vertex of the surface as shown on the middle and right of Figure 8.11.

Unlike the original definition of free-form deformations that relies on tensor-product Bézier functions of various degrees with cube domains, cage-based deformations can conform to the shape of the object. And while cage-based deformations are not typically thought of this way, these deformations are obviously S-patch deformations of degree 1. This insight means that we can construct volumetric cage-based deformations of higher degree as well. Unfortunately, the number of control points for these deformations grows quite rapidly. For example, the cage in Figure 8.11 has 51 control points as a linear S-patch. A quadratic S-patch would contain 1,326 control points. Moreover, the majority of those control points exist in the interior of the cage, which would make it difficult for a user to manipulate those points. Hence, higher degree cage-based deformations are not practical from a user-interface perspective. One alternative would be to use selective degree elevation [364] to avoid inserting more control points than desired as the degree of the patch is elevated. However, such a method is beyond the scope of this chapter, and we refer the interested reader to the corresponding reference.

CHAPTER 9

Generalized Triangulations

Pooran Memari

LIX, CNRS, École Polytechnique, Université Paris Saclay, Palaiseau, France

CONTENTS

The main concept of barycentric coordinates is intimately linked to the duality between primal and dual complexes. This chapter presents a general framework in $\mathbb{R}^d$ to define a family of compatible dual complexes for a given triangulation. We derive a combinatorial characterization theorem for triangulations that admit a compatible, possibly non-orthogonal dual complex. We show how the mesh optimization methods in geometric modeling can benefit from the parameterization of the space of generalized triangulations.

9.1 INTRODUCTION

A triangle mesh is a partition of a bounded domain with simplices (triangles in 2D, tetrahedra in 3D) so that any two of these simplices are either disjoint or sharing a face. The resulting triangulation (also called tetrahedralizations in 3D) provides a discretization of space through both its primal (simplicial) elements *and* its dual (cell) elements. Both types of element are crucial to a variety of numerical techniques. A growing trend in numerical simulation is the simultaneous use of primal *and* dual meshes: Petrov–Galerkin finite element/finite volume methods [24, 271, 298] and methods based on exterior calculus [65, 119, 176] use the ability to store quantities on both primal and dual elements to enforce (co)homological relationships, which for instance appear in Hodge theory. The choice of the dual, defined by the location of the dual vertices, is however not specified a priori. The barycentric dual, for which barycenters are used instead of circumcenters, is used for certain finite volume computations, but it fails to satisfy both the orthogonality and the convexity conditions on general triangulations. A very common dual to a triangulation in $\mathbb{R}^d$ is the cell complex that uses the circumcenters of each d-simplex as dual vertices. It corresponds to the Voronoi diagram in the case of Delaunay triangulations [137, 313]. The circumcentric Delaunay–Voronoi duality has been extensively used in diverse fields. Building on a number of results in algebraic and computational geometry, a more general primal-dual pair of complexes can be defined [274], as we briefly describe in the first sections of this chapter. A barycentric representation of primal-dual triangulations is then presented as a parameterization of this space of generalized triangulations.

Classical methods, such as *centroidal Voronoi tessellations* [127], lead to "good" simply connected domain dual elements, while *optimal Delaunay triangulations* [9] optimize the shape of primal elements. Nowadays, recent numerical methods on meshes can benefit from optimizations over *both* primal and dual complexes. At the end of this chapter, we present a unifying approach to mesh quality [289], which is based on the placement of primal *and* orthogonal dual elements with respect to each other, for instance, to make orthogonal barycentric duals.

9.2 GENERALIZED PRIMAL-DUAL TRIANGULATIONS

We start by recalling important notions on this topic. More detailed definitions can be found in [289] and [274].

Complex and triangulation

A *cell complex* in $\mathbb{R}^d$ is a set K of convex polyhedra (called cells) satisfying two conditions:

1. Every face of a cell in K is also a cell in K.
2. If C and C' are cells in K, their intersection is either empty or a common face of both.

A *simplicial complex* is a cell complex whose cells are all simplices.

For a set X of points $\{\boldsymbol{x}_i\}_{i=1,\dots,|X|}$ in $\mathbb{R}^d$, a triangulation of X is a simplicial complex K for which each vertex of K is in X, and the union of all the cells of K is the convex hull of X. Note that the triangulations we consider usually come from a point set and hence partition a simply connected domain in $\mathbb{R}^d$ (corresponding to the convex hull of the point set).

Also, in the definition of a triangulation of X, we do not require *all* the points of X to be used as vertices; a point $\boldsymbol{x}_i \in X$ is called *hidden* if it is not used in the triangulation.

Dual cell complexes

Two cell complexes K and K' in $\mathbb{R}^d$ are said to be dual to one another if there exists a duality map $*$ from K to K' that associates any cell σ in K of dimension i $(0 \leq i \leq d)$ to a cell $*\sigma$ in K' of dimension $d - i$ for all $0 \leq i, j \leq d$, $\sigma_i \subset \sigma_j \leftrightarrow *\sigma_i \supset *\sigma_j$.

9.2.1 Some classical examples

Power diagrams and weighted Delaunay triangulations

For a given point $\boldsymbol{p} \in \mathbb{R}^d$ and a scalar $\omega \in \mathbb{R}$, the power distance from any point $\boldsymbol{x} \in \mathbb{R}^d$ to the so-called weighted point $(\boldsymbol{p}, \omega)$ is defined by an additive modification of the Euclidean metric as $\|\boldsymbol{x} - \boldsymbol{p}\|^2 - \omega$ (also called the *Laguerre distance*).

In terms of the power distance, the bisector separating any pair of weighted points $(\boldsymbol{p}, \omega_p)$ and $(\boldsymbol{q}, \omega_q)$ is a hyper-plane with points $\boldsymbol{x}$ satisfying $\|\boldsymbol{x} - \boldsymbol{p}\|^2 - \omega_p = \|\boldsymbol{x} - \boldsymbol{q}\|^2 - \omega_q$. Therefore, the difference of weights acts as a shift of the bisector towards the smaller weight. This leads to the definition of a generalized Voronoi diagram called *power diagram*. Let $(X, \boldsymbol{\omega}) = \{(\boldsymbol{x}_i, \omega_i)\}_{i \in I}$ be a weighted point set in $\mathbb{R}^d$. To each $\boldsymbol{x}_i$ we associate its weighted Voronoi region

$$V(\boldsymbol{x}_i, \omega_i) = \{\boldsymbol{x} \in \mathbb{R}^d : \|\boldsymbol{x} - \boldsymbol{x}_i\|^2 - \omega_i \leq \|\boldsymbol{x} - \boldsymbol{x}_j\|^2 - \omega_j \forall j\}.$$

Note that drastic differences between weights may induce empty power cells and thereby isolate sites with no neighbors. We denote sites associated with zero volume cells as hidden sites [20]. Also, when the weights are all equal, the power diagram coincides with the Euclidean Voronoi diagram of X. While many other generalizations of Voronoi diagrams exist, they do not form straight-edge and convex polytopes and are thus not relevant here.

The *dual* of the power diagram of $(X, \boldsymbol{\omega})$ is the weighted Delaunay triangulation of $(X, \boldsymbol{\omega})$. This triangulation contains a k-simplex with vertices $\boldsymbol{x}_{i_0}, \boldsymbol{x}_{i_1}, \ldots, \boldsymbol{x}_{i_k}$ in X, if and only if $V(\boldsymbol{x}_{i_0}, \omega_{i_0}) \cap V(\boldsymbol{x}_{i_1}, \omega_{i_1}) \cap \cdots \cap V(\boldsymbol{x}_{i_k}, \omega_{i_k}) \neq \emptyset$, $\forall k \geq 0$.

Regular triangulations

Note that for a simply connected domain (that is, the convex hull of $X \subset \mathbb{R}^d$), the weighted Delaunay triangulation coincides with the so-called regular triangulation that is defined as the projection of the lower hull of the lifted point set $F(X)$, for some lifting F of X to $\mathbb{R}^{d+1}$. As it is shown in [113], this equivalence is no longer true for domains with non-trivial topology.

Weighted circumenters

For any simplex σ of the weighted Delaunay triangulation, let $\mathbf{c}(\sigma)$ denote the *weighted circumcenter* of σ, that is, the unique intersection of (the mutually orthogonal affine spaces supporting) the primal simplex σ^k and its weighted dual $*\sigma^k$. The positions of weighted circumcenters can be computed explicitly from the positions of vertices and the weights of the corresponding simplex [289, Equation 7].

Of particular importance are the weighted circumcenters of the d-simplices for a mesh $\mathcal{T}$ in $\mathbb{R}^d$: these form the vertices of its dual complex $\mathcal{D}$. Any point in the affine space supporting a d-simplex σ can be the weighted circumcenter of σ for a choice of weights on the corresponding vertices. As we will see, this can be seen as a parameterization of the space of dual complexes.

Computation

Algorithms to construct power diagrams make use of the same strategies as in the case of Voronoi diagrams, where the Euclidean distances are replaced by power distances. Therefore, the worst time complexity to generate a power diagram with n weighted points is $O(n \log n)$ in $\mathbb{R}^2$ and $O(n^{\frac{d}{2}})$ in $\mathbb{R}^d$. Efficient implementations of these algorithms, such as in the CGAL library [86], have reported that the complexity bound is typically reached in cases with coplanar sites, while the general case performs close to linear in the number of sites.

9.2.2 Primal-dual triangulations (PDT)

Compatible dual complexes of triangulations

We now show that the notion of mesh duality can be extended, so that the dual complex is defined geometrically and independently from the triangulation—while the combinatorial compatibility between the triangulation and its dual is maintained.

Definition 9.1 (Simple cell complex)**.** A cell complex K in $\mathbb{R}^d$ is called *simple* if every vertex of K is incident to $d+1$ edges. K is called *labeled* if every d-dimensional cell of K is assigned a unique label; in this case, we write $K = \{C_1, \ldots, C_n\}$, where n is the number of d-dimensional cells of K, and C_i is the i-th d-dimensional cell.

Definition 9.2 (Compatible dual complex)**.** Let T be a triangulation of a set $X = \{\boldsymbol{x}_1, \ldots, \boldsymbol{x}_n\}$ of points in $\mathbb{R}^d$ and let $K = \{C_{i_1}, \ldots, C_{i_n}\}$ be a labeled simple cell complex, that is, there is a one-to-one correspondence between $\boldsymbol{x}_p$ and C_{i_p}. We then call K a *compatible dual complex* of T if, for every pair of points $\boldsymbol{x}_p$ and $\boldsymbol{x}_q$ that are connected in T, the cells C_{i_p} and C_{i_q} share a face.

This compatibility between K and T is purely combinatorial, that is, it simply states that the connectivity between points induced by K coincides with the one induced by T. Notice that the cell C_{i_p} associated to the point $\boldsymbol{x}_p$ does not necessarily contain $\boldsymbol{x}_p$ in its interior. Moreover, the edge $[\boldsymbol{x}_p, \boldsymbol{x}_q]$ and its dual $C_{i_p} \cap C_{i_q}$ are not necessarily *orthogonal* to each other, unlike most conventional geometric dual structures. Consequently, we can generalize the notion of mesh duality.

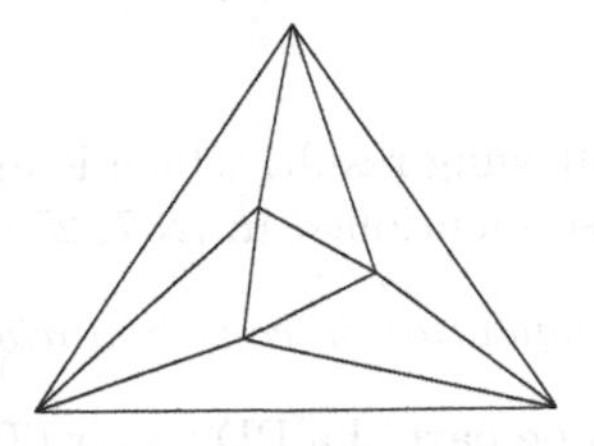
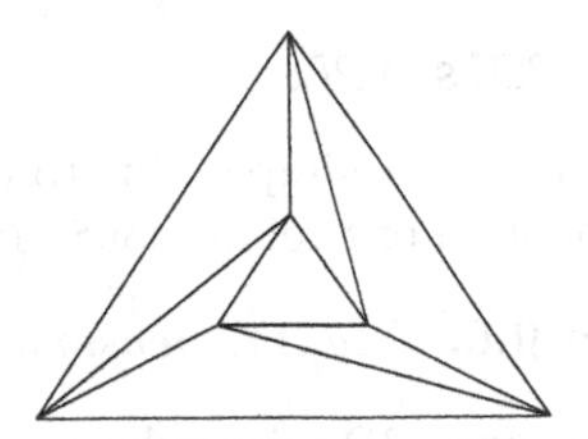

Figure 9.1 A regular triangulation (left), once deformed (right), becomes a combinatorially regular triangulation that is *not* regular itself. (Image taken from [237].)

Definition 9.3 (Primal-dual triangulation (PDT)). A pair (T, K) is said to form a d-dimensional *primal-dual triangulation* if T is a triangulation in $\mathbb{R}^d$ and K is a compatible dual complex of T. If every edge $[\boldsymbol{x}_p, \boldsymbol{x}_q]$ and its dual $C_{i_p} \cap C_{i_q}$ are orthogonal to each other, then the pair (T, K) is said to form an *orthogonal primal-dual triangulation.*

9.3 A CHARACTERIZATION THEOREM

Let us consider a set of points X in $\mathbb{R}^d$. For any set of weights $\boldsymbol{\omega}$, the power diagram of $(X, \boldsymbol{\omega})$ and the corresponding dual regular triangulation form a PDT of the convex hull of X. In other words, any regular triangulation admits a compatible dual complex and can be part of a PDT. Moreover, in [274] we showed that for simply connected domains, any triangulation that admits a compatible dual complex is indeed regular up to a perturbation. For a more formal statement we need the following definitions.

9.3.1 Combinatorially regular triangulations (CRT)

Definition 9.4 (Combinatorial equivalence). Two triangulations T and T' are *combinatorially equivalent* if there exists a labeling that associates to each point $\boldsymbol{x}_i$ in T a point $\boldsymbol{x}'_i$ in T' so that the connectivity between the $\boldsymbol{x}_i$'s induced by T matches the connectivity between the $\boldsymbol{x}'_i$'s induced by T'.

Definition 9.5 (Combinatorially regular triangulations (CRT)). A triangulation T of a d-dimensional point set X is called a *combinatorially regular triangulation* if there exists a d-dimensional point set X' admitting a regular triangulation T', such that T and T' are combinatorially equivalent.

These CRT triangulations have been introduced in [237] under the name of *weakly* regular triangulations, since a displacement of their vertices suffices to make them regular. Figure 9.1 shows an example of a combinatorially regular triangulation that is not regular itself.

Existence of PDTs in 2D.

The 2D case is rather simple, due to the following result, which is mainly based on a classical theorem of Steinitz [368] and also mentioned in [237, 274].

Proposition 9.6. *Any 2-dimensional triangulation is combinatorially regular.*

Therefore, every 2D triangulation T can be part of a PDT pair (T, K). However, in higher dimensions ($d \geq 3$), the situation is rather different [274].

Proposition 9.7. *For $d \geq 3$, there exist d-dimensional triangulations that do not admit any compatible dual complex.*

This is equivalent to the fact that there exist triangulations that are not combinatorially regular. The simplest non-combinatorially regular examples are the Brucker sphere and the Barnette sphere [141].

9.3.2 Equivalence between PDT and CRT

As shown in [274], combinatorially regular triangulations are *the only triangulations that admit compatible dual complexes.* The proof revolves around a theorem due to Aurenhammer:

> Every simple cell complex in $\mathbb{R}^d$, $d \geq 3$, is dual to a regular triangulation.

This theorem was proved in [18] through an iterative construction, which is valid in any dimension $d \geq 3$. In [274], we used this theorem to prove the following theorem, which surprisingly implies that in higher dimensions there are triangulations that do *not* admit a dual complex.

Theorem 9.8 (PDT characterization). *A d-dimensional triangulation T admits a compatible (not necessarily orthogonal) dual complex* if and only if *T is combinatorially regular.*

9.3.3 Parametrization of primal-dual triangulations

We established that for simply connected domains, the space of primal-dual triangulations covers all dual complexes in $d \geq 3$. It also covers all 2D triangulations, but only combinatorially regular triangulations in $d \geq 3$. The proof of Theorem 9.8 leads naturally to a parameterization of all triangulations that admit a compatible dual complex [274].

Definition 9.9. A *parameterized primal-dual triangulation* is a primal-dual triangulation parameterized by a set of triplets $(\boldsymbol{x}_i, \omega_i, \boldsymbol{v_i})$, where $\boldsymbol{x}_i$ is the *position* in $\mathbb{R}^d$ of the i-th node, ω_i is a real number called the *weight* of $\boldsymbol{x}_i$, and $\boldsymbol{v_i}$ is a d-dimensional vector called the *displacement vector* of $\boldsymbol{x}_i$. The triangulation associated with the triplets $(\boldsymbol{x}_i, \omega_i, \boldsymbol{v_i})$ is defined such that its dual complex K is the *power diagram of weighted points* $(\boldsymbol{p_i}, \omega_i)$, where $\boldsymbol{p_i} = \boldsymbol{x}_i + \boldsymbol{v_i}$.

The dual complex K can be seen as the generalized Voronoi diagram of the $\boldsymbol{x}_i$'s for the distance $d(\boldsymbol{x}, \boldsymbol{x}_i) = \|\boldsymbol{x} - \boldsymbol{x}_i - \boldsymbol{v}_i\|^2 - \omega_i$. If the vectors $\boldsymbol{v}_i$ are all zero, then the parameterized primal-dual triangulation T is regular, thus perpendicular to its dual K, and the pair (T, K) forms an orthogonal primal-dual triangulation. This proves that weighted Delaunay triangulations are sufficient to parameterize the set of all orthogonal primal-dual triangulations of a simply connected domain (see also [166]). The displacement vectors extend the type of triangulations and duals we can parameterize.

Characterizing the classes of *equivalent* triplets parameterizing the same PDT and completed with constraints for the parameters in order to avoid redundancy between equivalent triplets, the following theorem was proved in [274].

Theorem 9.10 (PDT parameterization). *There is a bijection between all primal-dual triangulations in* $\mathbb{R}^d$ *and sets of triplets* $(\boldsymbol{x}_i, \omega_i, \boldsymbol{v}_i)$, $1 \leq i \leq n$, *where* $\boldsymbol{x}_i, \boldsymbol{v}_i \in \mathbb{R}^d$, $\omega_i \in \mathbb{R}$ *with* $\sum_i \omega_i = 0$, $\sum_i \boldsymbol{v}_i = 0$, *and* $\sum_i \|\boldsymbol{x}_i + \boldsymbol{v}_i\|^2 = \sum_i \|\boldsymbol{x}_i\|^2$.

Using this parameterization, the particular case of Delaunay and Voronoi PDT of a set of points $\{\boldsymbol{x}_i\}_{i=1,\ldots,n}$ is naturally parameterized by the triplets $(\boldsymbol{x}_i, 0, \mathbf{0})$.

We now employ this parameterization to derive a novel discretization of calculus that verifies structural geometric properties, such as the linear accuracy for the gradient and Laplacian operators that are useful in geometry processing and graphics applications (see also [115]).

9.4 DISCRETE REPRESENTATION USING GENERALIZED TRIANGULATIONS

9.4.1 Discrete exterior calculus framework

The theory of discrete exterior calculus (see Section 5.2.1) provides one possible discretization of differential geometry, which has proven to be highly successful in improving and analyzing finite element methods.

The geometric principles on which these methods of computations are built are very intuitive. In this framework, discrete differential forms are considered as continuous forms integrated over each discrete element (simplices and cells). Once discrete forms and vector fields are defined, a calculus can be developed by defining discrete operators.

For example, the classical operations of gradient, divergence, and curl, as well as the theorems of Green, Gauss, and Stokes can all be expressed concisely in terms of two differential forms: the discrete exterior derivative (d) and the Hodge star ($\star$). More details on the discrete exterior calculus framework can be found in a variety of sources [118, 192], as well as in a recent Siggraph course [109].

Here, we are particularly interested in the discrete Hodge star $\star$, which takes geometry into account by encoding how integrated measures over primal elements are transferred to their associated dual elements. More concretely, the discrete diagonal Hodge star for discrete p-forms is defined as a diagonal matrix with entries

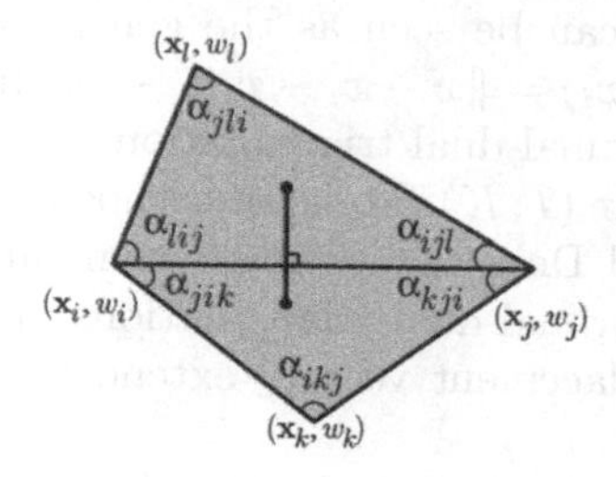

Figure 9.2 Notations for the cotangent formula.

per primal element σ of dimension p given as $\star_\sigma = |*\sigma|/|\sigma|$, where $|\sigma|$ indicates the signed measure of σ.

In the classical case of two-dimensional meshes and their circumcentric duals, this quantity is the ratio between the dual and the primal edge length. As explained in Section 5.3.6, this quantity coincides with the well-known cotangent formula [262]: $(\cot(\alpha_{ikj}) + \cot(\alpha_{jli}))/2$ (see Figure 9.2).

In the general case of two-dimensional orthogonal primal-dual meshes, we make use of the closed-form expressions for primal and dual volumes to define a weighted version of the discrete Hodge star operator, leading to

$$\frac{1}{2}\left(\cot\alpha_{ikj} + \cot\alpha_{jli} + (\omega_i - \omega_k)\frac{\cot\alpha_{kji}}{\|\boldsymbol{x}_i - \boldsymbol{x}_j\|^2} + (\omega_j - \omega_k)\frac{\cot\alpha_{jik}}{\|\boldsymbol{x}_i - \boldsymbol{x}_j\|^2}\right.$$
$$\left. + (\omega_i - \omega_l)\frac{\cot\alpha_{ijl}}{\|\boldsymbol{x}_i - \boldsymbol{x}_j\|^2} + (\omega_j - \omega_l)\frac{\cot\alpha_{lij}}{\|\boldsymbol{x}_i - \boldsymbol{x}_j\|^2}\right).$$

Observe that the dual structure for constant weights corresponds to the circumcentric dual, and therefore the Hodge star for one-forms reduces to the cotangent formula in this case of constant weights.

This way, most of the discrete exterior calculus methods in graphics (including the literature on Laplacian, Laplace–Beltrami, and discrete conformal parameterizations) can be directly adapted by including the contribution due to the weights.

9.4.2 Applications in mesh optimization

In the last sections we derived a natural parameterization of all primal-dual structures of simply connected domains in $\mathbb{R}^d$. This parameterization can in particular improve mesh optimization algorithms as it provides a convenient way to explore a large space of primal-dual structures. We already provided a first step in this direction by designing pairs of primal-dual structures that optimize accuracy bounds on differential operators using our parameterization [289], thus extending variational approaches designed to improve either primal (optimal Delaunay triangulations [397]) or dual (centroidal Voronoi tessellations) structures. In that work,

Hodge-optimized triangulations (HOT) were introduced, as a family of well-shaped primal-dual pairs of complexes designed for fast and accurate computations in computer graphics. Other existing work most commonly employs barycentric or circumcentric duals. While barycentric duals guarantee that the dual of each simplex lies within the simplex, circumcentric duals are often preferred due to the induced orthogonality between primal and dual complexes. This new approach instead promotes the use of weighted duals (power diagrams) and allows for much greater flexibility in the location of dual vertices while keeping primal-dual orthogonality. This provides an invaluable extension to the usual choices of duals by only adding one additional scalar per primal vertex. Furthermore, in that work, we introduce a family of functionals on pairs of complexes that we derive from bounds on the errors induced by diagonal Hodge stars, commonly used in discrete computations. The minimizers of these functionals, called HOT meshes, are shown to be generalizations of centroidal Voronoi tessellations and optimal Delaunay triangulations, and to provide increased accuracy and flexibility for a variety of computational purposes [289].

Following this framework, minor changes can be accommodated seamlessly in existing mesh optimization codes in order to allow for much greater flexibility. Weights can be, for instance, optimized (with fixed connectivity or not) to locally "displace" dual vertices onto an immersed boundary [26] through a least-squares fit. Vertices can be optimized as well, for instance, in applications requiring local or global remeshing to maintain good numerical analysis.

To some extent, this approach can even help to deal with situations where the primal triangulation is given and cannot safely be altered. For instance, moving vertices and/or changing the connectivity of a triangle mesh in $\mathbb{R}^3$ is potentially harmful, as it affects the surface shape. Still, the ability to optimize weights to drive the selection of the dual mesh is very useful. We can easily optimize primal-dual triangulations (meshes) by minimizing a functional (energy) with respect to the weights. The connectivity is kept intact, regardless of the weights—only the position and shape of the compatible dual is optimized.

As is shown in [289], even an optimized Delaunay triangulation with exceptionally high-quality tetrahedra can be made significantly better centered (with dual vertices closer to the inside of their associated primal simplex), using a simple weight optimization by this method (see Figure 9.3).

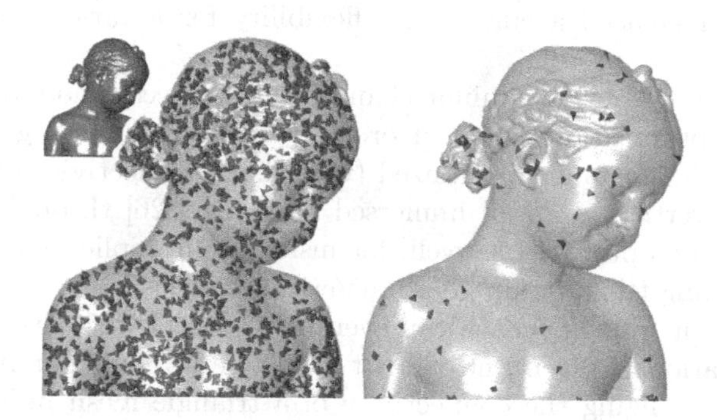

Figure 9.3 In a high-quality optimized Delaunay mesh of Bimba con Nastrino, we show the set of tetrahedra for which the distance between weighted circumcenter and barycenter is greater than a threshold, before (left) and after (right) the HOT optimization [289]. The primal triangulations are the same, and this significant optimization is done only on the dual, using the weights. See color insert.

CHAPTER 10

Self-Supporting Surfaces

Etienne Vouga

University of Texas, Austin, USA

CONTENTS

THE STUDY of masonry structures has fascinated architects, geometers, and engineers since antiquity. Artifacts like the Pantheon in Rome and Gothic cathedrals, which have stood for centuries without the benefit of modern mechanical engineering and materials science research, stand testament to the power and elegance of masonry construction techniques. This chapter sketches how the mechanics of masonry structures, and smooth and discrete geometry, are inseparably entwined, beginning with a historical overview of how stable structures were designed, covering some fundamental theory, and then surveying recent works in computer graphics on computational design and analysis of masonry structures. Weight functions defining generalized barycentric coordinates appear as a key component of this analysis, where they represent the force flowing through the structure.

10.1 INTRODUCTION AND HISTORICAL OVERVIEW

To study whether an arch or dome stands or falls is to study geometry. This is the key insight underlying *equilibrium analysis* or limit analysis, whose principles have been exploited for centuries by architects seeking to understand the mechanics of masonry structures.

10.1.1 What is a masonry structure?

Masonry structures are man-made structures composed of unreinforced concrete, brick, or stone, such as cathedrals, domes, and vaults. In particular, we consider *freeform* masonry structures, which though assembled of volumetric blocks, have a thin dimension (so that they resemble a surface). Figure 10.1 (left) shows an example of freeform masonry.

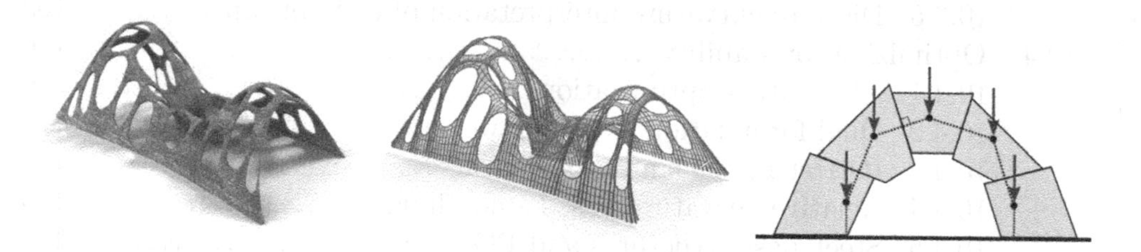

Figure 10.1 Left: A freeform masonry structure, composed of rigid, unreinforced blocks. (Image courtesy of Mathias Höbinger, used by permission.) Middle: The thrust network that certifies that this masonry structure is *stable* under its own weight. Each edge radius is proportional to the magnitude of compressive force carried by the edge. Right: An arch in 2D, with the self-load, and the thrust line certifying its stability, both illustrated. See color insert.

Following Heyman [189], we make the following assumptions about the masonry structures we consider:

1. The material has effectively infinite compressive strength (reasonable since the yield stress of stone far exceeds the working stress in practical structures).

2. In contrast, the structure has *no* tensile strength: mortar may be present at block interfaces, but this mortar is weak and cracks readily due to settling and weathering, so that it cannot be assumed to have any structural strength.

3. Failure of the structure by sliding is impossible (prevented either by the orientation of the block interfaces, or the presence of sufficient friction). Failure can occur only due to hinging apart of the structure.

The assumptions about compressive and tensile strength are critical ingredients in the geometric analysis of masonry, since they remove concerns about *material* failure: when a network of blocks crumbles, it is not due to buckling or crumpling of the blocks, but rather, due to *geometric* failure of the arrangements of the blocks alone. We thus need make no further assumptions about the composition, density, size, or other physical characteristics of the blocks.

10.1.2 Heyman's safe theorem

Heyman, building on the ideas of Kooharian [226], applied limit theorems from the theory of plasticity to formulate the *safe theorem* of 2D masonry structures* for masonry structures:

> If a line of thrust can be found which is in equilibrium with the external loads and lies wholly within the masonry, then the structure is safe.

A *thrust line* is a geometric representation of the forces flowing through the masonry structure. More specifically, consider a network of rigid blocks (see Figure 10.1, right). To each block i we associate an external load $\boldsymbol{F}_i \in \mathbb{R}^2$; typically this load is the block's self-weight and so is parallel to the direction of gravity. We also classify blocks as *anchored* boundary blocks if they are planted into the ground, and thus can support arbitrary loads from arbitrary directions, and *free* boundary blocks or *interior* blocks if they must be supported.

A thrust line is a graph G dual to the block network (so that to each block i corresponds a vertex $\boldsymbol{v}_i$ of G), and an associated set of weights $w_{ij} > 0$ on the edges of G (with $w_{ij} = w_{ji}$), so that the weighted sum of all edge vectors coming in to an interior or free boundary vertex $\boldsymbol{v}_i$ exactly balances the external load:

$$\boldsymbol{F}_i = \sum_{j \sim i} w_{ij}(\boldsymbol{v}_i - \boldsymbol{v}_j), \tag{10.1}$$

where the sum is over all vertices j incident to i.

*Note that this "theorem" only holds for structures obeying the material assumptions of Section 10.1.1: the restriction on sliding is particularly subtle, as discussed by Bagi [23].

Figure 10.2 Hanging nets and weights used by Gaudí to design the form of the Sagrada Família church in Barcelona. (Image courtesy of Flickr user *ralmonline alm* and reproduced under the CC-BY-2.0 license; colors have been converted to grayscale and white-balanced.)

Intuitively, the thrust line can be interpreted as a network of infinitesimal *beams* connected by pin joints at the vertices of G, with the weighted length of the edges $w_{ij}\|\boldsymbol{v}_j - \boldsymbol{v}_i\|$ the magnitude of the compression force within each beam. The safe theorem guarantees that if such a beam network exists that is in equilibrium under the external loads acting on a masonry structure, and the beam network is contained entirely within the volume of the structure, then the structure itself is stable. This ability of the safe theorem to reduce the problem of determining stability by one dimension has made thrust line analysis tremendously valuable as a practical technique for analyzing masonry arches, vaults, and domes [190, 203].

10.1.3 Gaudí and hanging nets

Although analyses of domes and vaults have been performed by considering thrust lines in 2D cross-sections, the safe theorem also applies to three-dimensional block structures, with (10.1) unchanged, with vertices of G (now a *thrust network*) and load vectors elements of $\mathbb{R}^3$.

The thrust network can again be viewed as a network of beams connected by pin joints, and under equilibrium when subject to the external load. This analogy has been exploited by architects and engineers like Frei Otto, Heinz Isler, and Antoni Gaudí, who used hanging chain and membrane models to find architectural forms [228]. Gaudí's masterpiece, the Sagrada Família, for instance, was designed using hanging nets and weights (see Figure 10.2). Notice that the thrust network equilibrium equations (10.1), when negated, describe the equilibrium of networks of string under *tension*; the shape of the hanging net, when turned upside down, thus yields a masonry structure that is stable by the safe theorem. This observation dates back at least as far as Hooke (who published in his famous anagram the observation that "as it hangs in a continuous flexible form, so it will stand contiguously rigid when inverted," and who used hanging nets and weights to design the dome of St Paul's Cathedral [210]).

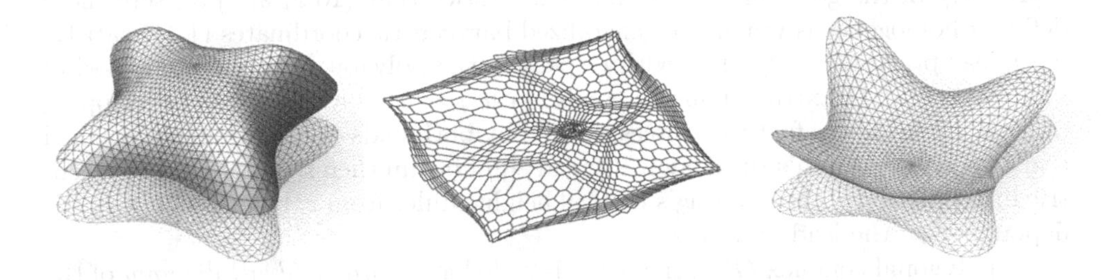

Figure 10.3 Left: A thrust network, inspired by the shape of the top of the Lilium tower, and its top view. Middle: The top view admits a reciprocal diagram, certifying horizontal equilibrium. Right: The discrete Airy stress surface S corresponding to this reciprocal diagram.

While existence of a thrust network within a masonry structure is sufficient to guarantee the structure's stability, the thrust line does *not* necessarily encode the true forces flowing through the structure: many possible thrust networks can exist within a masonry volume, and any one of these certifies the structure's stability. Consider for instance a four-legged stool: the stool stands with any one of its legs removed, and the true force flowing through the four legs cannot be determined by examining the geometry of the stool alone. In practice such *static indeterminacy* is commonplace.

10.1.4 Maxwell's reciprocal diagrams

Given a graph G and vertex loads $\boldsymbol{F}_i$, when do there exist edge weights that place the thrust network in equilibrium? To address this question, we begin by making the additional assumption about the structure that the external loads are purely vertical, that is, parallel to the $\hat{\boldsymbol{z}}$ axis (this is reasonable, since the self-weight of the structure usually dominates most other external loads such as those due to wind, etc.). We can thus write $\boldsymbol{F}_i = F_i\hat{\boldsymbol{z}}$ for scalar loads F_i. It will also be useful to project the vertices $\boldsymbol{v}_i$ of G into the xy-plane, to yield a *top view* $\bar{G}$ with vertices $\bar{\boldsymbol{v}}_i$, to which we associate heights z_i (see Figure 10.3). The equilibrium equations (10.1) then *split* into the pair

$$0 = \sum_{j\sim i} w_{ij}(\bar{\boldsymbol{v}}_j - \bar{\boldsymbol{v}}_i) \tag{10.2}$$

$$-F_i = \sum_{j\sim i} w_{ij}(z_j - z_i) \tag{10.3}$$

of *horizontal* and *vertical* equilibrium equations for each interior or free boundary vertex i. Notice that only the vertical equations involve the loads or heights of the vertices of G; existence of weights that satisfy the horizontal equations is purely a property of the geometry of the top view. Notice that (10.2) are precisely those defining homogeneous weights of generalized barycentric coordinates (1.11) satisfying linear precision [149] for $\bar{\mathbf{v}}_i$ with respect to the polygon P that is composed of $\bar{\mathbf{v}}_i$'s one-ring. The extra symmetry condition $w_{ij} = w_{ji}$ on weights in overlapping one-rings reflects the fact that forces exerted by both ends of a beam must be equal and opposite (Newton's third law). Vertical equilibrium then requires that barycentric interpolation of the one-ring's vertical heights differ from z_i by an amount F_i/W depending on the load at vertex i.

A polygonal complex $\bar{G}^*$ in the plane is called a *reciprocal (dual) diagram* of the top view $\bar{G}$ if it is a dual complex of $\bar{G}$, and every edge of $\bar{G}^*$ is orthogonal to its dual edge on $\bar{G}$ (see Figure 10.3).† Reciprocal diagrams and their relationship to static equilibrium have been studied extensively, starting with James Maxwell [270], who observed the following fact.

Theorem 10.1. *Given a top view $\bar{G}$, edge weights w_{ij} exist satisfying horizontal equilibrium* (10.2) *if $\bar{G}$ possesses a reciprocal diagram, and only if every simply connected subcomplex of $\bar{G}$ possesses a reciprocal diagram.*

Indeed, observe that if weights exist satisfying (10.2), the edges surrounding a vertex $\bar{\mathbf{v}}_i$ can be scaled by w_{ij} and rotated by ninety degrees to form a dual face (sometimes called the *force polygon*) on the reciprocal diagram, and similarly, given a dual face, edge weights can be computed by taking edge length ratios.

For simplicity of exposition, in what follows we assume that a global reciprocal diagram can always be constructed from edge weights; the fact that the reciprocal diagram can only be pieced together locally when $\bar{G}$ has higher genus is not important in practice.

Finally, notice that the compression-only constraint $w_{ij} \geq 0$ has a natural geometric analogue on the reciprocal diagram: none of the dual edges in $\bar{G}^*$ may be inverted.

10.2 SMOOTH THEORY

Before we continue analyzing the stability of thrust networks, we take a detour to study the properties of the *continuum limit* of a surface-like masonry structure made of infinitesimally small blocks, so that the geometry of the masonry structure can be represented as a smooth surface M embedded in $\mathbb{R}^3$. In this setting, external load magnitude (we again assume loads are only vertical) is represented by a scalar density function $F\colon M \to \mathbb{R}$ on the surface, and static equilibrium of M under these loads can be characterized using the classical theory of elasticity [178]. To simplify the analysis we will assume that M has no overhangs, that is, it is a graph over the plane, though such overhangs do occur in some practical masonry structures (such as the oculus in the dome of the Pantheon).

†Some authors define the reciprocal dual so that the dual edges are parallel to the primal edges; this notion of reciprocality differs from $\bar{G}^*$ only by an unimportant ninety-degree rotation.

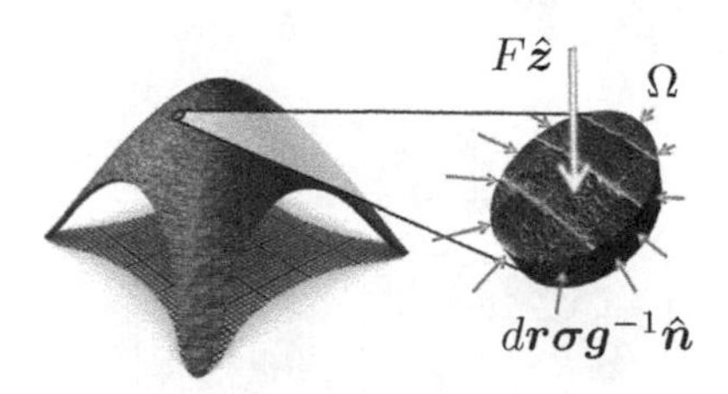

Figure 10.4 Balance of forces on a small patch.

10.2.1 Equilibrium equations

Consider a point $\boldsymbol{x}$ on M, and Ω a small neighborhood of $\boldsymbol{x}$. As above, we project Ω onto the xy-plane to yield a *top view* $\bar{\Omega} \subset \mathbb{R}^2$, and parameterize Ω by a height function $z\colon \bar{\Omega} \to \mathbb{R}$. The embedding of the patch Ω is then given by $\boldsymbol{r}(x, y) = (x, y, z(x, y))$ and the induced metric on $\bar{\Omega}$ is $\boldsymbol{g} = \boldsymbol{I} + d\boldsymbol{z} \otimes d\boldsymbol{z}$.

Internal forces acting through the boundary of Ω are characterized by the stress tensor $\boldsymbol{\sigma}\colon \bar{\Omega} \to GL(2, \mathbb{R})$; for a tangent vector $\boldsymbol{v}$ at a point on the top view, $\boldsymbol{\sigma v}$ measures the internal force that would result from making a small cut in Ω in the tangent direction perpendicular to $d\boldsymbol{rv}$. Balance of forces on Ω means that the internal forces integrated along $\partial\Omega$ must exactly cancel the total external load on Ω (see Figure 10.4),

$$\int_{\bar{\Omega}} F\hat{\boldsymbol{z}}\, d\bar{A} = \int_{\partial\bar{\Omega}} d\boldsymbol{r}\boldsymbol{\sigma}\boldsymbol{g}^{-1}\hat{\boldsymbol{n}}\, ds,$$

where $d\bar{A} = \sqrt{\det \boldsymbol{g}}\, dx\, dy$ is the area element on Ω, $\hat{\boldsymbol{n}}$ is the outward-pointing unit normal (with respect to the metric $\boldsymbol{g}$) on the boundary $\partial\bar{\Omega}$, and ds is the differential on $\partial\Omega$. Applying Stokes's theorem and contracting Ω to a point yields the pointwise equilibrium condition

$$F\hat{\boldsymbol{z}}\sqrt{\det \boldsymbol{g}} = \nabla \cdot \left[(\sqrt{\det \boldsymbol{g}}) \boldsymbol{g}^{-1} \boldsymbol{\sigma} \nabla \boldsymbol{r} \right],$$

where the divergence operator $\nabla\cdot$ is understood to act on matrices by producing a vector whose entries are the divergences of the columns of the matrix. Equilibrium conditions also exist on the free boundaries: the traction on these boundaries must be zero, that is, $\boldsymbol{\sigma g}^{-1}\hat{\boldsymbol{n}} = 0$, where $\hat{\boldsymbol{n}}$ is the boundary normal on $\partial\bar{M}$.

Since the stress tensor is self-adjoint with respect to $\boldsymbol{g}$, $\boldsymbol{S} = \left(\sqrt{\det \boldsymbol{g}}\right)\boldsymbol{g}^{-1}\boldsymbol{\sigma}$ is a symmetric matrix; the fact that M supports only compressive, and not tensile, internal forces becomes the constraint that $\boldsymbol{S}$ is positive semidefinite. Therefore M is in static equilibrium if there exists a section of matrices $\boldsymbol{S} \succeq 0$ over $\bar{\Omega}$ with

$$F\hat{\boldsymbol{z}}\sqrt{\det \boldsymbol{g}} = \nabla \cdot [\boldsymbol{S}\nabla \boldsymbol{r}] \tag{10.4}$$

at every point on the interior of $\bar{\Omega}$.

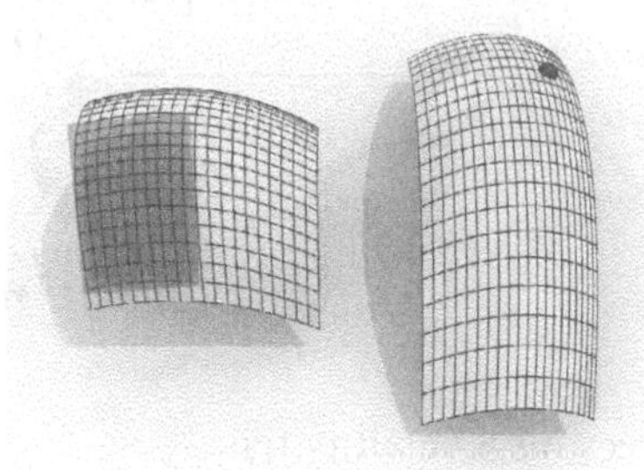

Figure 10.5 A function $z(x, y)$ and its polar dual. Tangent planes map to points.

10.2.2 Airy stress potential

We can again decompose the equilibrium equation (10.4) into horizontal and vertical components,

$$\mathbf{0} = \nabla \cdot \boldsymbol{S}$$
$$F\sqrt{\det \boldsymbol{g}} = \nabla \cdot [\boldsymbol{S}\nabla z].$$

The first set of equations, together with symmetry of $\boldsymbol{S}$, are precisely the local integrability conditions required for existence of a scalar function, the *Airy stress potential* s, with

$$\boldsymbol{S} = \begin{bmatrix} s_{yy} & -s_{xy} \\ -s_{xy} & s_{xx} \end{bmatrix} = \boldsymbol{J}^T[\mathbf{H}_s]\boldsymbol{J},$$

where $\mathbf{H}_s$ is the Hessian of s and $\boldsymbol{J}$ is the ninety-degree rotation matrix.

In direct analogy to the discussion above on reciprocal diagrams, if $\bar{M}$ is simply connected, s exists globally, and we have the following theorem.

Theorem 10.2. *The surface M with simply connected top view $\bar{M}$ is in compressive static equilibrium if there exists a convex function $s\colon \bar{M} \to \mathbb{R}$ with*

$$F\sqrt{\det \boldsymbol{g}} = \nabla \cdot \boldsymbol{J}^T[\mathbf{H}_s]\boldsymbol{J}\nabla z$$

at every point on the interior of $\bar{M}$, and

$$[\mathbf{H}_s]\boldsymbol{J}\hat{\boldsymbol{n}} = \mathbf{0}$$

on the free boundary, where $\hat{\boldsymbol{n}}$ is the normal to $\partial\bar{M}$.

10.2.3 Relative curvatures

The standard Gauss map ξ maps every point on a surface $z(x, y)$ to the corresponding point on the unit sphere with matching tangent plane. This idea can be generalized: given a second, convex surface $s(x, y)$, we define the *relative* Gauss map $\boldsymbol{\xi}^s : \mathbb{R}^2 \to \mathbb{R}^2$ as the function mapping every point on z to the point[‡] on s with identical tangent plane,

$$[\nabla z](x, y) = [\nabla s](\boldsymbol{\xi}^s(x, y)).$$

From this relative Gauss map we can define a relative shape operator $d\boldsymbol{\xi}^s(x, y) = [\mathbf{H}_z(x, y)][\mathbf{H}_s(\boldsymbol{\xi}^s(x, y))]^{-1}$, and mean and Gaussian curvatures of z *relative to* s by $\frac{1}{2}\operatorname{tr}(d\boldsymbol{\xi}^s)$ and $\det(d\boldsymbol{\xi}^s)$.

The isotropic polar *dual* z^* of a surface $z(x, y)$ is formed by mapping

$$[x, y, z(x, y)] \mapsto [z_x(x, y), z_y(x, y), -z + xz_x(x, y) + yz_y(x, y)]; \tag{10.5}$$

notice that (a) this mapping is an involution when z is strictly convex; (b) it maps planes to points; and (c) it maps points with parallel tangent plane to points that are *vertically offset*, that is, are coincident when projected onto the xy-plane [310]. See Figure 10.5 for an example of a surface and its dual. The Gauss map, expressed on the dual surfaces z^* and s^*, is thus simply the identity: it maps points on z^* to points on s^* with identical (x, y) coordinates. Dual relative curvatures thus have the particularly simple formulas

$$K_{z^*}^{s^*} = \frac{\det(\mathbf{H}_z)}{\det(\mathbf{H}_s)},$$

$$H_{z^*}^{s^*} = \frac{1}{2}\operatorname{tr}\left([\mathbf{H}_s]^{-1}[\mathbf{H}_z]\right).$$

Of special note is the Maxwell hyperboloid $h(x, y) = \frac{1}{2}(x^2 + y^2)$, whose dual is simply a reflection about the xy-plane, and for which $K_{z^*}^{h} = \det(\mathbf{H}_z)$; curvatures relative to h arise naturally in *isotropic geometry*, where the hyperboloid plays the role of the canonical unit sphere [310].

10.2.4 Curvature interpretation of equilibrium

As suggested by the above notation, the Airy stress function $s(x, y)$ of a surface in equilibrium can be interpreted as a height field defining a convex surface. This yields a *geometric* characterization of equilibrium in terms of the dual relative curvatures of the surface z and its Airy stress potential s; to see this, note that $\boldsymbol{J}^T[\mathbf{H}_s]\boldsymbol{J} = \det(\mathbf{H}_s)[\mathbf{H}_s]^{-1}$; therefore

$$H_{z^*}^{s^*} = \frac{1}{2\det(\mathbf{H}_s)}\operatorname{tr}\left(\boldsymbol{J}^T[\mathbf{H}_s]\boldsymbol{J}[\mathbf{H}_z]\right) = \frac{1}{2\det(\mathbf{H}_s)}\nabla \cdot \boldsymbol{J}^T[\mathbf{H}_s]\boldsymbol{J}\nabla z$$

and the surface M is in equilibrium if and only if there exists a convex surface $s(x, y)$, defined on the same top view $\bar{M}$ as z, with

$$2K_{s^*}^{h} H_{z^*}^{s^*} = F\sqrt{\det \boldsymbol{g}} \tag{10.6}$$

at every interior point of $\bar{M}$, and $[\mathbf{H}_s]\boldsymbol{J}\hat{\boldsymbol{n}} = 0$ on the free boundary [405].

‡We ignore issues of existence and uniqueness of such a point.

The importance of this relation is that it measures equilibrium purely in terms of the geometry of two height fields z and s; we will see that this relationship can be directly ported to the discrete setting, where these two surfaces become piecewise-affine meshes. And the discretization reconnects with the reciprocal diagrams of Maxwell and Gaudí.

10.3 DISCRETE THEORY

We now turn to the vertical discrete equilibrium equation (10.3). This equation takes the very familiar form of a discrete Laplace–Beltrami operator; we can rewrite it as

$$-F_i/A_i = \frac{1}{A_i}\sum_{j\sim i} w_{ij}(z_j - z_i) = (L_w z)_i, \tag{10.7}$$

where A_i is the dual Voronoi area[§] of $\bar{\boldsymbol{v}}_i$ (the area of the region in the plane nearer to $\bar{\boldsymbol{v}}_i$ than to any other vertex of $\bar{G}$) and L_w denotes the discrete Laplace–Beltrami operator with w_{ij} taking the place of the usual cotangent weights (see Section 1.2.2). Horizontal equilibrium with $L_w \boldsymbol{v} = 0$ corresponds to the linear precision property (see Section 1.1.2); together with the compression-only constraint $w_{ij} \geq 0$, horizontal equilibrium guarantees that this so-called *stress Laplacian* L_w is a *perfect* discrete Laplace–Beltrami operator, in the sense of satisfying the four properties discussed in Section 5.3.7, as originally investigated by Wardetzky et al. [410]. We thus have yet another characterization of equilibrium: a thrust network is in equilibrium under loads F_i if and only if there exists a perfect Laplace–Beltrami operator L_w on $\bar{G}$, such that the heights satisfy the Poisson equation (10.7) (with Neumann boundary conditions at the free boundaries, and Dirichlet at the anchored boundaries). The right-hand side F_i/A_i can be interpreted as a *load density* (with respect to area on the top view) that will be applied to the masonry structure.

10.3.1 Thrust network analysis

Philippe Block, in his seminal work on *Thrust Network Analysis* [52, 54], exploited the above observations to formulate an algorithm for designing equilibrium thrust networks. The form-finding proceeds in four steps:

1. First, a top view $\bar{G}$ is constructed in the silhouette of the desired form.
2. Next, a reciprocal diagram is found for $\bar{G}$. For regular quadrilateral grids, this can be done in a marching fashion. As discussed below, a reciprocal diagram is rarely unique, and later work [53, 273] explored the large space of possible diagrams to increase design flexibility.
3. Edge weights w_{ij} are extracted from the reciprocal diagram.

§Some authors prefer to define the discrete Laplace–Beltrami operator in terms of barycentric dual areas, or a Galerkin mass matrix; these alternatives are also suitable.

4. Given external loads, the Poisson equation (10.7) is solved to determine the heights z of the network vertices.

The user can explore the space of possible forms by interactively changing the top view, reciprocal diagram, and/or applied external loads.

10.3.2 Thrust networks as block networks

The safe theorem certifies that if a block network contains a thrust network in equilibrium within its volume, then the block network is itself stable. This relationship can be reversed: once a thrust network in equilibrium has been constructed, a masonry structure can be designed enclosing the thrust network, which is stable by the safe theorem. There are two subtleties with this approach in practice: first, the loads on the structure must be the same as those that were imposed on the thrust network (although scaling all of the loads by a constant factor is allowed, since the equilibrium equations (10.1) are scale-invariant, a fact exploited since ancient times to analyze full-sized masonry structures using scale models). Second, masonry blocks have only finite coefficients of friction and resistance to sliding, so the orientation of the block interfaces cannot be entirely neglected.

Nevertheless, every thrust network G in equilibrium corresponds to some set of blocks in frictionless equilibrium [356]. First, compute an *intrinsic* Voronoi diagram on G (so that the cell corresponding to vertex $\boldsymbol{v}_i$ is piecewise-affine and contains pieces of all faces adjacent to $\boldsymbol{v}_i$). Each cell can be "fattened" by normal extrusion to give a network of blocks that completely enclose the thrust network, with block interfaces perpendicular to the dual edges of the thrust network. Each block can now be modified (by adding or removing material, while keeping it in contact with all of its neighbors) so that its total mass equals the load F_i imposed on the thrust network, and so that its centroid is at $\boldsymbol{v}_i$.

The thrust network and external loads can then be interpreted as a network of contact and external forces, respectively, acting on each block's centroid; equilibrium of the thrust network guarantees balance of forces and moments for each block, and thus that the block network is stable [425].

10.3.3 Discrete Airy stress potential

In the smooth setting, we were able to characterize equilibrium in terms of a scalar Airy stress potential over the top view; so far, in the discrete setting, we have edge weights, or equivalently reciprocal diagrams, or equivalently perfect Laplace–Beltrami operators. We now discretize the Airy stress potential [405], and show how it ties to the other characterizations described above.

Given a top view $\bar{G}$, we call a polygonal 2-complex S in $\mathbb{R}^3$ a *discrete Airy stress surface* with respect to $\bar{G}$ if:

- the top view $\bar{S}$ of S (formed by projecting all vertices of S into the plane) is identical to $\bar{G}$;
- the faces of S are planar;

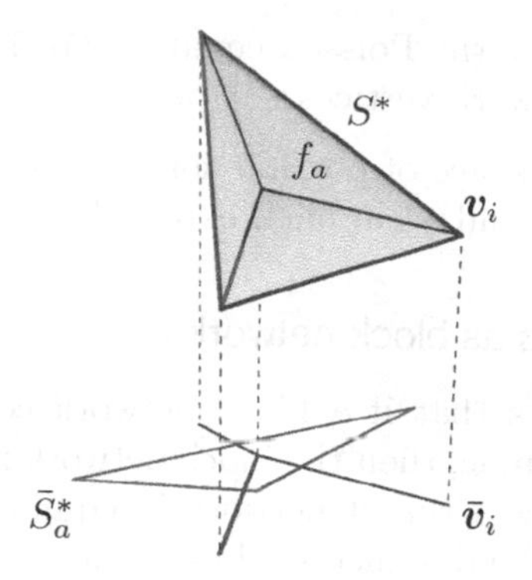

Figure 10.6 Relationship between S^* and reciprocal diagrams of $\bar{G}$.

- when interpreted as a function over $\bar{G}$, S is convex.

See Figure 10.3 (right) for an example of such an S. A discrete Airy stress surface, like the thrust network itself, can be encoded by an assignment of heights $s_i \in \mathbb{R}$ to the vertices of $\bar{G}$; however, not all such assignments encode valid discrete Airy stress surfaces, since the resulting points $(\bar{\boldsymbol{v}}_i, s_i) \in \mathbb{R}^3$ cannot always be interpolated by planar faces.

Applying isotropic polar duality (10.5) to S yields a dual point for every face of S; denote by $\bar{S}^*$ the top view of this dual discrete Airy stress surface. $\bar{S}^*$ and $\bar{S}$ are dual as polygonal complexes in the plane. Moreover, they are *reciprocal* duals:[¶] let $\bar{S}^*_a$ and $\bar{S}^*_b$ be two adjacent vertices of $\bar{S}^*$, and $f_a(x, y), f_b(x, y)$ be the affine height functions describing the corresponding faces on S. Then $\bar{S}^*_a = \nabla f_a$, and the dual edge connecting $\bar{S}^*_a$ and $\bar{S}^*_b$ is parallel to $\nabla f_a - \nabla f_b$. Since

$$\nabla f_a \cdot (\bar{\boldsymbol{v}}_i - \bar{\boldsymbol{v}}_j) = \nabla f_b \cdot (\bar{\boldsymbol{v}}_i - \bar{\boldsymbol{v}}_j) = s_i - s_j,$$

where the primal edge ij is the common edge to faces a and b, the corresponding edges of $\bar{S}$ and $\bar{S}^*$ are orthogonal (see Figure 10.6).

Therefore a reciprocal diagram can be constructed from any discrete Airy stress surface. The converse is also true: given a reciprocal diagram, assign arbitrary values of s to the vertices of an arbitrary face a. Then the *gradient* of all of the neighboring faces can be computed by adding to the gradient of S on face a the vector corresponding to the dual edge on the reciprocal diagram joining the dual vertices of the two faces. All of S can be constructed by gluing together faces in this way. Local integrability of S follows directly from closure of the force polygon on the reciprocal diagram,

$$\sum_i (\nabla f_{i+1} - \nabla f_i) = \sum_i (\bar{\boldsymbol{v}}^*_{i+1} - \bar{\boldsymbol{v}}^*_i) = 0,$$

[¶]Note that there are two distinct notions of duality at work here: polar duality of S and S^*, and reciprocal duality of their top views.

where the sums index over the faces circulating around a vertex of $\bar{G}$.

The discrete Airy stress surface is not unique, since the choice of the gradient of the first face of S was chosen arbitrarily. Different stress surfaces of the same diagram will differ by addition of a linear function to S. The above constructions also require $\bar{G}$ to be simply connected; otherwise S can only be assembled locally.

10.3.4 FEM discretization of Airy stress

By the above construction, lengths on the reciprocal diagram correspond to the magnitude of the *jump* in the gradient of S when moving between neighboring faces. This relationship has also been observed by Fraternali [156], who derived it in the context of finite element analysis of cracking in masonry vaults. Starting from the smooth setting, both the thrust surface $z(x, y)$ and the Airy stress potential $s(x, y)$ can be discretized using piecewise linear shape functions over the top view $\bar{G}$. Doing so, and discretizing horizontal equilibrium using Galerkin finite elements, yields the orthogonal relationship between gradient jumps of S and edges of $\bar{G}$.

10.3.5 Discrete curvature interpretation of equilibrium

Relative curvature of a polyhedral surface at a vertex with respect to a second polyhedral surface has been studied by Bobenko et al. [56] and formulated in terms of *mixed areas* of vertex neighborhoods. This concept can be extended to dual isotropic relative curvature [311] to yield formulas for the dual relative mean curvature at a vertex $\boldsymbol{v}_i$ of a discrete surface G,

$$H_{G^*}^{S^*}(\boldsymbol{v}_i) = \frac{A_i(G^*, S^*)}{A_i(S^*, S^*)},$$
$$A_i(G^*, S^*) = \frac{1}{4}\sum_j \left[\det(\bar{\boldsymbol{v}}_j^*, \bar{S}_j^*) + \det(\bar{S}_j^*, \bar{\boldsymbol{v}}_j^*)\right],$$

where the sum in the mixed area term A_i is over the faces in the one-ring of $\boldsymbol{v}_i$. It can be shown [405] that the mixed area in the numerator is

$$2A_i(G^*, S^*) = \sum_{j \sim i} w_{ij}(z_j - z_i),$$

where the weights w_{ij} are those corresponding to the reciprocal diagram constructed from S. Moreover, the dual Gaussian curvature relative to the Maxwell paraboloid is

$$K_{S^*}^{h}(\boldsymbol{v}_i) = \frac{A_i(S^*, S^*)}{A_i},$$

where A_i (with no arguments) is, as before, the Voronoi area of $\bar{\boldsymbol{v}}_i$ in the top view. Putting together all of the pieces and combining them with (10.3), we have, in direct analogy to (10.6), the following theorem.

Theorem 10.3. *A thrust network G with simply connected top view $\bar{G}$ is in equilibrium if and only if there exists a discrete Airy stress surface S over $\bar{G}$ with*

$$-F_i/A_i = 2K^h_{S^*}(\boldsymbol{v}_i)H^{S^*}_{G^*}(\boldsymbol{v}_i) \tag{10.8}$$

at interior vertices $\boldsymbol{v}_i$, with faces of S abutting consecutive pieces of free boundary coplanar.

Here, F_i/A_i, like $F\sqrt{\det \boldsymbol{g}}$ in (10.6), should be interpreted as a load density over the *top view* rather than the thrust surface itself.

10.4 OPTIMIZING FOR STABILITY

We now have the theoretical pieces necessary to formulate some algorithms for solving inverse problems related to stable masonry. In its simplest form, the inverse problem asks: given a thrust network M, is it in equilibrium? If not, what changes must be made to M so that it is now in equilibrium? Once such a network is found, a masonry structure can be built that is adapted to the shape of the thrust network, as described in Section 10.3.2.

The first question can be answered by looking for non-negative edge weights w_{ij} that satisfy equilibrium (10.1). If such weights do not exist, some modification must be made, either to the combinatorics of M, or to its geometric shape, in order for it to become stable.

10.4.1 Alternating optimization

If a thrust network is not stable, one can still look for weights that satisfy equilibrium as much as possible, in the sense of minimizing the *residual force* at each node of the thrust network,

$$\arg\min_{w_{ij}\geq 0} \sum_i \left\| \boldsymbol{F}_i - \sum_{j\sim i} w_{ij}(\boldsymbol{v}_i - \boldsymbol{v}_j) \right\|^2, \tag{10.9}$$

where the sum is taken over all interior and free boundary vertices of M. This optimization amounts to solving a least squares problem with nonnegativity constraints on the variables, for which there exist specialized solvers [234].

One could now fix the weights w_{ij} and solve for displacements of the vertices of the thrust network that minimize the force residual; since this is a square system of linear equations, it usually has a solution. While the resulting thrust network M' would be in equilibrium, there is no guarantee that M' is at all similar in shape to M, and in practice it is not, particularly if the initial network M was far from equilibrium. Vouga et al. [405] proposed an *alternating minimization* that repeatedly solves for the w_{ij}, while allowing the thrust network to slowly relax away from M:

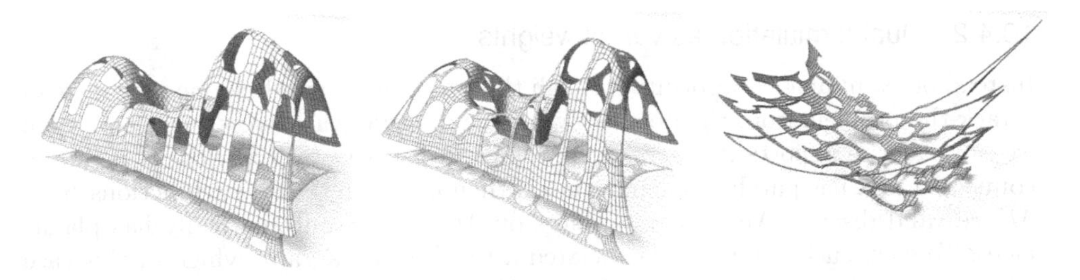

Figure 10.7 Left: The initial, unstable thrust network M and its top view. Middle: Several iterations of alternating optimizations bring the network to equilibrium. Right: The Airy stress surface S certifying stability. Since M is not simply connected, S exists only locally; it was cut to reduce its genus for visualization.

1. Begin iteration $k = 0$ with a guess $M_k = M$ for the stable thrust network.
2. Estimate the loads $\boldsymbol{F}_i$ on the vertices of M_k.
3. Fix M_k and solve for weights w_{ij} that minimize the force residuals; see (10.9).
4. Fix w_{ij} and improve the positions of the vertices M_k to yield a new thrust network M_{k+1} closer to equilibrium.
5. Repeat from Step 1 until convergence.

Recomputing the loads at every iteration is required since the load typically depends on the surface areas of the faces of M_k, which change in every iteration. In Step 4, the new thrust network M_{k+1} can be computed from M_k by solving a least squares problem with one objective term penalizing the force residual, and a second penalizing displacements of the vertices of M_{k+1} away from their positions in M_k; Vouga et al. suggest penalizing motion in the direction normal to M_k more severely than tangential sliding, since only the former affects the aesthetics of the final masonry structure.

The above algorithm, though simple and effective in practice (see Figure 10.7), has several limitations. First, it is not guaranteed to terminate: although both Steps 3 and 4 are guaranteed to strictly decrease the force residual, Step 2 can increase it. In practice, termination is not an issue when reasonable boundary conditions and loads are imposed. Second, the optimization modifies M by displacing its vertices, but does not consider improving equilibrium by remeshing M or performing other combinatorial edits. Third, the algorithm does not consider changing the boundary conditions (for instance, adding a supported pillar at the interior of an unstable region of the network) even though such operations are quite natural to human architects. Finally, the constrained optimization in Step 3 can be a performance bottleneck and a source of numerical instability: many solutions might exist to (10.9), some of which could have edge weights with extremely large or extremely small magnitudes.

10.4.2 Dual formulation as vertex weights

Instead of examining equilibrium through the lens of edge weights, one may instead attack the problem from the perspective of the discrete Airy stress surface S. Here we can make an important observation: when a thrust network top view $\bar{M}$ is simply connected and has purely triangular faces, then *all* discrete convex functions S on $\bar{M}$ are valid discrete Airy stress surfaces on $\bar{M}$, since S automatically has planar faces. To each such S must be associated a reciprocal diagram, which in this case is the *power diagram* [19] (see Section 9.2.1) with vertex weight r_i at $\bar{\boldsymbol{v}}_i$ given by

$$r_i = \frac{1}{2}\|\boldsymbol{v}_i\|^2 - \frac{1}{2}s_i.$$

Some sets of edge weights do not correspond to valid reciprocal diagrams, and hence Airy stress surfaces (if they do not satisfy (10.2)) and many sets of edge weights can give the same Airy stress surface. On the other hand, the vertex weights s_i form a *reduced basis* for the valid Airy stress surfaces, an observation that de Goes et al. [113] and Liu et al. [255] leveraged into efficient algorithms for finding equilibrium thrust networks M' approximating a designed network M.

Given $\bar{M}$, the algorithm seeks S_i and heights z_i satisfying the vertical equilibrium condition (10.3) while keeping the z_i as close as possible to their designed values; de Goes et al. [113] propose to penalize displacements of z_i more for vertices initially closer to equilibrium. The compression-only constraint, that S is convex, becomes a set of linear inequality constraints encoding that every edge of S is a convex edge. The algorithm can be easily modified to handle the more general case where $\bar{M}$ has holes, and S exists only locally. It can also be augmented to allow sliding of the top view vertices, using edge flips to keep $\bar{M}$ a regular triangulation [255].

10.4.3 Perfect Laplacian optimization

Recall that equilibrium of M is equivalent to the existence of a *perfect* Laplacian (10.2) on the top view $\bar{M}$ for which vertical heights of the network vertices satisfy a Poisson equation (10.3). Herholz et al. [187] studied perfect Laplacians on polygonal meshes and developed a different approach to finding an equilibrium thrust network M' given a guess M. The key insight is that a current guess for the weights w_{ij}, together with the specified supported boundary vertex positions and the external loads, uniquely determine heights z_i of a thrust network M' in equilibrium, with $\bar{M} = \bar{M}'$, via (10.3). Comparing edge lengths of M' to M suggests that the edge weights should be updated as

$$w_{ij}^{k+1} = w_{ij}^{k}\frac{\|\boldsymbol{v}_i^k - \boldsymbol{v}_j^k\|}{\|\boldsymbol{v}_i - \boldsymbol{v}_j\|},$$

where w_{ij}^k and $\boldsymbol{v}_i^k$ are the edge weights and vertex positions at the k-th iteration of the algorithm. This process is repeated until convergence of M'.

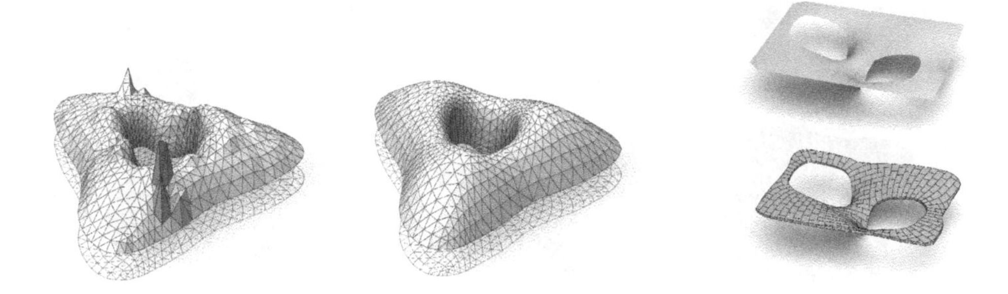

Figure 10.8 Left: Direct optimization of the thrust network can cause "spikes" to appear where the stresses change rapidly over the surface. Optimizing for constant dual-isotropic mean curvature yields a better-behaved thrust network. (Data for this image courtesy of Yang Liu and Hao Pan.) Right: Laying out blocks on a thrust surface by staggered grouping of quadrilateral faces, as a final step of the design process. (Image courtesy of Daniele Panozzo, used by permission.)

Remarkably, this optimization works well even in cases where the top view $\bar{M}$ has challenging geometry, such as sliver triangles or non-convex polygonal faces, which is far from admitting a perfect Laplacian. Moreover, the update rules for M and w_{ij} do not require the constrained optimization of the alternating algorithm in Section 10.4.1, instead needing only the solution of sparse linear systems. On the other hand, the forms M' to which this approach converges can differ substantially from the initial design M.

10.4.4 Relative-curvature-based smoothing

One drawback of the optimization methods presented so far is that they do not guarantee the quality of the final, equilibrium thrust network M'; indeed, a common problem is the presence of "spikes" in the equilibrium network, caused by edge weights that vary wildly over the star of a vertex (or equivalently, stresses that do not change smoothly over a small neighborhood). See Figure 10.8, left. Liu et al. [255] developed a method for addressing this problem by insisting that M has *constant dual-isotropic relative mean curvature* relative to its Airy stress surface S: $H^{S^*}_{M^*} = H$ for a constant H (taken to be the median value of $H^{S^*}_{M^*}$ in the input network). If this assumption were to hold, equilibrium, as understood in terms of curvatures in (10.8), would reduce to

$$-F_i = 2A_i K^h_{S^*} H = 2A_i(S^*, S^*)H,$$

where the dual cell areas $A_i(S^*, S^*)$ depend on the stress surface heights s_i. Liu et al. solve for these using Newton's method, taking care that S remains convex (this is done by computing the convex hull and projecting any violating vertices

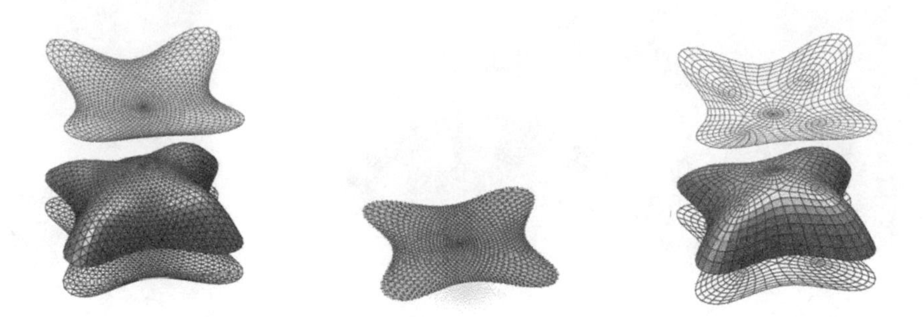

Figure 10.9 Left: Initial thrust network and Airy stress surface. Middle: Eigenvectors of $[\mathbf{H}_s]^{-1}[\mathbf{H}_z]$ trace out curve networks conjugate to both surfaces. Right: Simultaneous remeshing of both thrust network and stress surface as planar quadrilateral meshes. (Images courtesy of Mathias Höbinger, used by permission.)

down to the convex hull in between Newton iterations). This curvature-smoothing operation dramatically improves the fairness of the thrust network.

10.4.5 Steel-glass structures and PQ faces

A popular alternative to masonry for realizing freeform structures are steel-glass constructions, made out of glass panels supported inside a steel skeleton. Under the assumption that the glass panels are not structural and carry no load, equilibrium of steel-glass structures is identical to equilibrium of thrust networks, with the exception that the steel beams can carry only bounded compressive load, and do support some tension.

Steel-glass structures introduce new aesthetic considerations, however. Quadrilateral glass panels are preferred to triangular ones, as the nodes where beams meet in the former case have lower valence and are thus easier to manufacture, and the structure requires fewer steel beams, making the structure more permeable to light. *Planar* glass panels are much preferred to curved panels, as it is much easier to cut planar panels from a large glass sheet, than it is to hot-bend custom panel shapes.

We are thus left with a geometry problem: given a thrust network M in equilibrium, can it be approximated by a thrust network N with planar quadrilateral faces? Let S be the Airy stress surface certifying M's stability. One could simply remesh M, and then resample S to yield Ψ with $\bar{\Psi} = \bar{N}$. While Ψ would still be convex, there is no guarantee that the faces of Ψ are planar, and thus Ψ is no longer necessarily a valid stress surface. One thus needs to *simultaneously* remesh both M and S so that both remeshed surfaces have an identical top view, and planar quadrilateral faces (see Figure 10.9).

The edges of any planar quadrilateral mesh approximating M follow conjugate curve networks [347] on M. More specifically, if M approximates a smooth surface

$z\colon \bar{M} \to \mathbb{R}$, M can be remeshed as a planar quadrilateral mesh by first finding two families of curves $\boldsymbol{a}(s,t), \boldsymbol{b}(s,t)$ in the plane with $\dot{\boldsymbol{a}}^T[\mathbf{H}_z]\dot{\boldsymbol{b}} = 0$. Since the same must hold for S approximating s, we need families of curves satisfying

$$\dot{\boldsymbol{a}}^T[\mathbf{H}_z]\dot{\boldsymbol{b}} = \dot{\boldsymbol{a}}^T[\mathbf{H}_s]\dot{\boldsymbol{b}} = 0,$$

which we can find by taking $\boldsymbol{a}$ and $\boldsymbol{b}$ to be streamlines following the eigenvectors of $[\mathbf{H}_s]^{-1}[\mathbf{H}_z]$. A new top view can be traced along these lines, and then lifted to planar quadrilateral remeshings N and Ψ of M and S, using the procedure of Schiftner [347]. Since remeshing does not alter the surface's coarse geometry, Ψ is close to being convex and to satisfying the curvature equilibrium equation (10.8); N and Ψ can be further polished using Newton's method to exactly satisfy equilibrium and face planarity.

The unexpected consequence of the above is the following theorem [405].

Theorem 10.4. *If a surface $z(x,y)$ can be approximated by a thrust network M in compressive equilibrium, it can also be approximated by a thrust network N in compressive equilibrium and with planar quadrilateral faces.*

However, there is no guarantee about the quality of the faces, which can vary dramatically in scale across the surface.

10.4.6 Block layouts from stable surfaces

Once a thrust network in equilibrium has been found, a masonry structure can be constructed around it that is guaranteed to be stable by the safe theorem, as discussed in Section 10.3.2. However, a practical issue remains: blocks arranged directly by dualizing the thrust network tend to lie in long "chains" or "strips" that easily come apart under weathering or horizontal load. Panozzo et al. [301] developed a post-processing algorithm that remeshes the thrust network into blocks that are rectangular and staggered, much as is seen in a regular brick wall. The algorithm first remeshes the surface into a quadrilateral mesh. If the quad mesh has no singularities, one can find a staggered block pattern by simply deleting every other "vertical" edge. Generally speaking, a nontrivial freeform surface has complex geometry with several singularities present after remeshing, so edges are instead removed by solving a 2-coloring problem, which guarantees that blocks are staggered and are composed of at most two quads (see Figure 10.8, right).

10.5 CONCLUSION AND OPEN PROBLEMS

Although computational analysis and form-finding of stable freeform masonry structures is by now a thoroughly studied topic, there remain several open questions that have not been addressed by any previous work. Answering them will require significantly expanding our understanding of the mathematics of stable masonry, and its connections to discrete and smooth geometry.

10.5.1 Sensitivity analysis of masonry structures

In the above we assumed that the loads are known (either prescribed, or specified as a function of the surface area of the masonry structure). We know multiple characterizations of when the structure is stable under these known loads. However, will the structure remain standing if the loads change (as can easily happen during, for instance, a snowstorm, wind storm, or earthquake)? Can bounds be proven guaranteeing that the structure remains standing, if the loads do not change by more than some tolerance? How must the structure be modified to improve its resilience to additional loads?

The main challenge to addressing changes in load is that there is no simple relationship between the loads and the internal forces (or the w_{ij}), thanks to static indeterminacy. Instead, if the structure can bear additional load, it usually can do so in infinitely many ways. A more complete understanding of the space of possible stresses that can bear a given load, and how that *space* changes as the loads change, remains elusive.

10.5.2 Progressive stable structures

The analysis presented in this chapter only applies to complete masonry structures. Of course, these must still be built one brick at a time, and the stability of the final structure does not in any way speak to the stability of the structure in a partial state (consider that an arch is unstable up until the keystone is placed). When can a masonry structure be built progressively, so that at every point the structure is stable? Can a design be modified to eliminate, or at least mitigate, the need for scaffolding or formwork? Some work in a similar vein [121] has looked at supporting blocks from above using chains attached to fixed vertical anchors, and where to place the anchors to minimize the number of chains that must be added or removed.

III

Applications in Computational Mechanics

CHAPTER 11

Applications of Polyhedral Finite Elements in Solid Mechanics*

Joseph E. Bishop

Sandia National Laboratories, Albuquerque, USA

CONTENTS

THE FINITE element method has revolutionized structural analysis since its inception over 50 years ago, by enabling the computer analysis of geometrically complex structures. The main requirement of the finite element method is that an appropriate partition, or mesh, of the structure be created first. The elements of the partition typically have standard shapes, such as the hexahedron, pentahedron,

*Research support of Sandia National Laboratories is gratefully acknowledged. Sandia National Laboratories is a multimission laboratory managed and operated by National Technology and Engineering Solutions of Sandia, LLC., a wholly owned subsidiary of Honeywell International, Inc., for the U.S. Department of Energy's National Nuclear Security Administration under contract DE-NA-0003525.

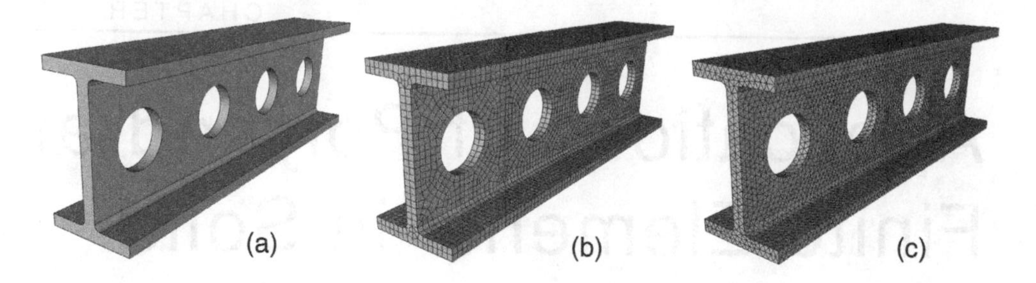

Figure 11.1 Example structure (a) with finite element meshes composed of standard element shapes: (b) hexahedral mesh, (c) tetrahedral mesh.

and tetrahedron. While this small library of standard element shapes is sufficient for many applications, there is a growing need for more general polyhedral shapes, ones that can have an arbitrary number of vertices, edges, and faces, and ones that can be non-convex. In this chapter, we discuss current and possible future applications of polyhedral finite elements in solid mechanics. These applications include rapid engineering analysis through novel meshing and discretization techniques, and fracture and fragmentation modeling. Several finite element formulations of general polyhedra have been developed. In this chapter we use a polyhedral formulation based on the use of harmonic shape functions. Harmonic shape functions are one example of several possible generalized barycentric coordinates, as discussed in Chapter 1.

11.1 INTRODUCTION

The analysis of stress and strain in a mechanical part or structure is critically important to the overall engineering design and development process. The finite element method has revolutionized analysis in solid mechanics by enabling the computer analysis of geometrically complex, nonlinear structures subjected to general loading conditions. Additionally, the finite element method can incorporate general material constitutive models, interface models, and constraints, making it an extremely versatile engineering analysis tool [41]. The main requirement of the finite element method is that an appropriate partition, or mesh, of the domain be created first, as shown in Figure 11.1. This partition of the domain typically consists of standard finite element shapes, such as the hexahedron and tetrahedron, as shown in Figures 11.2(a) and 11.2(b), respectively. Once the finite element mesh is created, analysis of the response of the structure to applied loads can be obtained using standard computational techniques, starting with the weak form of the governing field equations of solid mechanics [41]. Figure 11.3 shows the stress field (von Mises) resulting from the application of torsional tractions to the I-beam shown in Figure 11.1. This type of structural analysis is now an indispensable tool in engineering practice.

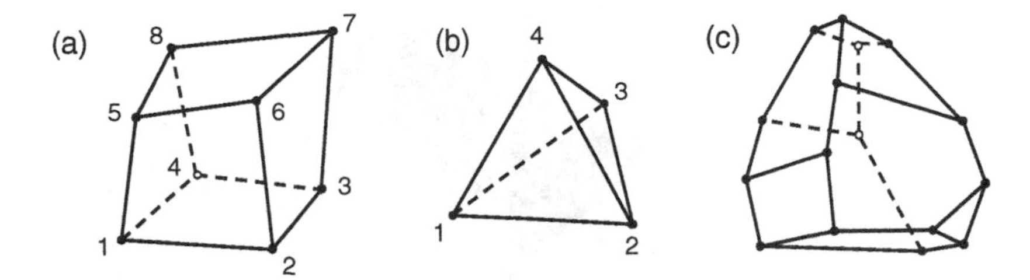

Figure 11.2 Finite element shapes: (a) hexahedron, (b) tetrahedron, (c) general polyhedron (possibly non-convex with warped faces).

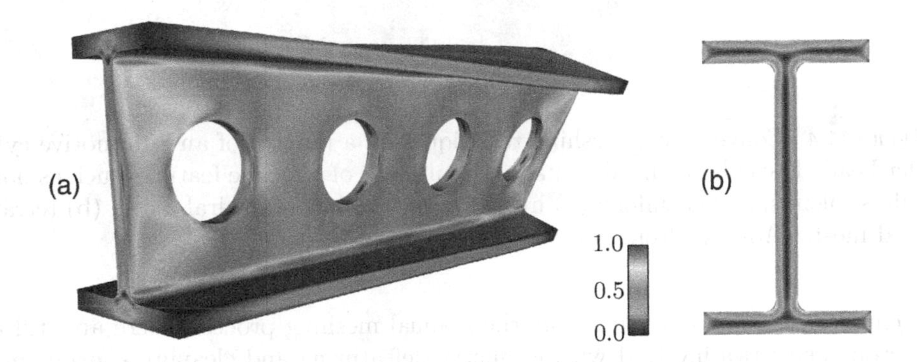

Figure 11.3 (a) Stress field (von Mises) resulting from a finite element analysis of the I-beam shown in Figure 11.1 with applied torsional tractions, (b) cross-section view. (Reproduced with permission from [48].) See color insert.

While the small library of element shapes consisting of hexahedra, pentahedra, and tetrahedra is sufficient for many applications, there is a growing need to have access to finite element formulations for general polyhedral shapes, ones that can have an arbitrary number of vertices, edges, and faces (see Figure 11.2(c)). For additional flexibility, it is also desirable to have formulations that can handle warped polygonal faces and allow for non-convex shapes. (Note that the conventional formulation of the hexahedral finite element, using trilinear shape functions mapped from a parent element, allows for warped faces.) While several finite element formulations of general polyhedra now exist (see, for example, [27, 47, 322]), in this chapter a formulation based on harmonic shape functions is used. Harmonic shape functions are one example of several possible generalized barycentric coordinates, as discussed throughout this book. A different formulation for polyhedra, one that is based on the virtual element method, is given in Chapter 15.

A driving application for general polyhedral finite elements in solid mechanics is rapid engineering analysis. Despite several decades of research on mesh generation, the meshing process for geometrically complex shapes, though much advanced, is still time consuming and labor intensive. While automated tetrahedral meshing

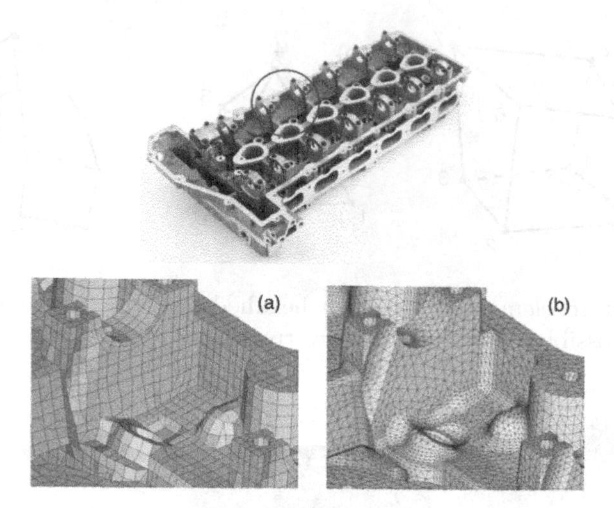

Figure 11.4 Conventional meshing techniques on a portion of an automotive cylinder head. The cylinder head contains a multitude of machine features such as holes, fillets, blends, and chamfers: (a) hybrid hexahedral/pentahedral mesh, (b) tetrahedral mesh. (Adapted from [45].)

technology seemingly circumvents the manual meshing process, there are still numerous heuristics involved with geometry defeaturing and cleanup, *a priori* mesh adaptation for minimization of solution error, and dimensional representations (e.g., beam versus shell versus continuum elements). The challenges in meshing realistic domains is shown in Figure 11.4. The automotive cylinder head has several geometric scales and numerous machine features such as holes, fillets, blends, and chamfers. Two examples of conventional meshing techniques are shown. Several possible uses of polyhedral finite elements to speed up the meshing process will be discussed in Section 11.4.

Another application of polyhedral finite elements is modeling pervasive fracture and fragmentation processes using dynamically changing mesh connectivity without remeshing [46, 49]. In this approach, fracture surfaces are allowed to nucleate and propagate only along inter-element faces. This is demonstrated in Figure 11.15 in which a quasi-brittle column impacts a rigid surface at an oblique angle. This technique for modeling fracture typically results in a type of mesh-dependency. The use of random polyhedral meshes can alleviate this mesh-dependency and restore a notion of convergence with mesh refinement. Details of this application and simulation are given in Section 11.5.

Additional advantages of polyhedral finite elements have been explored in modeling fluid dynamics using finite-volume techniques [207] and for topology optimization [390]. These applications will not be discussed here. The application of polyhedral finite elements for extremely large deformations in solid mechanics is discussed in Chapter 12.

This chapter is organized as follows. The governing equations of solid mechanics are reviewed in Section 11.2. Section 11.3 provides an overview of a finite element formulation based on harmonic shape functions. Several approaches of rapid meshing/analysis are discussed in Section 11.4. The use of polyhedral elements in modeling fracture and fragmentation is discussed in Section 11.5. A summary of this chapter and an outlook are provided in Section 11.6.

11.2 GOVERNING EQUATIONS OF SOLID MECHANICS

In this section, the governing equations of solid mechanics are briefly reviewed. Consider the motion of a body $\mathcal{B}$ with interior domain Ω and boundary Γ subjected to a body force $\boldsymbol{b}$ and applied surface tractions $\boldsymbol{t}$. A Lagrangian description of the motion of $\mathcal{B}$ is used. The initial configuration of the body is denoted by Ω_0 with boundary Γ_0. In the initial configuration, the position vector of a material point is denoted by $\boldsymbol{X}$. In the deformed configuration, the position of a material point is denoted by $\boldsymbol{x}$. The displacement vector $\boldsymbol{u}$ is given by $\boldsymbol{u} = \boldsymbol{x} - \boldsymbol{X}$. The (material) derivative of the current position $\boldsymbol{x}$ with respect to the original position $\boldsymbol{X}$ is given by $\partial \boldsymbol{x}/\partial \boldsymbol{X}$. This vector derivative is called the deformation gradient, denoted by $\boldsymbol{F}$, and in terms of the displacement is given by

$$\boldsymbol{F} = \frac{\partial \boldsymbol{u}}{\partial \boldsymbol{X}} + \boldsymbol{I},$$

where $\boldsymbol{I}$ is the identity tensor. From the deformation gradient, one can define measures of strain, for example, engineering strain, logarithmic strain, Lagrangian strain, and Eulerian strain [63]. The Lagrangian strain tensor $\boldsymbol{E}$ is defined as

$$\boldsymbol{E} := \frac{1}{2}(\boldsymbol{C} - \boldsymbol{I}),$$

where

$$\boldsymbol{C} := \boldsymbol{F}^T \boldsymbol{F}$$

is the right Cauchy–Green tensor.

In order to describe the kinetics of the material, one needs to derive measures of stress and the constitutive relations between stress and strain. In the context of hyperelasticity, one posits a potential energy function W of the strain tensor,

$$W = W(\boldsymbol{E}).$$

Certain restrictions are placed on W to obtain a constitutive model that is independent of the observer (objectivity) and to obtain a stable material. The first Piola–Kirchhoff stress tensor $\boldsymbol{P}$ is defined to be work conjugate to the derivative of W with respect to the deformation gradient $\boldsymbol{F}$,

$$\boldsymbol{P} = \frac{\partial W}{\partial \boldsymbol{F}} \tag{11.1}$$

so that

$$dW = \boldsymbol{P} : d\boldsymbol{F}.$$

An idealized material model useful for demonstrating large deformation behavior of a finite element formulation is the isotropic neo-Hookean model. For this material model the potential energy is of the form [63]

$$W(\boldsymbol{F}) = \frac{\mu}{2}(\boldsymbol{C} : \boldsymbol{I}) - \mu \log J + \frac{\lambda}{2}(\log J)^2, \tag{11.2}$$

where μ and λ are material constants, and $J^2 = \det \boldsymbol{C}$. Substituting (11.2) into (11.1) results in the following constitutive relationship between the first Piola–Kirchhoff stress $\boldsymbol{P}$ and the deformation gradient $\boldsymbol{F}$,

$$\boldsymbol{P} = \left[\mu(\boldsymbol{F}\boldsymbol{F}^T - \boldsymbol{I}) + \lambda(\log J)\boldsymbol{I}\right] \boldsymbol{F}^{-T}. \tag{11.3}$$

This material model is used in the simulation results shown in Figure 11.6. In the small deformation limit, (11.3) becomes the standard linear elasticity equation with μ and λ the shear modulus and the Lamé parameter, respectively. Hyperelastic formulations also exist for large-deformation plasticity [139], where now the stress tensor is a function of the entire history of deformation.

Finally, the governing equilibrium equations (conservation of linear momentum) are required. In Section 11.3, generalized barycentric coordinates are used to form the finite element shape functions directly in the initial configuration of the domain. Additionally, the weak form of the equilibrium equations will be integrated in the reference configuration. Therefore, a *total*-Lagrangian formulation of the governing equations is appropriate (as opposed to *updated*-Lagrangian). The conservation of linear momentum is then given by [41]

$$\frac{\partial \boldsymbol{P}}{\partial \boldsymbol{X}} : \boldsymbol{I} + \rho_0 \boldsymbol{b} = \rho_0 \ddot{\boldsymbol{u}} \tag{11.4}$$

with displacement boundary conditions

$$\boldsymbol{u} = \bar{\boldsymbol{u}} \quad \text{on} \quad \Gamma_u \tag{11.5}$$

and traction boundary conditions

$$\boldsymbol{P} \cdot \boldsymbol{N} = \boldsymbol{t}_0 \quad \text{on} \quad \Gamma_{\text{t}}, \tag{11.6}$$

where ρ_0 is the initial density, $\boldsymbol{b}$ is the body force vector per unit mass, $\ddot{\boldsymbol{u}}$ is the acceleration vector, $\boldsymbol{N}$ is the outward unit normal on Γ_0, $\boldsymbol{t}_0$ is the traction vector per unit initial area, and $\overline{\Gamma_u \cup \Gamma_{\text{t}}} = \Gamma_0$ and $\Gamma_u \cap \Gamma_{\text{t}} = \emptyset$.

11.3 POLYHEDRAL FINITE ELEMENT FORMULATION

The formulation of general polyhedral finite elements, ones with an arbitrary number of vertices and faces, both convex and non-convex, is an active area of research. A finite element formulation for polyhedra with planar faces, applicable to nonlinear solid mechanics, has been achieved by Rashid and Selimotic [322]. Shape functions were constructed by using a polynomial basis optimized for smoothness, while satisfying other constraints necessary for mesh convergence. The shape functions, however, are only weakly compatible at element faces. More recently, Rashid and Sadri [321] have proposed a polygonal finite element formulation for use in solid mechanics in which the discrete values of both the shape functions and their derivatives at the quadrature points are determined based only on the satisfaction of linear consistency, integration consistency, and a combined smoothness and compatibility measure. The recent virtual element method (VEM) [27, 161] provides a type of mean-gradient formulation for polyhedra with necessary stabilization for convergence with mesh refinement [81]. This method is discussed in Chapter 15. Recently, the author has developed a displacement-based finite element formulation for general polyhedra based on the use of harmonic shape functions for applications in nonlinear solid mechanics [47]. This latter formulation is briefly reviewed here. Other types of generalized barycentric coordinates, such as those described in Chapter 1, could possibly be used as well.

11.3.1 Weak form of governing equations

The weak form of (11.4) through (11.6) is given by the following variational problem [41]: find $\boldsymbol{u} \in \boldsymbol{H}^1(\Omega_0)$, where $\boldsymbol{H}^1(\Omega_0) := [H^1(\Omega_0)]^d$, with $\boldsymbol{u} = \bar{\boldsymbol{u}}$ on Γ_u such that

$$\int_{\Gamma_t} \mathbf{t}_o \cdot \delta\boldsymbol{u}\, d\Gamma_0 + \int_{\Omega_0} \rho_0 \mathbf{b} \cdot \delta\boldsymbol{u}\, d\Omega_0 - \int_{\Omega_0} \mathbf{P} : (\partial(\delta\boldsymbol{u})/\partial\boldsymbol{X})\, d\Omega_0 - \int_{\Omega_0} \rho_0 \ddot{\mathbf{u}} \cdot \delta\boldsymbol{u}\, d\Omega_0 = 0 \tag{11.7}$$

for all *test* functions $\delta\boldsymbol{u} \in \boldsymbol{H}_0^1(\Omega_0)$. Here, $H^1(\Omega_0)$ is the Sobolev function space of degree one containing functions that possess square-integrable weak derivatives, and the Sobolev space $\boldsymbol{H}_0^1(\Omega_0) = \{\boldsymbol{u} \in \boldsymbol{H}^1(\Omega_0) : \boldsymbol{u} = 0 \text{ on } \Gamma_u\}$. Any $\boldsymbol{u} \in \boldsymbol{H}^1(\Omega_0)$ is referred to as a *trial* function. The Bubnov–Galerkin procedure for obtaining an approximate solution to (11.7) uses a finite dimensional approximation to $\boldsymbol{H}^1(\Omega_0)$, denoted by $\boldsymbol{V}^h$, with $\boldsymbol{V}^h \subset \boldsymbol{H}^1(\Omega_0)$ and $\boldsymbol{V}_0^h = \{\boldsymbol{u}^h \in \boldsymbol{V}^h : \boldsymbol{u}^h = 0 \text{ on } \Gamma_u\}$. Let $\{\psi_I, I = 1, \ldots, N\}$ be a basis for $\boldsymbol{V}^h$ so that any $\boldsymbol{u}^h \in \boldsymbol{V}^h$ may be written as $\boldsymbol{u}^h(\boldsymbol{X}) = \sum_{I=1}^N \boldsymbol{u}_I \psi_I(\boldsymbol{X})$. A finite element approximation entails choosing basis functions with local support defined by a mesh. Let Ω_e represent the domain of a finite element with vertex (nodal) coordinates $\{\boldsymbol{v}_i, i = 1, \ldots, n\}$. The basis functions $\psi_I(\boldsymbol{X})$ are decomposed into element shape functions denoted as $\{\phi_i(\boldsymbol{X}), i = 1, \ldots, n\}$, where n is the number of vertices in the polyhedral element.

11.3.2 Shape functions

For a finite element formulation of polyhedra, several types of barycentric coordinates can be adopted for the element shape functions including harmonic [215], maximum entropy [202, 375], and reproducing kernel [46]. These barycentric coordinates are described in Chapter 1 along with several other types. Harmonic shape functions are used here. The shape functions $\phi_i(\boldsymbol{X})$ are defined in the reference (original) configuration, so there is no mapping to a parent coordinate system as is done with standard elements. Integration points within the element, used to in-

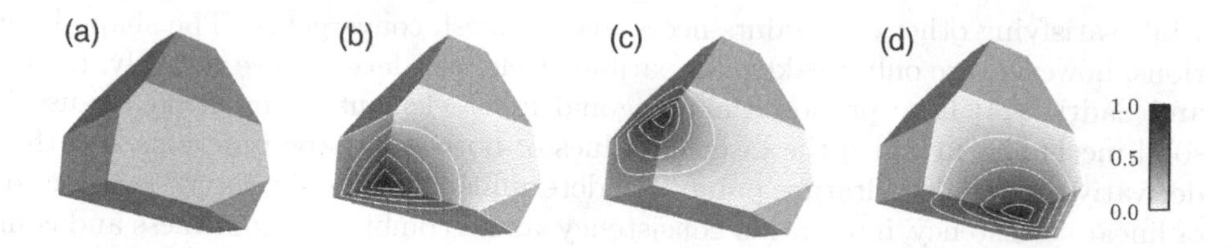

Figure 11.5 (a) Example polyhedron with element shape functions (b–d) based on harmonic coordinates. Shape functions for three vertices are shown. (Adapted from [47].)

tegrate the contribution to the volume integral in (11.7), are also defined in the reference configuration.

Harmonic functions satisfy the Laplace equation (1.17). For use in defining shape functions of a polyhedral finite element, solutions of (1.17) with appropriate boundary conditions possess the necessary properties of partition of unity (1.8) and linear reproduction (1.9). These properties result in an isoparametric displacement-based finite element formulation with an optimal rate of convergence, assuming that the weak form given by (11.7) is integrated in a consistent manner [47]. The shape functions also possess the Kronecker-delta property at the nodes. This property greatly simplifies the enforcement of displacement boundary conditions in the finite element solution.

The appropriate boundary conditions for the solution of (1.17) on a polyhedral finite element are developed hierarchically. Equation (1.17) is first applied to each edge, with appropriate end-point boundary conditions (zero or one). Equation (1.17) is then solved for each face, with the previous edge solutions applied as boundary conditions. Equation (1.17) is then solved in the interior of the polyhedron, with the previous face solutions applied as boundary conditions. Example shape functions are shown in Figure 11.5. In [47], solutions to (1.17) were approximated using the finite element method on a sub-tetrahedral mesh of the element. The construction of this sub-tetrahedral mesh was facilitated by assuming that the element shape was star-convex with respect to the vertex-averaged centroid. These approximate solutions still provide the necessary properties of finite element shape functions described previously.

11.3.3 Element integration

The integration points of the polyhedral finite element are defined based on the approach of Rashid and Selimotic [322]. In this approach, integration points are placed in so-called tributary volumes associated with each polyhedral vertex. Thus, the number of integration points is equal to the number of vertices. In practice, this integration method is sufficient to prevent any zero-energy modes of deformation. However, this integration scheme has only linear precision (can integrate linear functions exactly). Since the harmonic shape functions are non-polynomial in

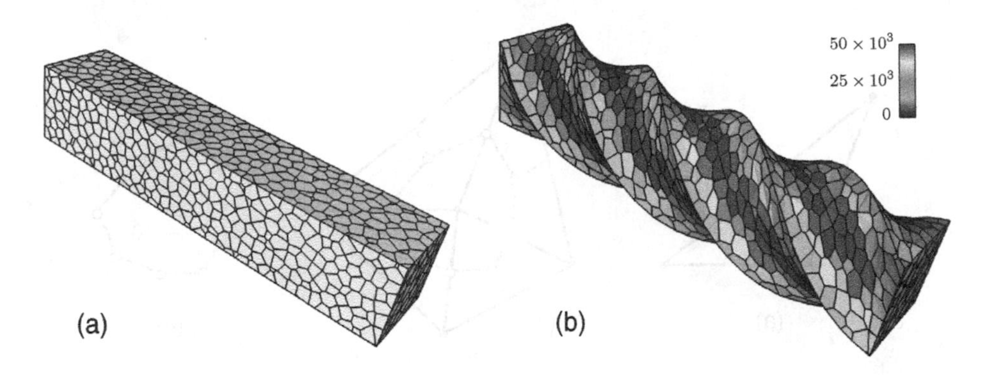

Figure 11.6 Large deformation example, a hyperelastic beam subjected to a torsion loading. One end is fixed while a full 360° rotation is applied to the opposite end. (a) original polyhedral mesh (Voronoi tessellation with maximal Poisson seeding), (b) deformed shape and von Mises stress field. See color insert.

nature, significant error is introduced with this low-order integration scheme. Furthermore, this error results in a failure to pass the patch test (a test for polynomial completeness in the element formulation). To circumvent this difficulty, the derivatives of the harmonic shape functions are modified slightly to satisfy the discrete divergence theorem while maintaining other necessary consistency properties [47]. This derivative correction has been further elucidated in [393].

A large deformation example using this polyhedral finite element formulation is shown in Figure 11.6. A beam of dimensions $100 \times 100 \times 500$ is fixed at one end while the opposite end is subjected to a full 360° rotation. The hyperelastic material model defined in (11.3) is used with $\lambda = 57.7 \times 10^3$ and $\mu = 38.4 \times 10^3$. A Voronoi tessellation with random seeding, described in Section 11.5, is used to mesh the beam with polyhedral elements.

11.4 RAPID ENGINEERING ANALYSIS

As noted in Section 11.1, for engineered mechanical structures the process of meshing realistic domains can be extremely challenging. This is also true for geological [154, 219], biological [68, 365, 434], and micromechanical structures [238, 249, 363]. Several recent discretization paradigms are focused on reducing the meshing task in solid mechanics. These include (1) meshfree methods [91, 245], (2) isogeometric methods [108], and (3) fictitious-domain methods (also called embedded domain and immersed boundary methods) [45, 132, 348, 349].

Additionally, there are a number of possible ways of using polyhedral finite elements to significantly decrease the burden of meshing. One approach to direct polyhedral mesh generation is to use Voronoi constructions [295]. While straightforward on simple convex domains using ghosting techniques [60], the generation of Voronoi meshes on geometrically complex structures is challenging. A more practi-

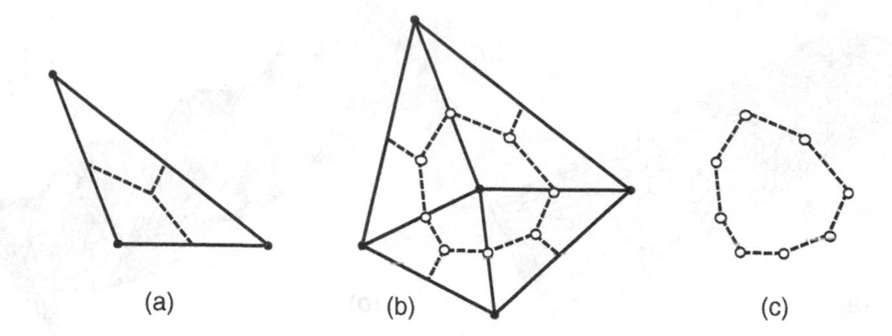

Figure 11.7 Example of polygonal element construction using the barycentric subdivision of a triangular mesh: (a) triangle element with barycentric subdivision using edge midpoints and the element centroid, (b) barycentric subdivision of a patch of triangle elements, (c) aggregation of sub-quadrilaterals into a polygonal element.

cal approach is to use barycentric subdivision of tetrahedral elements into hexahedra. In this approach, each tetrahedral element is subdivided into four hexahedral elements using edge midpoints and the element centroid. Figure 11.7(a) shows a two-dimensional example. This approach of all hexahedral meshing typically results in highly distorted hexahedral elements that may be unsuitable for use as individual elements within a finite element analysis. However, a polyhedral element can be created by aggregating the set of hexahedral elements attached to the original nodes of the tetrahedral mesh. This process is demonstrated in Figures 11.7(b,c) in two dimensions for an interior vertex. A three-dimensional example is shown in Figure 11.8. Furthermore, the sub-hexahedra that form the polyhedron provide a convenient structure for element integration and shape function construction.

Another approach to polyhedral meshing is to first create a highly refined tetrahedral mesh that captures all of the geometric detail, but is too refined to be computationally practical. The tetrahedra can then be adaptively aggregated into larger polyhedral elements producing a mesh that is computationally feasible yet still captures the geometric detail. Standard domain-decomposition techniques could be used to form the aggregates. This approach has been studied in the context of discontinuous Galerkin finite elements [25]. However, aggregates of a large number of tetrahedral elements will result in a polyhedral element with a large number of vertices. The number of vertices could be reduced by creating a "virtual mesh" with reduced geometrical representation. Discretization techniques based on primitive triangle meshes but using maximum-entropy coordinates defined on larger cell aggregates have recently been investigated [281].

In an approach similar to fictitious-domain methods, cut-cell methods embed the geometrically complex domain within a structured hexahedral mesh of a bounding box. The desired mesh of the geometrically complex domain is then obtained by

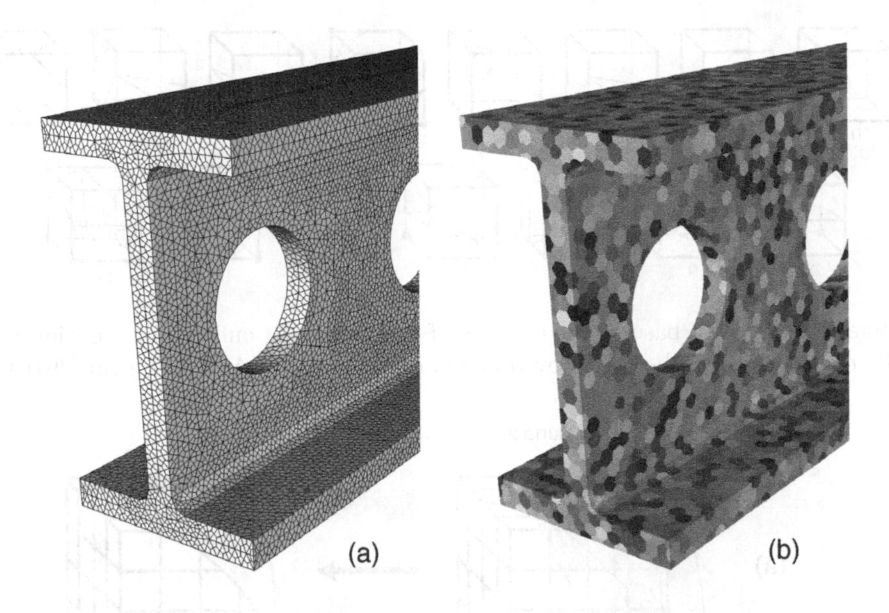

Figure 11.8 Example of polyhedral meshing using barycentric subdivision (see Figure 11.7(a)): (a) Barycentric dual of the tetrahedral mesh shown in Figure 11.1(c); (b) Each polyhedral cell is represented as a different color. See color insert.

"simply" cutting hexahedral elements that intersect the domain boundary to form polyhedra. Hexahedra are then left on the interior and exterior. The exterior hexahedra and polyhedra are discarded. This technique has been used in computational fluid dynamics with finite-volume discretizations [162, 207, 430]. It has also been investigated for use in solid mechanics by Rashid and Selimotic [322], and more recently by Sohn and coworkers [224, 365, 366], with promising results. Song and coworkers [256, 387] have developed a similar approach using the "scaled boundary" finite-element formulation of the resulting cut-cell polyhedra. In practice the "cutting" of a hexahedron by an arbitrary surface is nontrivial, particularly since the cutting operation needs to produce a continuous surface and preserve the topology of the original domain.

One approach to simplifying the computational-geometry problem is to use the classical marching cubes algorithm [261, 291]. This algorithm is commonly used to construct isosurfaces of field quantities. In this approach, each vertex of the hexahedron is identified as either interior or exterior to the given domain. For a hexahedron, there are $2^8 = 256$ possible cases, which can be reduced to 15 base configurations by rotation and symmetry operations. These 15 base configurations are shown in Figure 11.9. With this categorization of vertices, the edges of the hexahedron that intersect the surface of the domain are then known. The actual intersection points within an edge can be determined by linear interpolation if a scalar field is assigned to each vertex, such as a signed distance function. A polyhedral element can then

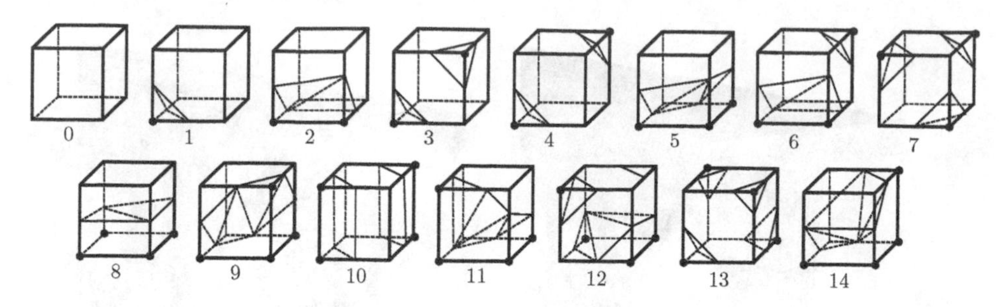

Figure 11.9 The 15 base configurations of the marching cubes triangulation table. The other cases can be found by rotation or symmetry. (Adapted from [291].)

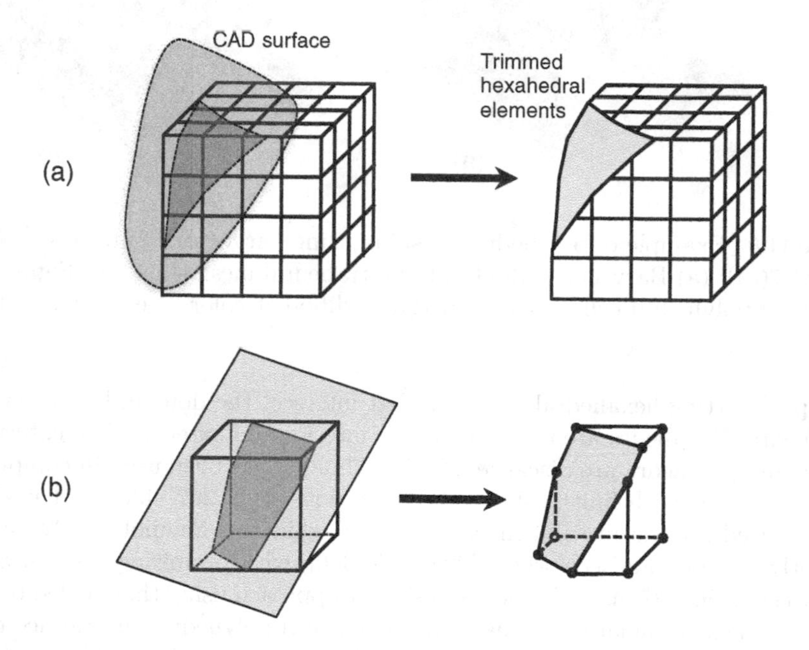

Figure 11.10 Trimmed hexahedral elements using the cut-cell process: (a) cutting process, before and after, (b) trimmed hexahedral element with nodes at the vertices. (Adapted from [224].)

be constructed from this triangulated surface within each hexahedron as shown in Figure 11.10. In practice, the created polyhedra may be non-convex. Also, tolerancing issues due to small edges are a primary concern, and topology issues may require disambiguation techniques [283, 401]. These techniques are extensible to adaptively refined (quadtree in 2D or octree in 3D) overlay meshes [358, 424].

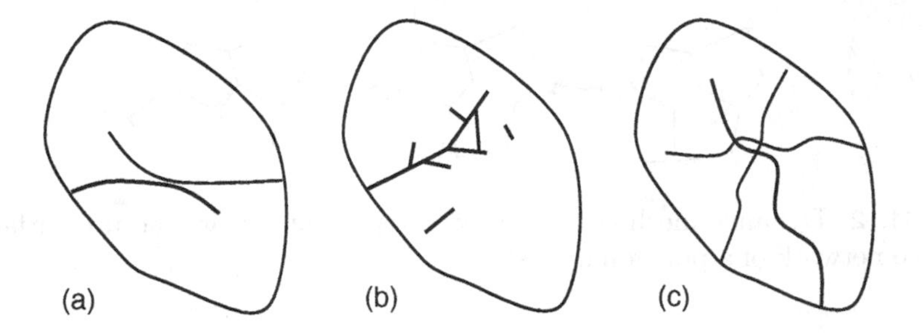

Figure 11.11 Examples of fracture processes in a continuum in which unrestricted fracture growth could lead to modeling difficulties: (a) two fractures propagate from the surface and touch at an arbitrarily small angle, (b) fractures initiate and branch arbitrarily, (c) three fractures propagate from the surface and intersect, creating an arbitrarily small fragment.

11.5 FRAGMENTATION MODELING

The modeling of failure and fracture of a solid body is one of the major challenges in theoretical and computational solid mechanics. The extent of fracturing in a solid body is termed pervasive when a number of cracks nucleate, propagate, branch, and coalesce. The highly nonlinear, multiscale problem of pervasive fracture is challenging to simulate. The simulation of pervasive fracture is further complicated by the extensive self-contact that can accompany the generation of new crack surfaces, especially during fragmentation.

Several computational fracture methods strive to model unrestricted fracture growth in a continuum using an explicit representation of the fracture surfaces, e.g., explicit meshing [290], extended finite element method [374], and embedded discontinuities [213]. However, the goal of unrestricted fracture growth quickly becomes computationally infeasible when considering pervasive fracture processes in which multiple fractures nucleate, branch, and intersect. Figure 11.11 gives three examples of unrestricted fracture processes in a continuum. Figure 11.11(a) shows two fractures initiating from a surface and approaching each other at an arbitrarily small angle. Figure 11.11(b) shows one fracture initiating from a surface with arbitrary branching, self-intersection, and initiation away from the surface. Figure 11.11(c) shows three fractures initiating from a surface, then intersecting and forming an arbitrarily small fragment. In order to enable robust simulations of pervasive fracture processes, the allowable fracture paths must be restricted to some degree. This may be thought of as regularizing the pervasive fracture problem. One approach to this regularization is to *a priori* discretize the space of all possible fracture paths with constraints on minimum fragment size, bifurcation locations, etc.

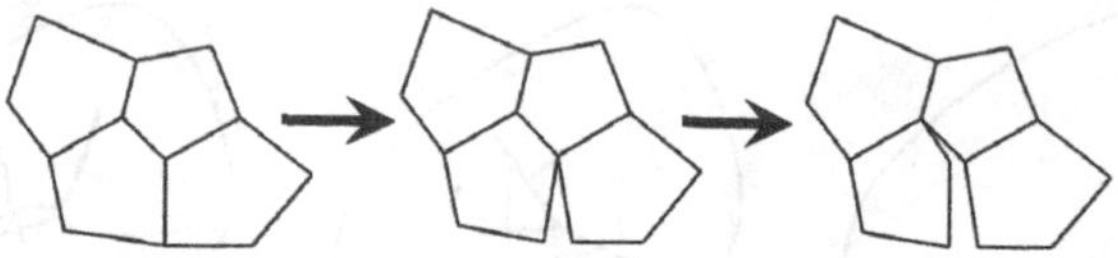

Figure 11.12 Dynamic mesh connectivity to represent a new fracture surface along the face network of a polygonal mesh.

11.5.1 Fracture methodology

One approach for regularizing the pervasive fracture problem is to only allow fracture surfaces to form at the inter-element faces of a finite element mesh as shown in Figure 11.12. At the inception of material softening leading to fracture, the connectivity of the mesh is modified to reflect a new surface, and a cohesive traction with a softening behavior is dynamically inserted. The cohesive traction is of the *extrinsic* form, as opposed to an *intrinsic* traction-separation relation that is *a priori* seeded throughout the face network of the finite element mesh [303]. The restriction of only allowing new surfaces to form at inter-element faces provides a necessary regularization of the resulting domain and surface topologies to obtain a robust simulation. This approach to modeling pervasive fracture processes has been used by several researchers [46, 239, 299, 300, 339, 340]. Instead of inserting new fracture surfaces at the start of material softening, one could also use gradient damage or phase field computational methodologies [64, 277, 278, 331] to model the physics of material softening and then change mesh connectivity when the load-carrying capacity of the material has decreased below a certain threshold.

This technique for modeling fracture typically results in a type of mesh-dependency in the crack paths. The use of a random mesh with no preferred orientation restores, in a weak sense, simulation objectivity within this class of fracture methods. A certain type of random mesh advocated by the author is the random close-packed (RCP) Voronoi mesh [46]. This type of mesh is described in more detail in the following subsection. The RCP Voronoi mesh provides a random face network for representing fracture surfaces. The polyhedral cells of the RCP Voronoi mesh are formulated as finite elements. The broader question of what type of mesh is optimal for this fracture methodology is an area of active research [302, 304, 325].

Within this computational approach for modeling pervasive fracture, there are a number of possibilities for controlling fracture initiation and growth. For example, at one extreme new fracture surfaces can be restricted to nucleate only at an existing fracture tip. At the other extreme new fracture surfaces can be allowed to nucleate at any element face. For quasi-static applications, the approach of only allowing new fractures to form from existing fracture surfaces seems more realistic. However, this restriction may be overly restrictive by not allowing new fracture surfaces to form in front of or near an existing fracture. Allowing fractures surfaces to form in front of or near an existing fracture would help ameliorate the artificial toughness induced by constraining crack growth to predefined surfaces. In effect, a damage band is formed around the crack front, in analogy to subscale damage mechanisms

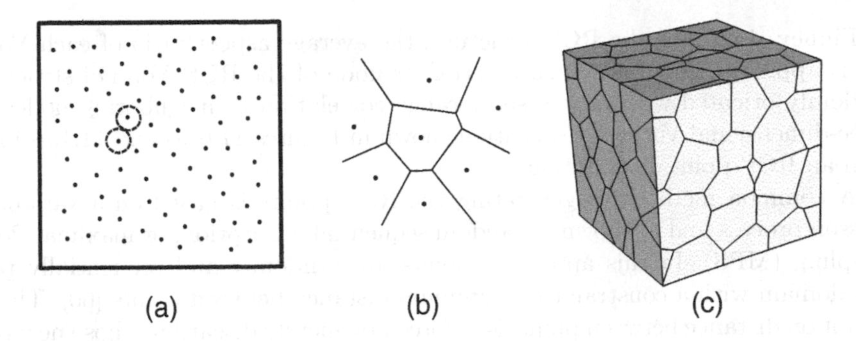

Figure 11.13 Example showing the generation of a Voronoi tessellation, or mesh, starting with a random close-packed arrangement of points. (a) Points are randomly assigned inside the domain with a constraint on minimum distance (red circles). (b) A Voronoi cell corresponding to the construction defined by (11.8) and a subset of points. (c) A three-dimensional example. The domain boundaries are reproduced by reflecting interior points about the domain boundary. (Reproduced with permission from [48].)

in the material. In the simulations presented here, fracture initiation and growth are allowed at any facet at any time during the simulation.

11.5.2 Random Voronoi meshes

Voronoi tessellations, or tilings, have a rich history in mathematics and science and have a number of interesting properties [295]. Given a finite set of points $\boldsymbol{x}_i$ (sites), the Voronoi tessellation is defined as the collection of regions, or cells, V_i where

$$V_i = \bigcap_{i \neq j} \{\boldsymbol{x} \in \Omega \mid d(\boldsymbol{x}_i, \boldsymbol{x}) < d(\boldsymbol{x}_j, \boldsymbol{x})\}. \tag{11.8}$$

Here, Ω represents the given domain, and the function $d(\boldsymbol{x}, \boldsymbol{y})$ is the Euclidean metric (distance) between the two points $\boldsymbol{x}$ and $\boldsymbol{y}$. Each spatial point belonging to the Voronoi cell V_i is closer to site i than all other sites $j \neq i$. Each Voronoi cell is the intersection of a collection of half-spaces, and therefore forms a convex polyhedron. An example of a two-dimensional Voronoi cell is shown in Figure 11.13(b). An example of a three-dimensional Voronoi tessellation is shown in Figure 11.13(c).

While the Voronoi tessellation can be formed from any finite set of points, a special structure arises from the close packing of equi-sized hard spheres [431]. A classic experiment of dropping hard spheres into a relatively large container produces a structure known as random close packed (RCP). The face-centered cubic (FCC) structure possesses an optimal packing factor of 0.740, whereas the RCP structure exhibits a maximum packing factor of 0.637. An example of a two-dimensional random close-packing is shown in Figure 11.13(a). The statistical geometry aspects of RCP structures and their associated Voronoi diagrams have been studied by Finney [142]. For the RCP structure, the average aspect ratio of each Voronoi cell is approximately one. Each junction or node of the RCP Voronoi structure is randomly oriented with only a short range correlation to neighboring nodes. The three-dimensional Voronoi tessellation shown in Figures 11.6(a) and 11.13(c) result from an RCP point distribution.

A common method for generating the RCP points is based on a variant of a Poisson process and is known as random sequential adsorption, or maximal Poisson sampling (MPS). In this approach, points are randomly and sequentially placed in a domain with a constraint on minimum distance between points [60]. The constraint on distance between points is enforced by merely discarding those new points that violate the constraint. The seeding process stops when the maximum packing threshold is reached within tolerance. This approach to generating the RCP seeding can be inefficient, and does not have a quantifiable stopping time. Recently, a very efficient MPS method has been proposed with a finite stopping time [135].

In practice, the Voronoi tessellation can contain numerous relatively small edges that are inappropriate for finite element analysis, in particular for explicit-dynamics time integration. The Voronoi mesh can be regularized for finite element computation by simply deleting those edges whose size is below a relative tolerance. In three dimensions, this edge-collapse operation will produce non-planar faces, and may give tangled polyhedra if the regularization is too aggressive.

11.5.3 Fragmentation

To illustrate the fracture methodology, two fragmentation examples are presented: (1) a rigid sphere impacting a quasi-brittle beam, and (2) a three-dimensional quasi-brittle beam impacting a rigid surface at an oblique angle. These examples illustrate the dynamically evolving fracture network as well as the self-contact capability required for this class of problems. The Mohr–Coulomb fracture initiation model, the cohesive traction-separation model, and material properties used in [46] are also used in these examples. The cohesive model is representative of a quasi-brittle material with a relatively large crack-tip cohesive-zone size. For a fracture-initiation criterion, the stress field is first interpolated to each facet using a volume-weighted average of the stress in the two attached elements. The traction vector on the facet is then used to assess the fracture criterion.

Figure 11.14 shows the simulation of a rigid sphere impacting a quasi-brittle beam. The sphere has a diameter of 100 mm, a velocity of 100 m/s, and has the density of steel (7800 kg/m^3). The beam has dimensions 100 mm × 100 mm × 1000 mm. The RCP Voronoi mesh is similar to the one shown in Figure 11.6(a). The simulation shows fragmentation under the impact location along with bending-induced fractures away from the impact location.

Figure 11.15 shows the simulation of a quasi-brittle beam of dimensions 100 mm × 100 mm × 500 mm impacting a rigid surface at an angle of 45 degrees with a striking velocity of 70.7 m/s normal to the rigid plane. The RCP Voronoi mesh shown in Figure 11.6(a) was used in the simulation. The deformed state is shown at three instances in time. The initial fractures initiate from the beam sur-

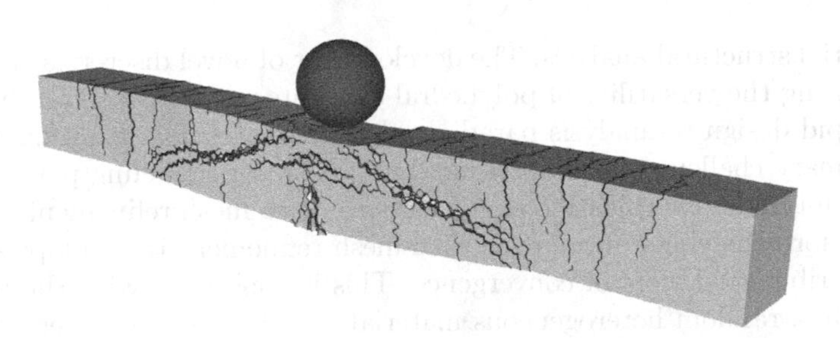

Figure 11.14 Pervasive-fracture simulation of a rigid sphere impacting a quasi-brittle beam. The Voronoi mesh used here is similar to the one shown in Figure 11.6(a).

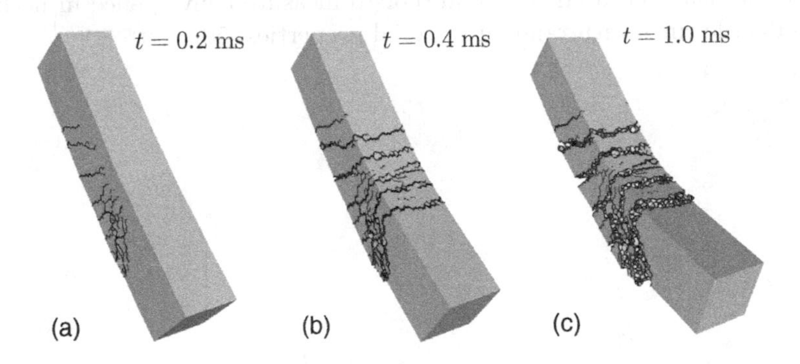

Figure 11.15 Pervasive-fracture simulation of a quasi-brittle beam impacting a rigid surface at an oblique angle. The fracture growth at various times is shown: (a) $t = 0.2$ ms, (a) $t = 0.4$ ms, (a) $t = 1.0$ ms. The Voronoi mesh shown in Figure 11.6(a) was used in the simulation.

face near the mid-section and then grow and coalesce into numerous fragments. This simulation shows the need for an efficient and robust fracture algorithm.

11.6 SUMMARY AND OUTLOOK

In this chapter, we discussed several applications of polyhedral finite elements in solid mechanics. These included rapid meshing of geometrically complex structures and pervasive-fracture modeling using random close-packed Voronoi meshes. Each of these applications requires a finite element formulation that can handle non-convex polyhedra and polyhedra with warped faces. A displacement-based finite element formulation based on the use of harmonic shape functions, a type of generalized barycentric coordinate, was reviewed for use in these applications.

With continuing research into polyhedral meshing tools and finite element formulations, the use of polyhedral finite elements is expected to become commonplace in industrial structural analysis. The development of novel discretization techniques for exploiting the versatility of polyhedral elements, e.g., cut-cell methods, will enable a rapid design-to-analysis paradigm that has heretofore been unobtainable.

A primary challenge in the presented approach for modeling pervasive fracture is the definition and verification of convergence with mesh refinement. One possible approach for achieving convergence with mesh refinement is to adopt a probabilistic or distributional view of convergence. This is consistent with the use of a random mesh, a random heterogeneous material, and the fact that pervasive fracture phenomena are inherently sensitive to initial conditions. Each pervasive-fracture simulation provides one realization, and the results of several simulations provide a sampling for approximating the true distribution of engineering quantities-of-interest such as fracture paths or fragment volumes. The concept of convergence in distribution has been used by the author to measure convergence in neck spacing of thin ductile rings with random material properties [50].

CHAPTER 12

Extremely Large Deformation with Polygonal and Polyhedral Elements*

Glaucio H. Paulino, Heng Chi
Georgia Institute of Technology, Atlanta, USA

Cameron Talischi
McKinsey & Company, Chicago, USA

Oscar Lopez-Pamies
University of Illinois at Urbana-Champaign, USA

CONTENTS

*Research support from the U.S. National Science Foundation through grant CMMI #1624232 (formerly #1437535) is gratefully acknowledged. The information presented in this chapter is the sole opinion of the authors and does not necessarily reflect the views of the sponsoring agency.

Soft organic materials are of great engineering interest, but challenging to model with standard finite elements. The challenges arise primarily because their nonlinear elastic response is characterized by non-convex stored-energy functions, they are incompressible or nearly incompressible, and oftentimes possess complex microstructures. Recently, polygonal finite elements have been found to possess several advantages over standard finite elements when studying the mechanics of such materials under finite deformations. This chapter summarizes a new approach, using polygonal and polyhedral elements in nonlinear elasticity problems involving extremely large and heterogeneous deformations. We present both displacement-based and two-field mixed variational principles for finite elasticity, together with the corresponding lower- and higher-order polygonal and polyhedral finite element approximations. We also utilize a gradient correction scheme that adds minimal perturbations to the gradient field at the element level in order to restore polynomial consistency and recover (expected) optimal convergence rates when the weak form integrals are evaluated using quadrature rules. With the gradient correction scheme, optimal convergence of the numerical solutions for displacement-based and mixed formulations with both lower- and higher-order displacement interpolants is confirmed by numerical studies of several boundary-value problems in finite elasticity. For demonstration purposes, we deploy the proposed polygonal discretization to study the nonlinear elastic response of rubber filled with random distributions of rigid particles considering interphasial effects. These physically motivated examples illustrate the potential of polygonal finite elements to simulate the nonlinear elastic response of soft organic materials with complex microstructures under finite deformations.

12.1 INTRODUCTION

Many organic materials are characterized by their ability to undergo large reversible deformations in response to a variety of stimuli, including mechanical forces, electric and magnetic fields, and temperature changes. While they have long been utilized in numerous engineering applications, modern advances in material science have demonstrated that soft organic materials, such as electro- and magneto-active elastomers, gels, and shape-memory polymers, hold tremendous potential to enable new technologies, and in particular, the new generation of sensors and actuators. From a mechanical perspective, although most of these materials can be approximated "simply" as nonlinear elastic, they more often than not contain highly heterogeneous, complex microstructures. The heterogeneity of the microstructures makes the underlying local deformations inherently complex.

Computational microscopic studies of nonlinear elastic materials, especially those with complex microstructures, help in gaining a quantitative understanding of the complex behavior of these materials. These studies are also essential in guiding the optimization of nonlinear elastic materials, so that they can enable new technological applications. It has long been recognized that standard finite elements, especially simplicial (triangular and tetrahedral) finite elements, are inadequate in simulating processes involving realistic, large deformations—an example is shown in Figure 12.1. By contrast, recent studies have demonstrated that polygonal elements possess significant potential in the study of nonlinear elastic materials under finite deformations. First, these elements are well-suited to model complex microstructures (e.g., particulate microstructures and microstructures involving different length scales) and to seamlessly incorporate periodic boundary conditions. Secondly, polygonal elements are found to be more tolerant to large localized deformations and to produce more accurate results in bending and shear than standard finite elements. With mixed formulations, lower-order mixed polygonal elements are also shown to be numerically stable on Voronoi-type meshes without the need for any additional stablization terms. Likewise, numerical stability of the polygonal finite elements is also demonstrated in topology optimization [160, 389, 392] and in fluid mechanics [394] problems.

One major challenge in the polygonal finite method is accurate numerical integration of the weak-form integrals. The finite element space for polygonal elements contains non-polynomial (e.g., rational) functions, and therefore, the use of existing quadrature schemes, typically designed for integration of polynomial functions, leads to non-vanishing consistency errors [268, 388] and consequently, sub-optimal or even non-convergent finite element solutions under mesh refinement. In practice, using a sufficiently large number of integration points can lower the consistency error for linear polygonal elements in two dimensions. A triangulation scheme with three integration points per triangle is shown to be sufficiently accurate for practical problems and mesh sizes [99, 388]. However, for higher-order polygonal elements, the number of integration points for such a scheme can become prohibitively large [100]. This is also the case for polyhedral elements in three dimensions [268, 393]. In practical applications, as the number of elements increases, the associated computational cost for a polyhedral discretization can become exorbitant.

Several attempts have been made in the literature to address the issue of numerical quadrature. For example, in the context of scalar diffusion problems, inspired by the virtual element method (VEM) [33, 72, 161], Talischi et al. [388] have proposed a polynomial projection approach to ensure the polynomial consistency of the bilinear form, and thereby ensure the satisfaction of the patch test and optimal convergence for both linear and quadratic polygonal elements. A similar approach was also adopted by Manzini et al. [268] to solve Poisson problems on polyhedral meshes. However, those approaches require the existence of a bilinear form, and therefore, extension to general nonlinear problems is non-trivial and is still an open question. Borrowing the idea of pseudo-derivatives in the meshfree literature [229],

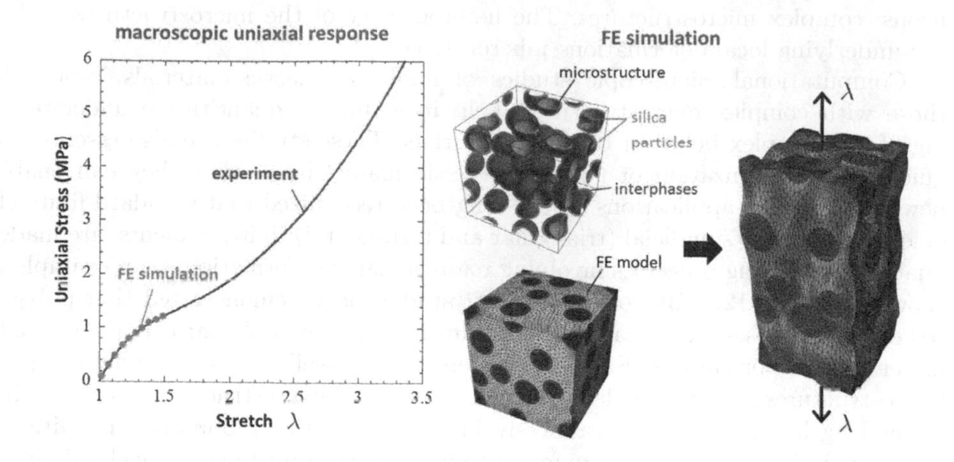

Figure 12.1 An illustration of the limited capability of standard tetrahedral elements to undergo large deformation. Experimentally, a synthetic rubber filled with 20% volume fraction of randomly distributed silica particles can be stretched more than four times its original length ($\lambda = 4$) without internal damage under uniaxial tension [315]. By contrast, a finite element model, based on standard 10-node hybrid elements with linear pressure, is only able to deform to a macroscopic stretch of $\lambda = 1.5$ [174]. See color insert.

Bishop has proposed an approach to correct the derivatives of the shape functions to enforce the linear consistency property on general polygonal and polyhedral meshes [46, 47]. With this correction, the linear patch test is passed and optimal convergence is achieved. Although applicable for general nonlinear scenarios, the extension to higher-order cases (for example, quadratic polygonal finite elements) is not straightforward, More recently, Talischi et al. [393] proposed a general gradient correction scheme that is applicable to both linear and nonlinear problems on arbitrary-order polygonal and polyhedral elements. By requiring a minimum order of accuracy in the numerical quadrature, the correction scheme is shown to restore optimal convergence for both linear and nonlinear problems [393]. In this chapter, we apply the gradient correction scheme to polygonal and polyhedral finite elements for finite elasticity problems and demonstrate that the gradient correction scheme can effectively and efficiently render optimally convergent polygonal and polyhedral finite element methods.

12.2 FINITE ELASTICITY FORMULATIONS

This section briefly recalls the equations of elastostatics applied to hyperelastic materials. Two variational formulations are presented: (i) the standard displacement-based formulation [63, 294, 429], and (ii) a general mixed formulation involving the displacement field and a pressure-like scalar field as trial fields. Throughout this chapter, we adopt a Lagrangian description of the fields.

Consider a body in its stress-free, undeformed configuration that occupies an open domain $\Omega \subset \mathbb{R}^d$ with boundary $\partial\Omega$. On its boundary, it is subjected to a prescribed displacement field $\boldsymbol{u}^0$ on $\Gamma^{\boldsymbol{x}}$ and traction $\boldsymbol{t}$ (per unit undeformed surface) on Γ^{t}, such that $\Gamma^{\boldsymbol{x}} \cup \Gamma^{\boldsymbol{t}} = \partial\Omega$ and $\Gamma^{\boldsymbol{x}} \cap \Gamma^{\boldsymbol{t}} = \varnothing$. The body is also assumed to be subjected to a body-force $\boldsymbol{f}$ (per unit undeformed volume) in Ω. A stored-energy function W is used to characterize the constitutive behavior of the body, which is assumed to be an objective function of the deformation gradient $\boldsymbol{F}$. In terms of W, the first Piola–Kirchhoff stress tensor $\boldsymbol{P}$ at each material point $\boldsymbol{x} \in \Omega$ is thus given by the following relation:

$$\boldsymbol{P}(\boldsymbol{x}) = \frac{\partial W}{\partial \boldsymbol{F}}(\boldsymbol{x}, \boldsymbol{F}),$$

which is used as the stress measure of choice in this chapter.

12.2.1 Displacement-based formulation

The well-known principle of minimum potential energy asserts that the equilibrium displacement field $\boldsymbol{u}$ minimizes the potential energy Π among all displacement fields that are kinematically admissible:

$$\Pi(\boldsymbol{u}) = \min_{\boldsymbol{v} \in \mathcal{K}} \Pi(\boldsymbol{v}), \tag{12.1}$$

with

$$\Pi(\boldsymbol{v}) = \int_{\Omega} W(\boldsymbol{x}, \boldsymbol{F}(\boldsymbol{v}))\, d\boldsymbol{x} - \int_{\Omega} \boldsymbol{f} \cdot \boldsymbol{v}\, d\boldsymbol{x} - \int_{\Gamma^t} \boldsymbol{t} \cdot \boldsymbol{v}\, ds. \tag{12.2}$$

In the above expression, $\mathcal{K}$ stands for a sufficiently large set of displacements such that $\boldsymbol{v} = \boldsymbol{u}^0$ on $\Gamma^{\boldsymbol{x}}$.

The weak form of the Euler–Lagrange equations associated with the variational problem (12.1) reads as follows:

$$\begin{aligned} G(\boldsymbol{v}, \delta\boldsymbol{v}) = \int_{\Omega} \frac{\partial W}{\partial \boldsymbol{F}}(\boldsymbol{x}, \boldsymbol{F}(\boldsymbol{v})) : \nabla(\delta\boldsymbol{v})\, d\boldsymbol{x} - \int_{\Omega} \boldsymbol{f} \cdot \delta\boldsymbol{v}\, d\boldsymbol{x} \\ - \int_{\Gamma^t} \boldsymbol{t} \cdot \delta\boldsymbol{v}\, ds = 0 \quad \forall \delta\boldsymbol{v} \in \mathcal{K}^0, \end{aligned} \tag{12.3}$$

where the test displacement field $\delta\boldsymbol{v}$ is the variation of $\boldsymbol{v}$, and $\mathcal{K}^0$ denotes the set of all the kinematically admissible displacement fields that vanish on $\Gamma^{\boldsymbol{x}}$.

12.2.2 A general two-field mixed variational formulation

For nonlinear elastic materials that are nearly incompressible, the variational principle in (12.1)–(12.2) is known to perform poorly when approximated by standard finite element methods [204]. As a standard remedy, mixed variational principles that involve not only the deformation, but also a pressure field are utilized [17, 75, 87, 90, 99, 188, 362, 385]. Based on the multiplicative splitting of the deformation gradient, the mixed variational formulations are categorized into two classes, the $\boldsymbol{F}$-formulation and the $\overline{\boldsymbol{F}}$-formulation [99]. In this section, we recall the $\boldsymbol{F}$-formulation; for the $\overline{\boldsymbol{F}}$-formulation, the reader is referred to [17, 99, 188] and references therein.

The basic idea is to introduce a function $\widehat{W}$ such that $W(\boldsymbol{x}, \boldsymbol{F}) = \widehat{W}(\boldsymbol{x}, \boldsymbol{F}, J)$ when $J = \det \boldsymbol{F}$, and its partial Legendre transformation,

$$\widehat{W}^*(\boldsymbol{x}, \boldsymbol{F}, \widehat{q}) = \max_J \{\widehat{q}(J-1) - \widehat{W}(\boldsymbol{x}, \boldsymbol{F}, J)\}.$$

Provided that $\widehat{W}$ is convex in the argument J, the following duality relation holds:

$$\widehat{W}(\boldsymbol{x}, \boldsymbol{F}, J) = \max_{\widehat{q}} \{\widehat{q}(J-1) - \widehat{W}^*(\boldsymbol{x}, \boldsymbol{F}, \widehat{q})\}. \tag{12.4}$$

Direct substitution of relation (12.4) in the principle of minimum potential energy (12.1)–(12.2) renders the following alternative mixed variational principle:

$$\widehat{\Pi}(\boldsymbol{u}, \widehat{p}) = \min_{\boldsymbol{v} \in \mathcal{K}} \max_{\widehat{q} \in \mathcal{Q}} \widehat{\Pi}(\boldsymbol{v}, \widehat{q}), \tag{12.5}$$

with

$$\begin{aligned} \widehat{\Pi}(\boldsymbol{v}, \widehat{q}) &= \int_\Omega \{-\widehat{W}^*(\boldsymbol{x}, \boldsymbol{F}(\boldsymbol{v}), \widehat{q}) + \widehat{q}[\det \boldsymbol{F}(\boldsymbol{v}) - 1]\} \, d\boldsymbol{x} \\ &\quad - \int_\Omega \boldsymbol{f} \cdot \boldsymbol{v} \, d\boldsymbol{x} - \int_{\Gamma^t} \boldsymbol{t} \cdot \boldsymbol{v} \, ds. \end{aligned} \tag{12.6}$$

In the above expression, $\mathcal{Q}$ denotes the set of square-integrable scalar functions and the maximizing scalar field $\widehat{p}$ is directly related to the equilibrium Cauchy hydrostatic pressure field $p := \operatorname{tr} \boldsymbol{\sigma}$ via [99]

$$p = \widehat{p} - \frac{1}{3 \det \boldsymbol{F}} \frac{\partial \widehat{W}^*}{\partial \boldsymbol{F}}(\boldsymbol{x}, \boldsymbol{F}, \widehat{p}) : \boldsymbol{F}.$$

The weak form of the Euler–Lagrange equations associated with the variational principle (12.5)–(12.6) reads as

$$\begin{aligned} G^{\boldsymbol{v}}(\boldsymbol{v}, \widehat{q}, \delta\boldsymbol{v}) = \int_\Omega \left[-\frac{\partial \widehat{W}^*}{\partial \boldsymbol{F}}(\boldsymbol{x}, \boldsymbol{F}(\boldsymbol{v}), \widehat{q}) + \widehat{q} \operatorname{adj}\big(\boldsymbol{F}^T(\boldsymbol{v})\big) \right] : \nabla(\delta\boldsymbol{v}) \, d\boldsymbol{x} \\ - \int_\Omega \boldsymbol{f} \cdot \delta\boldsymbol{v} \, d\boldsymbol{x} - \int_{\Gamma^t} \boldsymbol{t} \cdot \delta\boldsymbol{v} \, ds = 0 \qquad \forall \delta\boldsymbol{v} \in \mathcal{K}^0, \end{aligned}$$

$$G^{\widehat{q}}(\boldsymbol{v}, \widehat{q}, \delta\widehat{q}) = \int_\Omega \left[\det \boldsymbol{F}(\boldsymbol{v}) - 1 - \frac{\partial \widehat{W}^*}{\partial \widehat{q}}(\boldsymbol{x}, \boldsymbol{F}(\boldsymbol{v}), \widehat{q}) \right] \delta\widehat{q} \, d\boldsymbol{x} = 0 \qquad \forall \delta\widehat{q} \in \mathcal{Q},$$

where the test field $\delta\widehat{q}$ is the variation of $\widehat{q}$, and $\operatorname{adj}(\cdot)$ stands for the adjugate operator.

12.3 POLYGONAL AND POLYHEDRAL APPROXIMATIONS

This section discusses the displacement and pressure spaces for the polygonal and polyhedral finite element approximations. In 2D, we present the constructions of conforming finite-dimensional displacement and pressure spaces, for both linear and quadratic polygonal finite elements. In 3D, we present the construction of linear displacement finite element space on convex polyhedra.

12.3.1 Displacement space on polygons in 2D

Let us begin in the 2D setting and consider a general polygon P with n edges and vertices, $\boldsymbol{v}_1, \ldots, \boldsymbol{v}_n$, in the undeformed configuration. We use generalized barycentric coordinates as the basis for a linear polygonal finite element space on P, denoted by $\mathcal{V}_1(P)$, with degrees of freedom associated with the vertices of P.

Quite a few barycentric coordinates can be found in the literature [15, 34, 149, 193, 215, 265, 275, 372] (see Chapter 1 for details), among which the mean value coordinates [144] are adopted in this chapter. The mean value coordinates, denoted by $\phi_1^{(1)}, \ldots, \phi_n^{(1)}$, can be constructed on arbitrary polygons. Contour plots of the mean value coordinates defined on non-convex polygons are shown in Figure 12.2(a).

A second-order space on P, denoted as $\mathcal{V}_2(P)$, can be constructed using pairwise products of the generalized barycentric coordinates [320]. $\mathcal{V}_2(P)$ is a $2n$-dimensional space having the degrees of freedom at each vertex of P, as well as at the mid-point of each edge. Its interpolants $\phi_1^{(2)}, \ldots, \phi_{2n}^{(2)}$ are expressed as:

$$\phi_i^{(2)}(\boldsymbol{x}) = \sum_{j=1}^{n} \sum_{l=1}^{n} c_{jl}^i \phi_j^{(1)}(\boldsymbol{x}) \phi_l^{(1)}(\boldsymbol{x}),$$

where $\phi_j^{(1)}$ are the mean value coordinates. The coefficients c_{jl}^i are computed such

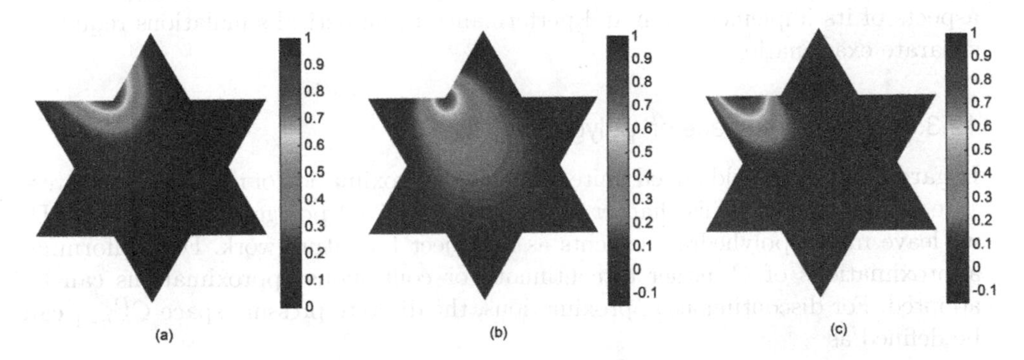

Figure 12.2 (a) Contour plot of a mean value basis $\phi_i^{(1)}$ over a concave polygon. (b) Contour plot of quadratic basis $\phi_i^{(2)}$ over a concave polygon associated with a vertex. (c) Contour plot of quadratic basis $\phi_i^{(2)}$ over a concave polygon associated with a mid-side node. See color insert.

that any quadratic function can be interpolated exactly by $\phi_i^{(2)}$:

$$q(\boldsymbol{x}) = \sum_{i=1}^{n} \left[q(\boldsymbol{v}_i)\phi_i^{(2)}(\boldsymbol{x}) + q(\hat{\boldsymbol{v}}_i)\phi_{i+n}^{(2)}(\boldsymbol{x})\right] \quad \forall q \in \mathcal{P}_2(E),$$

and the Kronecker-delta property

$$\phi_i^{(2)}(\boldsymbol{v}_j) = \phi_{i+n}^{(2)}(\hat{\boldsymbol{v}}_j) = \delta_{ij}, \quad \phi_i^{(2)}(\hat{\boldsymbol{v}}_j) = \phi_{i+n}^{(2)}(\boldsymbol{v}_j) = 0 \quad \forall i, j = 1, \ldots, n,$$

is satisfied, where $\hat{\boldsymbol{v}}_i = (\boldsymbol{v}_i + \boldsymbol{v}_{i+1})/2$ denotes the mid-side nodes.

Although the construction above is derived assuming that the polygonal elements are strictly convex [320], we also find that it is valid on concave polygons when constructed from the mean value coordinates. Basis function plots for a vertex node and a mid-side node of a concave polygon are shown in Figures 12.2(b) and 12.2(c), respectively.

12.3.2 Displacement space on polyhedra in 3D

Let $P \in \mathbb{R}^3$ be a polyhedron whose boundary consists of planar polygonal faces in its undeformed configuration with its n vertices located at $\boldsymbol{v}_1, \ldots, \boldsymbol{v}_n$. As in the 2D case, one can define generalized barycentric coordinates $\phi_1^{(1)}, \ldots, \phi_n^{(1)}$ as the basis for a linear polyhedral finite element space on P using vertex degrees of freedom.

This chapter considers the Wachspress coordinates, as defined in [413] and analyzed in [147], which are valid for simple polyhedra. This means that the collection of faces that include $\boldsymbol{v}$ for each $i = 1, \ldots, n$ consists of exactly three faces. As before, we denote by $\mathcal{V}_1(P) = \text{span}\{\phi_1^{(1)}, \ldots, \phi_n^{(1)}\}$ the finite element space on P, which by linear precision of the Wachspress coordinates includes $\mathcal{P}_1(P)$. We remark that the gradient correction scheme described later in this chapter is applicable to other types of coordinates on linear and quadratic polyhedrons as well, though certain aspects of its implementation and performance in numerical simulations require a separate examination.

12.3.3 Pressure space on polygons in 2D

Regarding the two-field mixed finite elements, approximation of the additional pressure field is needed. This chapter only considers mixed polygonal elements in 2D; we leave mixed polyhedral elements as a subject for future work. For conforming approximations of $\mathcal{Q}$, either discontinuous or continuous approximations can be adopted. For discontinuous approximations, the discrete pressure space $\mathcal{Q}_{h,k-1}^D$ can be defined as:

$$\mathcal{Q}_{h,k-1}^D = \{\widehat{q}_h \in \mathcal{Q} : \widehat{q}_h|_P \in \mathcal{P}_{k-1}(P)\ \forall P \in \Omega_h\},$$

where k is the order of the element. As implied by the above definition, the approximated pressure field may be discontinuous across element boundaries. This type

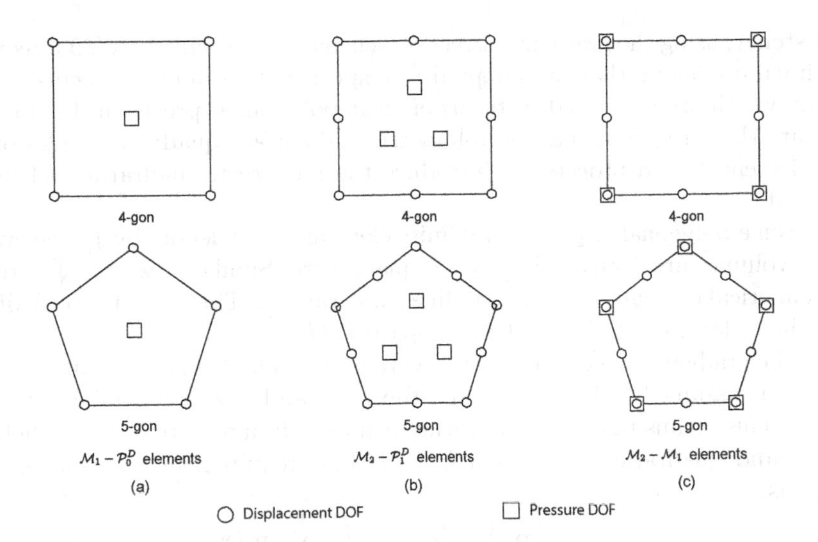

Figure 12.3 Illustration of the degrees of freedom of displacement and pressure fields for different mixed polygonal elements. (a) $\mathcal{M}_1 - \mathcal{P}_0^D$ elements, (b) $\mathcal{M}_2 - \mathcal{P}_1^D$ elements, and (c) $\mathcal{M}_2 - \mathcal{M}_1$ elements.

of mixed element is similar to the Crouzeix–Raviart (C–R) elements in fluid problems [110]. For the remainder of this chapter, we adapt the acronyms from [100] and denote this family of mixed elements as $\mathcal{M}_k - \mathcal{P}_{k-1}^D$ elements. Therefore, the pressure space of the $\mathcal{M}_1 - \mathcal{P}_0^D$ element consists of piecewise constant functions, which are constant over each element. Similarly, the pressure space of the $\mathcal{M}_2 - \mathcal{P}_1^D$ element contains piecewise linear functions that vary linearly over each element.

Alternatively, a continuous approximation of the pressure space can be defined in the following manner:

$$\mathcal{Q}_{h,k-1}^C = \{\widehat{q}_h \in C^0(\Omega) : \widehat{q}_h|_P \in \mathcal{M}_{k-1}(P)\ \forall P \in \Omega_h\}.$$

This class of elements resembles the Taylor-Hood (T–H) elements in fluid problems [395], and they are henceforth denoted as the $\mathcal{M}_k - \mathcal{M}_{k-1}$ elements ($k \geq 2$).

Here we consider both types of mixed polygonal finite elements up to quadratic order. As an illustration, the degrees of freedom (DOFs) of the displacement field and pressure field for these mixed polygonal elements are shown in Figure 12.3.

12.4 QUADRATURE RULES AND ACCURACY REQUIREMENTS

On general polygonal and polyhedral elements, generalized barycentric coordinates are usually non-polynomial functions. As discussed in [100, 388, 393], the evaluation of weak form integrals involving barycentric coordinates and their gradients by means of available quadrature schemes leads to errors that are persistent under mesh refinement. In order to control the quadrature errors and restore polynomial consistency, using the gradient correction scheme by Talischi et al. [393] requires the quadrature schemes that are adopted to respect certain minimal accuracy requirements, which are expressed in terms of their polynomial precision. In this section, we state these requirements for volumetric and surface quadrature rules on a general element P and proceed to introduce the numerical quadrature schemes that are used.

Given a polygonal or polyhedral finite element P, we denote by $\oint\!\!\!\!\!\!\int_P$ the evaluation of the volume (area) integral $\int_P$ using quadrature. Similarly, we use $\oint_{\partial P}$ to denote the numerical evaluation of surface (line) integral $\int_{\partial P}$. Throughout, we shall assume that the order k of the finite element space $\mathcal{V}_k(P)$ is fixed.

In the gradient correction scheme by Talischi et al. [393], the volumetric quadrature $\oint\!\!\!\!\!\!\int_P$ is assumed to be exact when the integrand is a polynomial field of order $2k-2$. This means that the quadrature can exactly integrate constant fields when $k = 1$ and quadratic fields when $k = 2$. This requirement guarantees that the integrals,

$$\int_P \boldsymbol{p} \cdot \nabla q \, d\boldsymbol{x}, \qquad \int_P q \nabla \cdot \boldsymbol{p} \, d\boldsymbol{x},$$

are exact when $\boldsymbol{p} \in [\mathcal{P}_{k-1}(P)]^d$ and $q \in \mathcal{P}_k(P)$. Additionally, an implicit assumption is made on the volume quadrature rule that it is sufficiently rich to eliminate the appearance of spurious elemental zero-energy modes that can compromise the stability of the resulting discretization scheme. For instance, for $k = 1$, a one-point rule consisting of an integration point at the centroid of E, with the volume $|P|$ being the weight, satisfies the above precision requirement, but leads to spurious zero-energy modes.

On the other hand, the boundary quadrature $\oint_{\partial P}$ is required to integrate polynomial fields of order $2k-1$ on each edge/face of the element. Therefore, 1-point $(k = 1)$ and 2-point $(k = 2)$ quadrature suffices for piecewise linear and cubic fields, respectively. This requirement ensures that integrals of the form

$$\int_{\partial P} q(\boldsymbol{p} \cdot \boldsymbol{n}) ds$$

are exact if $\boldsymbol{p} \in [\mathcal{P}_{k-1}(P)]^d$ and q are piecewise k-th order polynomials on ∂P with $\boldsymbol{n}$ being the unit normal vector to the boundary. For 2D elements, because the functions in $\mathcal{V}_k(P)$ have k-th order polynomial variation on ∂P, the assumed accuracy ensures exact integration:

$$\oint_{\partial P} v(\boldsymbol{p} \cdot \boldsymbol{n}) \, ds = \int_{\partial P} v(\boldsymbol{p} \cdot \boldsymbol{n}) \, ds,$$

for any $v \in \mathcal{V}_k(P)$ and $\boldsymbol{p} \in [\mathcal{P}_{k-1}(P)]^d$.

We next describe the quadrature schemes that are used in the numerical studies. For the 2D case, we adopt a triangulation scheme that divides each polygonal element into triangles by connecting the centroid to each vertex and apply the Dunavant rules [131] in each subdivided triangle. According to the above-stated

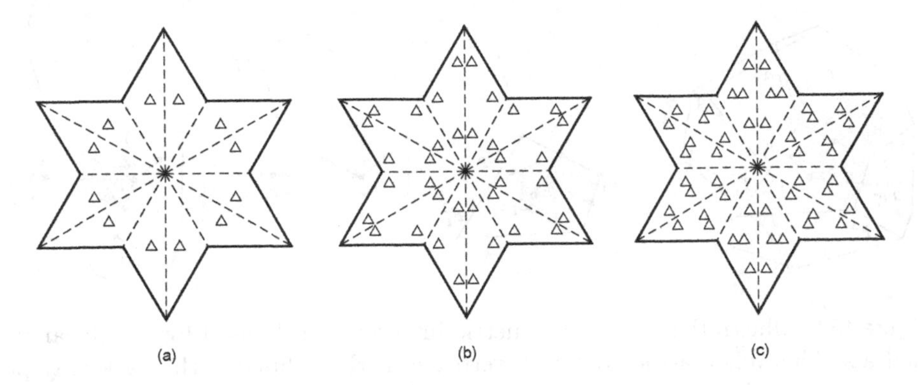

Figure 12.4 Illustration of the "triangulation" scheme for general polygons in the physical domain. (a) 1st order triangulation scheme, (b) 2nd order triangulation scheme, and (c) 3rd order triangulation scheme.

requirements, instead of using a one-point rule, the 1st order triangulation scheme with one integration point per subdivided triangle is used for linear elements to avoid spurious energy modes. For quadratic elements, the 2nd order triangulation scheme is employed, which contains three integration points per subdivided triangle. Furthermore, the 3rd order triangulation scheme is also used to investigate the effect of increasing integration orders on the accuracy of the results in the gradient correction scheme. Illustrations of these schemes are shown in Figure 12.4(a)–12.4(c). As a side note, the triangulation scheme requires polygonal elements to be star-convex with respect to their centroids, which is indeed the case for all the examples presented here. Other quadrature schemes can also be utilized in conjunction with the gradient correction scheme, such as those specifically designed for integrating polynomial functions over arbitrary polygonal domains [102, 285, 286], as long as they satisfy the requirements stated in this section.

For volumetric integration in the 3D case, one possibility is to split the element into tetrahedra and use the available quadrature schemes on these subdomains. A more economical alternative for the linear polyhedron that satisfies the polynomial precision requirement is proposed in [322] and utilized in [47, 393]. This scheme, which is illustrated in Figure 12.5, can exactly integrate constant and linear fields. Additional options are noted in [102], which is designed for general polyhedral domains.

Regarding the boundary quadrature, the one-dimensional Gauss–Lobatto quadrature rule is adopted in 2D, which uses two integration points per edge for linear elements and three integration points per edge for quadratic elements. As for the boundary quadrature in 3D, the integration on each face is carried out by means of a vertex quadrature rule that requires only the evaluation of the integrand at the vertices of the face. The face is split into quadrilateral regions associated with its vertices by connecting the centroid to the midpoint of the edges of the faces. The

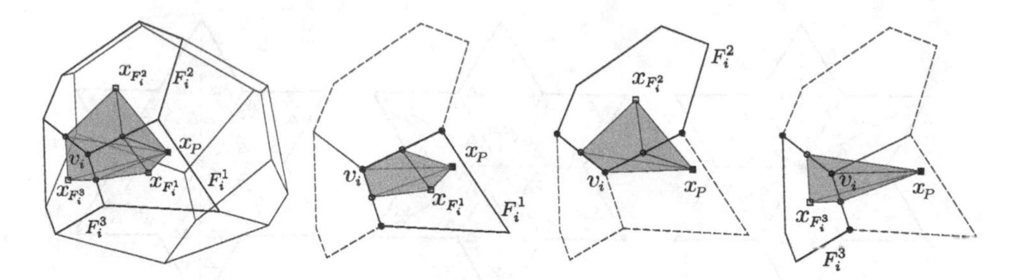

Figure 12.5 Illustration of the volumetric integration rule used for 3D linear polyhedrals. The weight associated with vertex $\boldsymbol{v}_i$ is the volume of the dark gray polyhedron formed by pyramids associated with the faces connected to $\boldsymbol{v}_i$. The pyramid associated with a face F_i^j is formed by the centroid of the element $\boldsymbol{x}_P$, centroid of the face $\boldsymbol{x}_{F_i^j}$, and the midpoint of the edges of F_i^j connected to $\boldsymbol{v}_i$. This figure is taken from [393].

quadrature weights are the areas of these quadrilateral regions. It can be shown that this scheme is exact for integration of linear fields (see Appendix of [268]).

12.5 GRADIENT CORRECTION SCHEME AND ITS PROPERTIES

In this section, the gradient correction scheme by Talischi et al. [393] is presented. First, we review the scheme for scalar problems and then show its extension to vectorial problems.

12.5.1 Gradient correction for scalar problems

Given an element P, the basic idea of the gradient correction scheme is to correct the gradient of functions in the local space $\mathcal{V}_k(P)$ with minimal perturbations that result in the satisfaction of a discrete divergence theorem. For scalar problems, the *corrected* gradient of $v \in \mathcal{V}_k(P)$, henceforth denoted by $\nabla_{P,k} v$, is taken to be a vector field over P that solves the optimization problem

$$\min_{\boldsymbol{\xi}} \oint_P \|\boldsymbol{\xi} - \nabla v\|^2 \, d\boldsymbol{x}, \tag{12.7}$$

subject to

$$\oint_P \boldsymbol{p} \cdot \boldsymbol{\xi} \, d\boldsymbol{x} = -\oint_P v \nabla \cdot \boldsymbol{p} \, d\boldsymbol{x} + \oint_{\partial P} v(\boldsymbol{p} \cdot \boldsymbol{n}) \, ds \quad \forall \boldsymbol{p} \in [\mathcal{P}_{k-1}(P)]^d. \tag{12.8}$$

The above constrained minimization is carried out over all sufficiently smooth vector fields on P such that the utilized quadrature makes sense. The corrected gradient

$$\nabla_{P,k} v = [(\nabla_{P,k} v)_x, (\nabla_{P,k} v)_y]^T \text{ in 2D}$$
$$\nabla_{P,k} v = [(\nabla_{P,k} v)_x, (\nabla_{P,k} v)_y, (\nabla_{P,k} v)_z]^T \text{ in 3D}$$

is thus the "closest"† field to $\nabla v = [(\nabla v)_x, (\nabla v)_y]^T$ (or $\nabla v = [(\nabla v)_x, (\nabla v)_y, (\nabla v)_z]^T$ in 3D) that satisfies the discrete divergence theorem against all polynomials of order k.

To assist the formal definition of the gradient correction scheme for scalar problems, we consider a basis $\{\zeta_1, \dots, \zeta_n\}$ for $[\mathcal{P}_{k-1}(P)]^d$ and replace the equality constraint (12.8) by the equivalent set of constraints:

$$\oint_P \zeta_a \cdot \boldsymbol{\xi}\, d\boldsymbol{x} = -\oint_P v\nabla \cdot \zeta_a\, d\boldsymbol{x} + \oint_{\partial P} v(\zeta_a \cdot \boldsymbol{n})\, ds, \quad a = 1, \dots, n$$

Introducing multipliers $\lambda_1, \dots, \lambda_n$, one obtains the Lagrangian:

$$\begin{aligned}\mathcal{L}(\boldsymbol{\xi}, \lambda_1, \dots, \lambda_n) = &\oint_P \|\boldsymbol{\xi} - \nabla v\|^2\, d\boldsymbol{x} \\ &+ \sum_{a=1}^{n} \lambda_a \left[\oint_P \zeta_a \cdot \boldsymbol{\xi}\, d\boldsymbol{x} + \oint_P v\nabla \cdot \zeta_a\, d\boldsymbol{x} - \oint_{\partial P} v(\zeta_a \cdot \boldsymbol{n})\, ds\right].\end{aligned}$$

The optimality of $\nabla_{P,k} v$ requires that for any variation $\boldsymbol{\eta}$,

$$D_{\boldsymbol{\xi}}\mathcal{L}(\nabla_{P,k} v, \lambda_1, \dots, \lambda_n)[\boldsymbol{\eta}] = 2\oint_P (\nabla_{P,k} v - \nabla v) \cdot \boldsymbol{\eta}\, d\boldsymbol{x} + \sum_{a=1}^{n} \lambda_a \oint_P \zeta_a \cdot \boldsymbol{\eta}\, d\boldsymbol{x} = 0,$$

and therefore

$$\oint_P \left(\nabla_{P,k} v - \nabla v + \sum_{a=1}^{n} \frac{1}{2}\lambda_a \zeta_a\right) \cdot \boldsymbol{\eta}\, d\boldsymbol{x} = 0.$$

This shows that the perturbation $\nabla_{P,k} v - \nabla v$ coincides with an element of $[\mathcal{P}_{k-1}(P)]^d$ at the location of the quadrature points, because the optimization problem (12.7)–(12.8) can only prescribe the values of $\nabla_{P,k} v$ at the quadrature points. Therefore, we can officially define the corrected gradient $\nabla_{P,k} v$ as the field satisfying the following two conditions:

- $\nabla_{P,k} v - \nabla v \in [\mathcal{P}_{k-1}(P)]^d$.
- For all $\boldsymbol{p} \in [\mathcal{P}_{k-1}(P)]^d$,

$$\oint_P \boldsymbol{p} \cdot \nabla_{P,k} v\, d\boldsymbol{x} = -\oint_P v\nabla \cdot \boldsymbol{p}\, d\boldsymbol{x} + \oint_{\partial P} v(\boldsymbol{p} \cdot \boldsymbol{n})\, ds. \tag{12.9}$$

We observe from these two conditions that $\nabla_{P,k}$ is a linear map, and so in practice only the action of $\nabla_{P,k}$ on the basis of $\mathcal{V}_k(P)$ must be computed. Here, we present the details on a systematic procedure for computing the corrected gradient of the basis functions for arbitrary k. We denote $\{\phi_1^{(k)}, \cdots, \phi_{n\mathcal{V}}^{(k)}\}$ as the basis for $\mathcal{V}_k(P)$,

---

†The distance here is with respect to the quadrature form of the L^2-metric.

where $n\mathcal{V}$ is the dimension of the space $\mathcal{V}_k(P)$. Our goal is to find a coefficient matrix $\boldsymbol{S}$ such that

$$\nabla_{P,k}\phi_i^{(k)} = \nabla\phi_i^{(k)} + \sum_{a=1}^{n\mathcal{P}} S_{ia}\boldsymbol{\zeta}_a \quad \forall i = 1,\ldots,n\mathcal{V}.$$

To explicitly compute $\boldsymbol{S}$, we further define matrices $\boldsymbol{R}$ of size $n\mathcal{V} \times n\mathcal{P}$, and $\boldsymbol{M}$ of size $n\mathcal{P} \times n\mathcal{P}$ with the following forms:

$$R_{ia} = \oint_{\partial P} (\boldsymbol{\zeta}_a \cdot \boldsymbol{n})\phi_i^{(k)}\, ds - \oint_P \phi_i^{(k)}\nabla \cdot \boldsymbol{\zeta}_a\, d\boldsymbol{x} - \oint_P \boldsymbol{\zeta}_a \cdot \nabla\phi_i^{(k)}\, d\boldsymbol{x},$$

and

$$M_{ab} = \oint_{\partial P} \boldsymbol{\zeta}_a \cdot \boldsymbol{\zeta}_b\, d\boldsymbol{x}.$$

Replacing v and $\boldsymbol{p}$ with ϕ_i^k and $\boldsymbol{\zeta}_b$ in (12.9) yields the linear system of equations

$$\sum_{a=1}^{n\mathcal{P}} S_{ia}M_{ab} = R_{ib} \quad \forall i = 1,\ldots,n\mathcal{V} \qquad \text{and} \qquad b = 1,\ldots,n\mathcal{P}.$$

Therefore, the coefficient matrix is obtained as $\boldsymbol{S} = \boldsymbol{R}\boldsymbol{M}^{-1}$.

In order to better understand the gradient correction scheme, let us denote by $\boldsymbol{p}_v$ the perturbation to ∇v that gives the corrected gradients, that is,

$$\nabla_{P,k}v = \nabla v + \boldsymbol{p}_v.$$

Inserting the above in (12.9), we get for any $\boldsymbol{p} \in [\mathcal{P}_{k-1}(P)]^d$,

$$\oint_P \boldsymbol{p}\cdot\boldsymbol{p}_v\, d\boldsymbol{x} = \left[-\oint_P v\nabla\cdot\boldsymbol{p}\, d\boldsymbol{x} + \oint_{\partial P} v(\boldsymbol{p}\cdot\boldsymbol{n})ds\right] - \oint_P \boldsymbol{p}\cdot\nabla v\, d\boldsymbol{x}. \tag{12.10}$$

Several observations can be made. First, the integral on the left-hand side of the above equation is exactly computed by the utilized quadrature (recall that we require the quadrature to be exact for polynomial functions up to order $2k-2$). Second, notice the right-hand side of the above expression represents the difference of the two quadrature approximations (the two quadrature approximations are the terms in the bracket and the last term, respectively) of $\int_P \boldsymbol{p}\cdot\nabla v\, d\boldsymbol{x}$. Therefore, (12.10) shows that the perturbations $\boldsymbol{p}_v$ capture the difference between two approximations of ∇v when integrated against polynomials of order $k-1$. $\boldsymbol{p}_v$ is, in fact, the "smallest" element in $[\mathcal{P}_{k-1}(P)]^d$ with respect to the L^2-norm that does so according to (12.7)–(12.8). Third, the smaller the error in the satisfaction of the discrete divergence theorem (i.e., right-hand side of (12.10)), the smaller the perturbation is to the gradient. This error depends on the accuracy of quadrature as well as the nature of the functions in the local space $\mathcal{V}_k(P)$. Last but not least, by setting $\boldsymbol{p} = \nabla_{P,k} q - \nabla q = \boldsymbol{p}_q$ in (12.10), we have

$$\begin{aligned}\int_P (\nabla_{P,k} q - \nabla q)^2 d\boldsymbol{x} &= -\fint_P q \nabla \cdot \boldsymbol{p}_q \, d\boldsymbol{x} + \oint_{\partial P} q(\boldsymbol{p}_q \cdot \boldsymbol{n}) \, ds - \fint_P \boldsymbol{p}_q \cdot \nabla q \, d\boldsymbol{x} \\ &= -\int_P q \nabla \cdot \boldsymbol{p}_q \, d\boldsymbol{x} + \int_{\partial P} q(\boldsymbol{p}_q \cdot \boldsymbol{n}) \, ds - \int_P \boldsymbol{p}_q \cdot \nabla q \, d\boldsymbol{x} \\ &= 0,\end{aligned}$$

which implies that the corrected gradient coincides with the exact gradient when applied to k-th order polynomials, i.e.,

$$\nabla_{P,k} q = \nabla q \quad \forall q \in \mathcal{P}_k(P). \tag{12.11}$$

This property of the correction scheme plays an important role in ensuring that the resulting discretizations are polynomially consistent and satisfy the patch test.

To aid in better understanding (12.11), we consider the case with linear space $\mathcal{V}_1(P)$, where an explicit expression for the corrected gradient can be readily obtained. Let $\boldsymbol{p} \in [\mathcal{P}_0(P)]^d$ and $v \in \mathcal{V}_1(P)$. We arrive at the following expression for the corrected gradient:

$$\nabla_{P,1} v = \nabla v + \frac{1}{|P|} \left(\oint_{\partial P} v \boldsymbol{n} \, ds - \fint_P \nabla v \, d\boldsymbol{x} \right),$$

which is similar to the corrected strain operator in the meshfree literature [296]. We can see from this expression that the perturbation is a constant field proportional to the error in satisfying the identity $\int_P \nabla v \, d\boldsymbol{x} = \int_{\partial P} v \boldsymbol{n} \, ds$ by the volumetric and surface quadrature schemes. This implies that if $v \in \mathcal{P}_1(P)$, then there is no correction to its gradient, i.e., $\nabla_{P,1} v = \nabla v$. Additionally, in the 2D case, recall that the variation of v on ∂P is piecewise linear, and the boundary quadrature is exact. In this case, $\oint_{\partial P} v \boldsymbol{n} \, ds = \int_{\partial P} v \boldsymbol{n} \, ds = \int_P \nabla v \, d\boldsymbol{x}$, and subsequently

$$\nabla_{P,1} v = \nabla v + \frac{1}{|P|} \left(\int_P \nabla v \, d\boldsymbol{x} - \fint_P \nabla v \, d\boldsymbol{x} \right). \tag{12.12}$$

The perturbation is simply the difference between the volume average of ∇v and its approximations through the quadrature. An immediate observation from (12.12) is that

$$\fint_P \nabla_{P,1} v \, d\boldsymbol{x} = \int_P \nabla v \, d\boldsymbol{x}.$$

This shows that the corrected gradient $\nabla_{P,1} v$ under the action of the quadrature behaves like ∇v under exact integration.

Generally, it is expected from the definition of the correction that

$$\fint_P \boldsymbol{\psi} \cdot \nabla_{P,k} v \, d\boldsymbol{x} - \int_P \boldsymbol{\psi} \cdot \nabla v \, d\boldsymbol{x} = O(h_P^k) \|\nabla v\|_{L^2(P)^d}$$

for $v \in \mathcal{V}_k(P)$ when $\boldsymbol{\psi}$ is a sufficiently smooth vector field. Here, h_P is the diameter of element P.

12.5.2 Gradient correction for vectorial problems

The gradient correction scheme defined in the preceding subsection for scalar problems can be readily extended to vectorial problems. For a vector field $\boldsymbol{v} = [v_x, v_y]^T$ (or $\boldsymbol{v} = [v_x, v_y, v_z]^T$ in 3D) $\in [\mathcal{V}_k(P)]^d$, the gradient correction scheme takes the form

$$\nabla_{P,k} \otimes \boldsymbol{v} = \begin{bmatrix} (\nabla_{P,k} v_x)^T \\ (\nabla_{P,k} v_y)^T \end{bmatrix} \text{ in 2D}, \quad \nabla_{P,k} \otimes \boldsymbol{v} = \begin{bmatrix} (\nabla_{P,k} v_x)^T \\ (\nabla_{P,k} v_y)^T \\ (\nabla_{P,k} v_z)^T \end{bmatrix} \text{ in 3D}.$$

Similar to the scalar case, the corrected gradient satisfies the discrete divergence theorem,

$$⨎_P \boldsymbol{p} : \nabla_{P,k} \otimes \boldsymbol{v}\, d\boldsymbol{x} = \int_{\partial E} (\boldsymbol{p} \cdot \boldsymbol{n}) \cdot \boldsymbol{v}\, ds - ⨎_P \boldsymbol{v} \cdot (\nabla \cdot \boldsymbol{p})\, d\boldsymbol{x} \quad \forall \boldsymbol{p} \in [\mathcal{P}_{k-1}(P)]^{d \times d},$$

and, moreover, for any sufficiently smooth second-order tensorial field $\boldsymbol{\psi}$, the element-level consistency error satisfies

$$⨎_P \boldsymbol{\psi} : \nabla_{P,k} \otimes \boldsymbol{v}\, d\boldsymbol{x} - \int_P \boldsymbol{\psi} : \nabla \boldsymbol{v}\, d\boldsymbol{x} = \mathcal{O}(h_P^k) \|\nabla \boldsymbol{v}\|_{L^2(P)^d}.$$

In the computational implementation of 2D problems, let us assume the set of basis functions $\{\boldsymbol{\phi}_1^{(k)}, \ldots, \boldsymbol{\phi}_{2n\mathcal{V}}^{(k)}\}$ of $[\mathcal{V}_k(P)]^2$ are of the form

$$\boldsymbol{\phi}_{2i-1}^{(k)} = [\phi_i^{(k)}, 0]^T, \quad \boldsymbol{\phi}_{2i}^{(k)} = [0, \phi_i^{(k)}]^T$$

for $i = 1, \ldots, n\mathcal{V}$, and where $n\mathcal{V}$ is the dimension of the space $\mathcal{V}_k(P)$. The correction scheme for vectorial problems in practice amounts to correcting each basis function as follows:

$$\nabla_{P,k} \otimes \boldsymbol{\phi}_{2i-1}^{(k)} = \begin{bmatrix} (\nabla_{P,k} \phi_i^{(k)})^T \\ \mathbf{0} \end{bmatrix}, \quad \nabla_{P,k} \otimes \boldsymbol{\phi}_{2i}^{(k)} = \begin{bmatrix} \mathbf{0} \\ (\nabla_{P,k} \phi_i^{(k)})^T \end{bmatrix}$$

for $i = 1, \ldots, n\mathcal{V}$, and where $\nabla_{P,k} \phi_i^{(k)}$ is the computed corrected gradient of the shape function according to the procedure outlined for the scalar problem.

Similarly, let us assume the set of basis functions for $\{\boldsymbol{\phi}_1^{(k)}, \ldots, \boldsymbol{\phi}_{3n\mathcal{V}}^{(k)}\}$ of the 3D space $[\mathcal{V}_k(P)]^3$ are of the form

$$\boldsymbol{\phi}_{3i-3}^{(k)} = [\phi_i^{(k)}, 0, 0]^T, \quad \boldsymbol{\phi}_{3i-1}^{(k)} = [0, \phi_i^{(k)}, 0]^T, \quad \text{and} \quad \boldsymbol{\phi}_{3i}^{(k)} = [0, 0, \phi_i^{(k)}]^T$$

for $i = 1, \ldots, n\mathcal{V}$. The gradient correction scheme for 3D vectorial problems corrects each basis function as

$$\nabla_{P,k} \otimes \boldsymbol{\phi}_{3i-2}^{(k)} = \begin{bmatrix} (\nabla_{P,k} \phi_i^{(k)})^T \\ \mathbf{0} \\ \mathbf{0} \end{bmatrix}, \quad \nabla_{P,k} \otimes \boldsymbol{\phi}_{3i-1}^{(k)} = \begin{bmatrix} \mathbf{0} \\ (\nabla_{P,k} \phi_i^{(k)})^T \\ \mathbf{0} \end{bmatrix},$$

$$\nabla_{P,k} \otimes \boldsymbol{\phi}_{3i}^{(k)} = \begin{bmatrix} \mathbf{0} \\ \mathbf{0} \\ (\nabla_{P,k} \phi_i^{(k)})^T \end{bmatrix} \quad (i = 1, \ldots, n\mathcal{V}).$$

12.6 CONFORMING GALERKIN APPROXIMATIONS

Consider Ω_h to be a finite element decomposition of the domain Ω into non-overlapping polygons or polyhedra, where h is the average element size. The boundary of the mesh, denoted as Γ_h, is assumed to be compatible with the applied boundary condition. In other words, Γ_h^t and Γ_h^x are both unions of edges of the mesh.

We define the numerical integration $\oint_{\Omega_h}$ on Ω_h as the summation of the contributions from numerical integrals $\oint_P$ from each element P following the standard assembly rule, namely, $\oint_{\Omega_h} = \sum_{P\in\Omega_h} \oint_P$, and $\oint_{\Gamma_h^t}$ as the numerical integration on Γ_h^t. Similarly, we define the gradient correction map on the global level, $\nabla_{h,k}\colon \mathcal{K}_{h,k} \to [L^2(\Omega_h)]^{d\times d}$, such that it coincides with the gradient correction map $\nabla_{P,k}$ at the element level,

$$(\nabla_{h,k}\boldsymbol{v}_h)|_P = \nabla_{P,k} \otimes (\boldsymbol{v}_h|_P) \quad \forall P\in\Omega_h \qquad \text{and} \qquad \boldsymbol{v}_h \in \mathcal{K}_{h,k}.$$

The Galerkin approximation of the displacement-based formulation then consists of finding $\boldsymbol{u}_h \in \mathcal{K}_{h,k}$, such that

$$G_h(\boldsymbol{u}_h, \delta\boldsymbol{v}_h) = 0 \quad \forall\delta\boldsymbol{v}_h \in \mathcal{K}_{h,k}^0,$$

where $\mathcal{K}_{h,k}^0 = \mathcal{K}_{h,k} \bigcap \mathcal{K}^0$ and $G_h(\boldsymbol{u}_h, \delta\boldsymbol{v}_h)$ is the discretization and quadrature evaluation of $G(\boldsymbol{u}, \delta\boldsymbol{v})$ in (12.3) with the exact gradient operator ∇ replaced by $\nabla_{h,k}$, which takes the form

$$G_h(\boldsymbol{u}_h, \delta\boldsymbol{v}_h) = \oint_{\Omega_h} \frac{\partial W}{\partial \boldsymbol{F}}(\boldsymbol{x}, \boldsymbol{I} + \nabla_{h,k}\boldsymbol{u}_h) : \nabla_{h,k}(\delta\boldsymbol{v}_h)\, d\boldsymbol{x} - \oint_{\Omega_h} \boldsymbol{f}_h \cdot \delta\boldsymbol{v}_h\, d\boldsymbol{x} - \oint_{\Gamma_h^t} \boldsymbol{t}_h \cdot \delta\boldsymbol{v}_h\, ds, \quad (12.13)$$

and terms $\boldsymbol{f}_h$ and $\boldsymbol{t}_h$ are the approximated body force and boundary traction, respectively.

For the two-field mixed formulation, if we introduce the additional finite element space $\mathcal{Q}_{h,k-1}^D$(or $\mathcal{Q}_{h,k-1}^C$)$\subseteq \mathcal{Q}$, the Galerkin approximation consists of finding $(\boldsymbol{u}_h, \widehat{p}_h) \in \mathcal{K}_{h,k} \times \mathcal{Q}_{h,k-1}^D$(or $\mathcal{Q}_{h,k-1}^C$), such that

$$G_h^{\boldsymbol{v}}(\boldsymbol{u}_h, \widehat{p}_h, \delta\boldsymbol{v}_h) = 0 \quad \forall\delta\boldsymbol{v}_h \in \mathcal{K}_{h,k}^0,$$
$$\widehat{G_h^q}(\boldsymbol{u}_h, \widehat{p}_h, \delta\widehat{q}_h) = 0 \quad \forall\delta\widehat{q}_h \in \mathcal{Q}_{h,k-1}^D \quad (\text{or } \mathcal{Q}_{h,k-1}^C),$$

with $G_h^{\boldsymbol{v}}(\boldsymbol{u}_h, \widehat{p}_h, \delta\boldsymbol{v}_h)$ and $\widehat{G_h^q}(\boldsymbol{u}_h, \widehat{p}_h, \delta\widehat{q}_h)$ being of the form

$$G_h^{\boldsymbol{v}}(\boldsymbol{u}_h, \widehat{p}_h, \delta\boldsymbol{v}_h) = \oint_{\Omega_h} \left[-\frac{\partial \widehat{W}^*}{\partial \boldsymbol{F}}(\boldsymbol{x}, \boldsymbol{I} + \nabla_{h,k}\boldsymbol{u}_h, \widehat{p}_h) + \widehat{p}_h \operatorname{adj}\big(\boldsymbol{I} + (\nabla_{h,k}\boldsymbol{u}_h)^T\big) \right] : \nabla_{h,k}(\delta\boldsymbol{v}_h)\, d\boldsymbol{x} - \oint_{\Omega_h} \boldsymbol{f}_h \cdot \delta\boldsymbol{v}_h\, d\boldsymbol{x} - \oint_{\Gamma_h^t} \boldsymbol{t}_h \cdot \delta\boldsymbol{v}_h\, ds, \quad (12.14)$$

$$\widehat{G_h^q}(\boldsymbol{u}_h, \widehat{p}_h, \delta\widehat{q}_h) = \oint_{\Omega_h} \left[\det(\boldsymbol{I} + \nabla_{h,k}\boldsymbol{u}_h) - 1 - \frac{\partial \widehat{W}^*}{\partial \widehat{p}}(\boldsymbol{x}, \boldsymbol{I} + \nabla_{h,k}\boldsymbol{u}_h, \widehat{p}_h) \right] \delta\widehat{q}_h \, d\boldsymbol{x}. \tag{12.15}$$

Notice that since we replace the gradient operators ∇ in both $\boldsymbol{u}_h$ and $\delta\boldsymbol{v}_h$ in (12.13), (12.14), and (12.15) with $\nabla_{h,k}$, the resulting approximations yield symmetric linearizations. Moreover, both of the above finite element approximations pass the first-order patch test [100]. This means that when the exact displacement field is a linear field, the above finite element approximations give the exact results.

12.7 NUMERICAL EXAMPLES

In this section, we present a series of numerical tests to assess the performance of the displacement-based and the two-field mixed polygonal and polyhedral finite element methods. For 2D problems, both displacement-based and mixed polygonal elements up to the second order are considered, whereas only first-order displacement-based polyhedral elements are studied for the 3D case. We demonstrate, through the convergence studies of two boundary-value problems, that the gradient correction scheme is capable of ensuring optimal convergence of polygonal and polyhedral elements in finite elasticity. For mixed polygonal elements, we also provide numerical evaluations and discuss inf-sup stability for different choices of pressure approximations.

Throughout this section, a neo-Hookean material model is considered. For a compressible neo-Hookean material, the stored-energy function characterizing the material behavior has the form

$$W(\boldsymbol{F}) = \frac{\mu}{2}[\boldsymbol{F} : \boldsymbol{F} - 3] - \mu(\det \boldsymbol{F} - 1) + \frac{3\kappa + \mu}{6}(\det \boldsymbol{F} - 1)^2,$$

where μ and κ denote the initial shear and bulk moduli of the material response. In the limit of incompressibility ($\kappa \to \infty$), the above compressible neo-Hookean material reduces to the standard neo-Hookean stored-energy function

$$W(\boldsymbol{x}, \boldsymbol{F}) = \begin{cases} \dfrac{\mu}{2}[\boldsymbol{F} : \boldsymbol{F} - 3], & \text{if } \det \boldsymbol{F} = 1, \\ +\infty, & \text{otherwise.} \end{cases} \tag{12.16}$$

The standard Newton–Raphson method is employed to solve the nonlinear system of equations, and we assume that convergence at each loading step is met if the norm of the residual falls below 10^{-8} times that of the initial residual.

12.7.1 Displacement-based polygonal and polyhedral elements

This subsection presents a convergence study of the displacement-based polygonal and polyhedral finite elements using the gradient correction scheme. In particular, we consider a boundary-value problem as shown in Figure 12.6(a), where a rectangular block of dimensions $1 \times 1 \times \pi$ is bent into a semi-circular shape. The corresponding analytical displacement solution $\boldsymbol{u}$ is given by

$$u_x = \; 1 + (1 + x)\cos(z) - x, \qquad u_y = 0, \qquad u_z = (1 + x)\sin(z) - z,$$

while the corresponding body force $\boldsymbol{f} = [f_x, f_y, f_z]^T$ reads as

$$f_x = -\frac{(1+x)\cos(z)(3\kappa - 2\mu)}{3}, \qquad f_y = 0, \qquad f_z = -\frac{(1+x)\sin(z)(3\kappa - 2\mu)}{3}.$$

Throughout, we set the initial shear and bulk moduli μ and κ as 1 and 10, and consider two error measures, the L^2-norm and the H^1-seminorm of the displacement error. More specifically, the L^2-norm and the H^1-seminorm, respectively, are given by

$$E_{0,\boldsymbol{u}} = \|\boldsymbol{u} - \boldsymbol{u}_h\| \qquad \text{and} \qquad E_{1,\boldsymbol{u}} = \|\nabla \boldsymbol{u} - \nabla \boldsymbol{u}_h\|,$$

where the L^2-norm, $\|\cdot\|$, is evaluated using a fifth order triangulation rule.

Since no deformation occurs in the u_y direction, this problem can be simplified into a plane strain problem as shown in Figure 12.6(b). We make use of structured

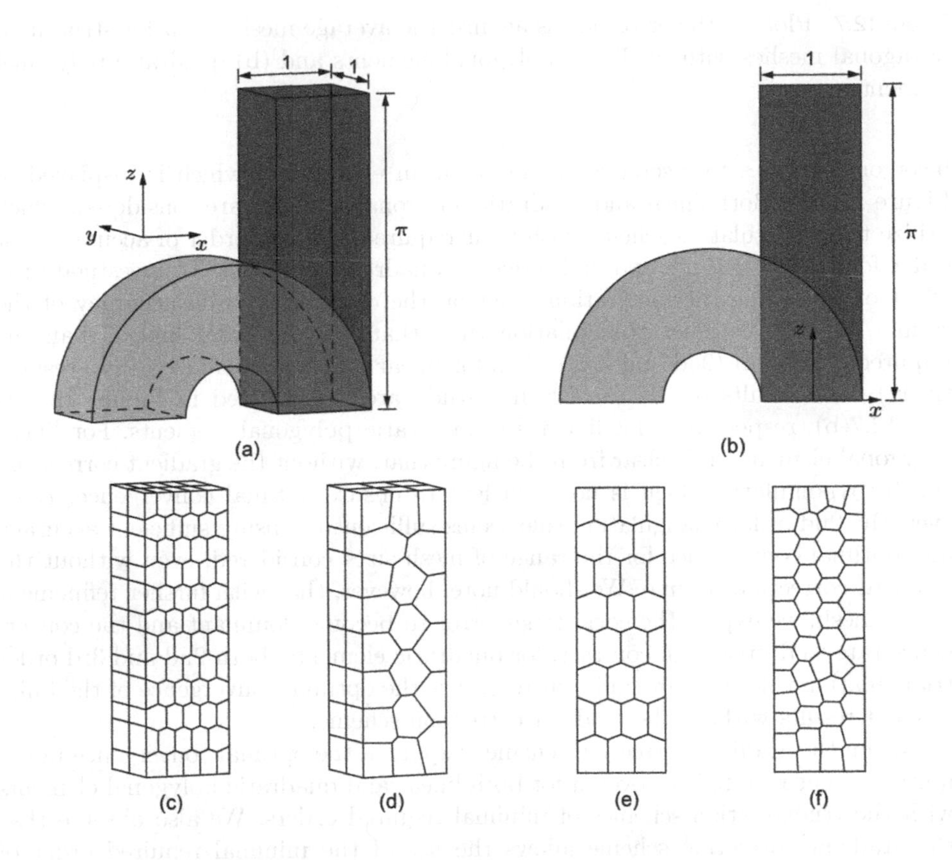

Figure 12.6 (a) An illustration of the 3D boundary-value problem. (b) Associated simplified 2D plane strain problem of (a). (c) An example of the extruded structured hexagonal mesh consists of 120 elements. (d) An example of the extruded CVT mesh with 40 elements. (e) An example of the 2D structured hexagonal mesh with 28 elements. (f) An example of the 2D CVT mesh with 50 elements.

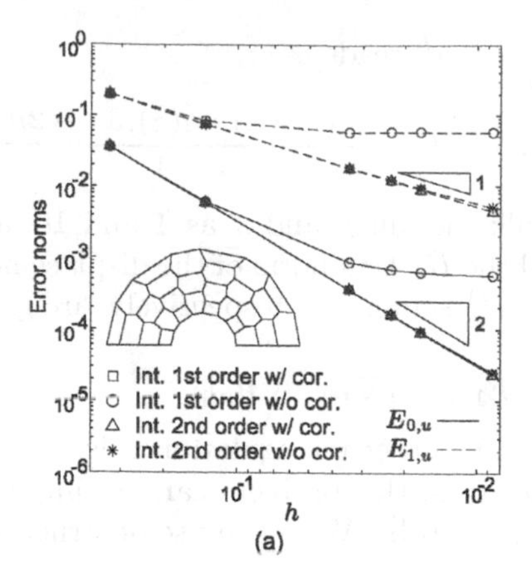

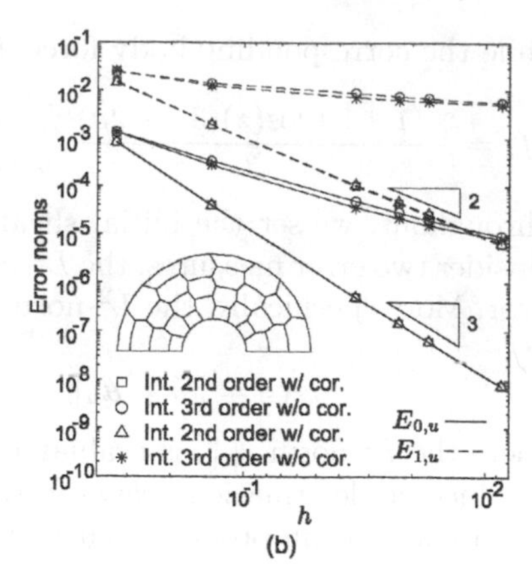

Figure 12.7 Plots of the error norms against the average mesh size h for structured hexagonal meshes with (a) linear polygonal elements and (b) quadratic polygonal elements.

hexagonal meshes to discretize the domain, an example of which is displayed in Figure 12.6(e). Both linear and quadratic polygonal elements are considered, which utilize the triangulation scheme with their required minimal order of accuracy (1st order for linear elements and 2nd order for quadratic element). To investigate the effect of increasing the integration order on the convergence and accuracy of the results, we also consider triangulation rules that are one order higher than the required minimum (2nd and 3rd orders for linear and quadratic elements, respectively). The results of the convergence study are summarized in Figure 12.7(a) and 12.7(b), respectively, for linear and quadratic polygonal elements. For linear polygonal elements, it is clear from the figure that, without the gradient correction, the 1st triangulation itself is not sufficient to ensure optimal convergence. However, the 2nd order triangulation rule seems sufficient to ensure sufficient accuracy and optimal convergence for the range of mesh sizes considered, even without the gradient correction scheme. We should note, however, that with further refinement of the mesh, we expect the consistency error to become dominant and the convergence rate to decrease. In contrast, for quadratic elements, both 2nd and 3rd order triangulation rules are not sufficient to ensure the optimal convergence of the finite element results without the gradient correction scheme.

When the gradient correction scheme is applied, the optimal convergence in the finite element results is recovered for both linear and quadratic polygonal elements with the triangulation schemes of minimal required orders. We also observe that the gradient correction scheme allows the use of the minimal required order of integration to achieve the same level of accuracy as with higher order integrations. If the gradient correction scheme is used, the displacement errors are identical for 1st and 2nd order triangulation rules for the linear polygonal elements and for 2nd and 3rd triangulation rules for the quadratic polygonal elements. This indicates

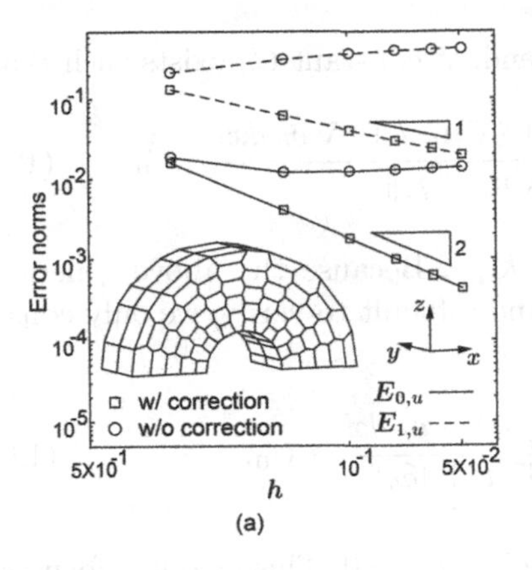

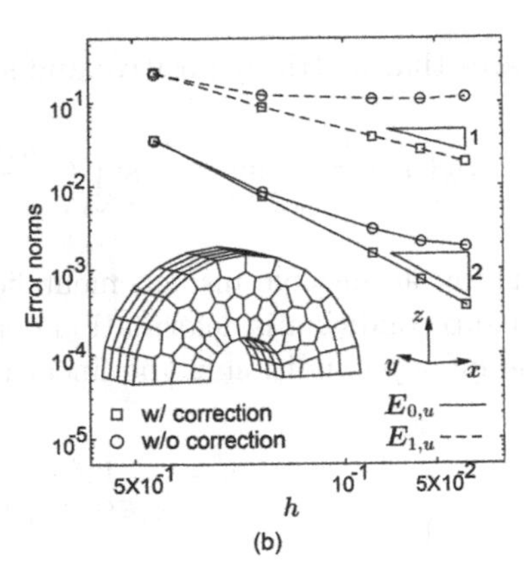

Figure 12.8 Plots of the error norms against the average mesh size h for (a) extruded structured hexagonal meshes and (b) extruded CVT meshes.

that the gradient corrections scheme can allow for a more efficient finite element implementation without sacrificing accuracy.

We also assess convergence for displacement-based polyhedral computations. To this end, we take the original 3D problem shown in Figure 12.6(a) and consider two types of polyhedral meshes, extruded structured hexagonal meshes and extruded centroid Voronoi tessellation (CVT) meshes, as shown in Figures 12.6(c) and 12.6(d), respectively. The convergence results for the displacement error norms as functions of the average mesh size are depicted in Figure 12.8. Each data point for the extruded CVT meshes is the mean of a five-mesh set. Again, the preliminary study indicates that the gradient correction scheme can ensure the optimal convergence for polyhedral elements.

12.7.2 Two-field mixed polygonal elements

In this subsection, together with the gradient correction scheme, the performance of mixed polygonal elements on inf-sup stability and convergence are numerically evaluated for the 2D case. In addition to the displacement error norms considered in the previous subsection, we define the L^2-norm of the pressure error

$$E_{0,\widehat{p}} = \|\widehat{p}_h - \widehat{p}\|,$$

which is evaluated using a fifth order quadrature scheme.

Inf-sup stability of polygonal discretizations

The satisfaction of the inf-sup stability condition is crucial to guarantee the convergence of mixed finite elements [58, 235, 293]. For finite elasticity problems, inf-sup stability is formally defined by the generalized inf-sup condition [235, 293], which states that a strictly positive and size-independent constant C_0 exists such that

$$\beta_h(\boldsymbol{u}_h) = \inf_{\widehat{q}_h \in \mathcal{Q}^{D \text{ or } C}_{h,k-1}} \sup_{\boldsymbol{v}_h \in \mathcal{K}_{h,k}} \frac{\int_\Omega \widehat{q}_h \, \mathrm{adj}\big(I + (\nabla \boldsymbol{u}_h)^T\big) : \nabla \boldsymbol{v}_h \, d\boldsymbol{x}}{\|\nabla \boldsymbol{v}_h\| \, \|\widehat{q}_h\|} \geq C_0. \tag{12.17}$$

for any admissible displacement field $\boldsymbol{u}_h \in \mathcal{K}_{h,k}$. Because the above generalized inf-sup condition is deformation dependent and difficult to verify, we only consider the inf-sup condition in the linear elasticity context,

$$\beta_h^0 = \inf_{\widehat{q}_h \in \mathcal{Q}^{D \text{ or } C}_{h,k-1}} \sup_{\boldsymbol{v}_h \in \mathcal{K}_{h,k}} \frac{\int_\Omega \widehat{q}_h \nabla \cdot \boldsymbol{v}_h \, d\boldsymbol{x}}{\|\nabla \boldsymbol{v}_h\| \, \|\widehat{q}_h\|} \geq C_0, \tag{12.18}$$

which essentially is a special case of (12.17) when $\boldsymbol{u}_h = 0$. This example focuses on the inf-sup stability of polygonal discretizations in 2D (similar to Figure 12.6(b)). It is well known that most of the lower order standard mixed finite elements in 2D are not inf-sup stable, whereas Beirão da Veiga et al. [30] have derived a geometrical condition to guarantee the satisfaction of (12.18), if every internal vertex node in the mesh is connected to no more than three edges. However, the analogous condition for polygonal discretizations with higher order mixed polygonal elements is still unknown. For 3D problems, even with lower order displacement and piecewise-constant pressure polyhedral elements, the inf-sup stability of polyhedral meshes is still an open question and needs further investigation. Here, we adopt the so called inf-sup test [88] to numerically evaluate the inf-sup stability of the linear and quadratic polygonal elements on several polygonal discretizations. We remark that while passing the inf-sup test only constitutes a necessary condition for the satisfaction of the inf-sup condition, the inf-sup test has been shown to reliably predict the stability properties of many well-known mixed finite elements [88, 391]. We consider a unit square domain with imposed boundary conditions as shown in Figure 12.9(a) and consider three Voronoi-type discretizations: structured hexagonal meshes, CVT meshes, and random Voronoi meshes, examples of which are provided in Figure 12.9(b)–(d). In general, the Voronoi-type discretizations with lower order elements satisfy the geometrical condition of Beirão da Veiga et al. [30], except for several limiting special cases such as the one where the Voronoi seeds are aligned in a Cartesian grid, forming a Cartesian mesh. In the inf-sup test, we compute the stability index β_h^0 as a function of the average mesh size h for all the families of discretizations considered in Figure 12.9(e)–(g). Each point for the CVT and random Voronoi meshes represents an average from five results. The figures indicate that all three polygonal discretizations are inf-sup stable with both discontinuous and continuous pressure approximations, at least in the linear elasticity regime. In comparison, while the standard linear and quadratic mixed elements with continuous pressure approximations (the T–H family) are unconditionally stable [58], most of those with discontinuous pressure approximations (the C-R family), such as lower-order mixed triangular and quadrilateral elements, are inf-sup unstable [58, 88, 204].

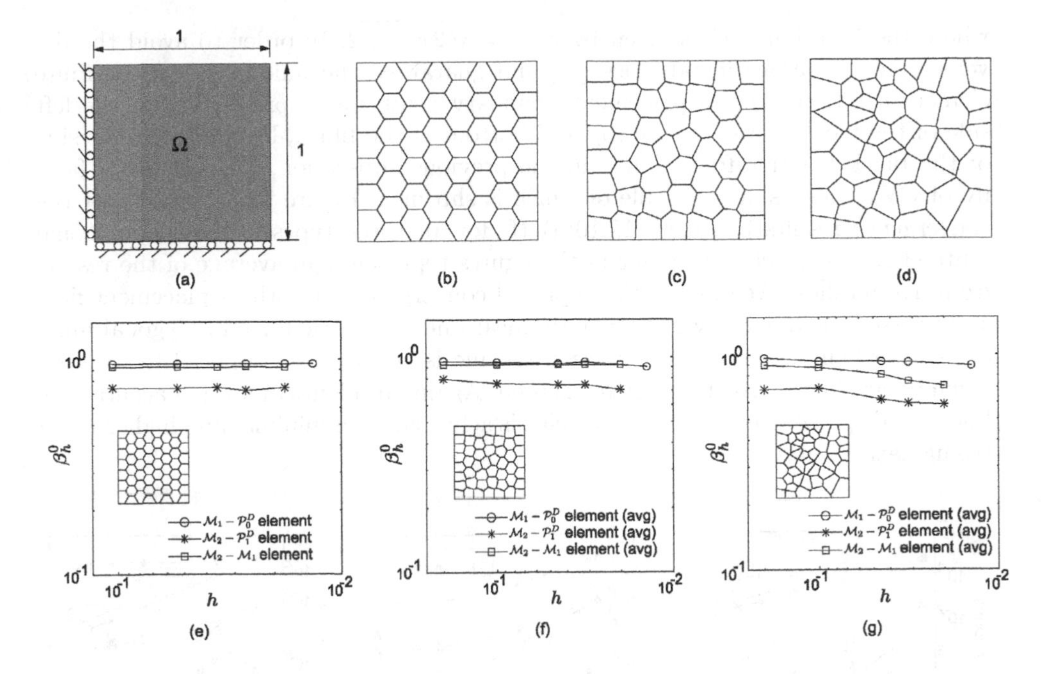

Figure 12.9 (a) Dimensions and boundary conditions adopted for the inf-sup test. (b) An example of the structured hexagonal dominant mesh with 56 elements. (c) An example of the randomly generated CVT mesh with 50 elements. (d) An example of the random Voronoi mesh with 50 elements. (e) Plot of the computed value of the stability index as a function of the average mesh size h for structured hexagonal dominant meshes. (f) Plot of the computed value of the stability index as a function of the average mesh size h for CVT meshes. (g) Plot of the computed value of the stability index as a function of the average mesh size h for random Voronoi meshes. This figure is adapted from [100].

Convergence study

Next we perform a convergence study of the two-field mixed elements through a 2D boundary-value problem, where an incompressible rectangular block with dimensions 1×1 is bent into a semicircle, as illustrated by Figure 12.6(b). The closed-form displacement and pressure solutions for such a problem read as

$$u_x = -r(-0.5) + r(x)\cos(y) - 0.5 - x, \qquad u_y = r(x)\sin(y) - y,$$

and

$$\widehat{p} = \frac{2\mu(4x^2 - 3)x - \sqrt{2}}{8x^2 - 4},$$

where the function $r(x)$ is given by $r(x) = \sqrt{2x + \sqrt{2}}$. In order to avoid the development of surface instabilities [44] and guarantee the uniqueness of the finite element solutions, the displacement boundary condition is prescribed on the left side of the block as well as the top and bottom of the block, leaving the right side of the block traction-free. Unlike in the previous subsection, we consider a family of CVT meshes, an example of which is shown in Figure 12.6(f), and plot the convergence results in Figure 12.10(a)–(f) for all three types of mixed polygonal finite elements. Each data point in the figures represents an average of the results from five meshes. We observe that optimal convergence in both displacement field and pressure field is recovered for both linear and quadratic mixed polygonal finite elements when the gradient correction scheme is adopted, as opposed to the case when no gradient correction scheme is used. Again, an identical level of accuracy in both the displacement and pressure fields is achieved with minimal and higher-order triangulation rules.

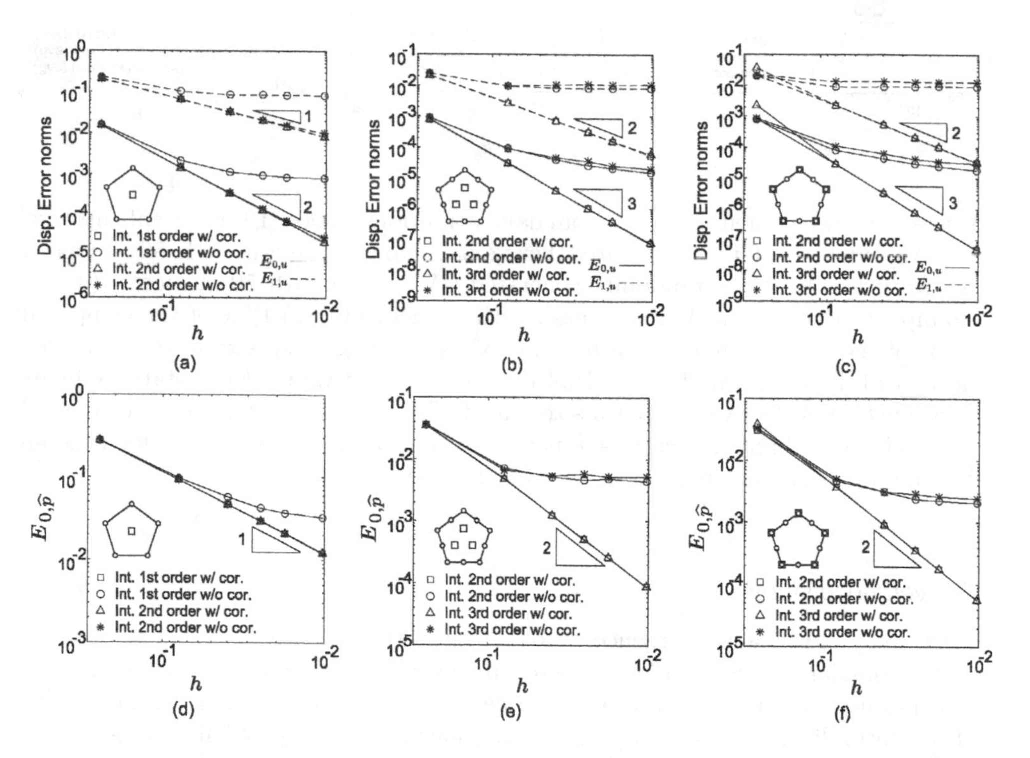

Figure 12.10 The displacement errors for the CVT meshes with (a) $\mathcal{M}_1 - \mathcal{P}_0^D$ elements, (b) $\mathcal{M}_2 - \mathcal{P}_1^D$ elements, and (c) $\mathcal{M}_2 - \mathcal{M}_1$ elements. The pressure errors for the CVT meshes with (d) $\mathcal{M}_1 - \mathcal{P}_0^D$ elements, (e) $\mathcal{M}_2 - \mathcal{P}_1^D$ elements, and (f) $\mathcal{M}_2 - \mathcal{M}_1$ elements.

12.8 APPLICATION TO THE STUDY OF FILLED ELASTOMERS

Experiments have demonstrated that typical filled elastomers contain stiff "interphases" or layers of stiff "bound rubber" around their inclusions [236]. The presence of such interphasial regions can significantly affect the macroscopic response of the filled elastomer when the fillers are submicron in size [175, 260]. In this section, the nonlinear elastic response of an incompressible elastomer reinforced with a random, isotropic distribution of circular rigid particles, bounded through finite-size interphases, is studied by means of polygonal finite elements. We aim at demonstrating the ability of polygonal finite elements to model the complex behavior of nonlinear elastic materials with extremely large localized deformations. Throughout this section, we consider each particle to be infinitely rigid and model them with the variational formulation proposed by Chi et al. [98]. This variational formulation treats the presence of each particle as a set of kinematic constraints on the displacement DOFs.

We approximate the truly random and isotropic distribution of particles by the periodic repetition of a unit cell in the $\boldsymbol{e}_1$ and $\boldsymbol{e}_2$ directions. The unit cell contains a total of 50 monodisperse rigid particles at an area fraction of $c_p = 25\%$. Each particle is bounded to the matrix by an interphase that is ten times stiffer than the matrix and whose thickness is assumed to be 20% of the radius of the particle, resulting in a total area fraction of $c_i = 11\%$. The mixed $\overline{\boldsymbol{F}}$-formulation is employed in this example and polygonal meshes with both linear and quadratic mixed elements are considered as shown in Figure 12.11(c) and (d). For comparison purposes, a finite element mesh with the commonly used 6-node hybrid quadratic triangular elements (termed CPE6MH) is included, as depicted in Figure 12.11(b). In order to make a fair comparison, all the meshes are generated such that they have a similar number of nodes in the matrix phase.

In the following study, we consider periodic boundary conditions [100, 260] for two loading conditions: (i) pure shear with average deformation gradient $\langle \boldsymbol{F} \rangle = \lambda \boldsymbol{e}_1 \otimes \boldsymbol{e}_1 + \lambda^{-1} \boldsymbol{e}_2 \otimes \boldsymbol{e}_2$ and (ii) simple shear with $\langle \boldsymbol{F} \rangle = \boldsymbol{I} + \gamma \boldsymbol{e}_1 \otimes \boldsymbol{e}_2$ where $\lambda \geq 1$ and $\gamma \geq 0$ denote the applied stretch and amount of shear, respectively. We remark that the periodicity in the pressure field cannot in general be enforced when employing mixed finite elements. At any rate, even when we use the T–H mixed elements where the pressure field is continuous, as shown in [100], the enforcement of periodic pressure boundary conditions only exerts a negligible influence on the solution. Moreover, the discontinuities of shear moduli between the matrix, interphase, and inclusions generate a discontinuous pressure field, which, in general, cannot be captured by mixed elements with continuous approximations for the pressure field, i.e., T–H elements. However, [100] has shown that using T-H elements (with minor levels of discontinuities (e.g., $\mu_i = 10\mu_m$) between the matrix and interphase) can still produce sufficiently accurate solutions, provided that the stiff inclusions are treated independently using the rigid-body constraint formulation by Chi et al. [98].

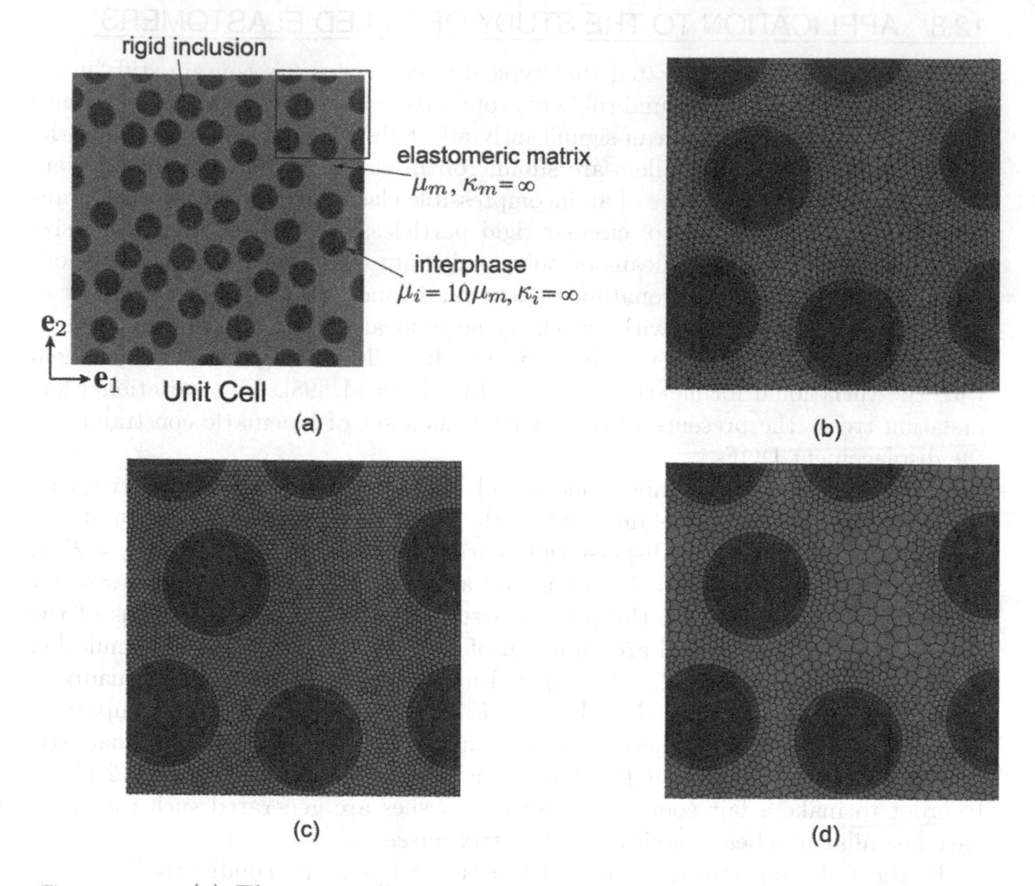

Figure 12.11 (a) The unit cell considered in the problem. (b) The quadratic triangular mesh comprised of 58,814 CPE6MH elements and 118,141 nodes in total, with 71,179 nodes in the matrix phase. (c) The linear polygonal mesh comprised of 39,738 elements and 76,020 nodes in total, with 66,095 nodes in the matrix phase. (d) The quadratic polygonal mesh comprised of 20,205 elements and 95,233 nodes in total, with 67,325 nodes in the matrix phase. This figure is adapted from [100]. See color insert.

12.8.1 Results for filled neo-Hookean elastomers

In this subsection, the matrix is assumed to be an incompressible neo-Hookean material, described by the stored-energy function (12.16). The initial shear modulus μ_m of the matrix is taken to be $\mu_m = 1$ MPa, and that of the interphase is set as $\mu_i = 10\,\mu_m = 10$ MPa. Figure 12.12(a)–(d) plots the deformed configurations of the unit cell obtained by $\mathcal{M}_2 - \mathcal{M}_1$, $\mathcal{M}_2 - \mathcal{P}_1^D$, $\mathcal{M}_1 - \mathcal{P}_0^D$, and CPE6MH elements, at their respective maximum global stretches, $\lambda_{\max} = 2.9132, 2.6456, 2.2515$ and 1.4308, under loading condition (i). The fringe scales in the deformed configurations correspond to the maximum principal stretch of each element (representing the level of local deformation), with those having a value of 8 or larger plotted in red. The maximum global and local stretches reached by each mesh are also summarized in Table 12.1. As an additional quantitative comparison, the relevant components of the macroscopic first Piola–Kirchhoff stress $\langle \boldsymbol{P} \rangle$ as functions of the applied global stretch λ or shear γ are shown in Figure 12.13(a)–(b) for loading conditions (i) and (ii). The simulated macroscopic responses are compared with the available analytical solution in the literature for neo-Hookean elastomers, reinforced with a random and isotropic distribution of polydisperse circular particles, which considers the interphasial effect [175, 258].

Several observations and conclusions can be made. First, the macroscopic responses obtained by all three types of meshes under both loading conditions (i) and (ii) agree well with the analytical solutions. Second, the results from the polygonal meshes reach significantly larger global deformation states than the ones from the quadratic triangular mesh. Since the effective volume fraction of the filled elastomer considered in this example is high ($c_p + c_i = 36\%$), resulting in extremely heterogeneous and large localized deformation fields, a remeshing strategy is likely to be ineffective in stretching the triangular mesh farther [284]. Third, the quadratic polygonal meshes exhibit better performance than linear meshes, as they can reach larger global deformation states, especially under loading condition (i). Partially, this is because the quadratic polygonal elements have richer approximating capabilities, allowing them to better capture the curvatures in the highly stretched regions (for example, see the region in Figure 12.12). Furthermore, while producing almost identical macroscopic responses, the quadratic polygonal mesh with the $\mathcal{M}_2 - \mathcal{M}_1$ element in this example appears to reach a larger global stretch than the one with $\mathcal{M}_2 - \mathcal{P}_1^D$ elements under loading condition (i) ($\lambda_{\max} = 2.91$ as compared to $\lambda_{\max} = 2.65$).

12.8.2 Results for a filled silicone elastomer

This subsection studies the large deformation response of a filled elastomer with both its matrix and interphase being a typical silicone rubber, characterized by the

	Max. global deformation		Max. local deformation	
	Pure shear	Simple shear	Pure shear	Simple shear
$\mathcal{M}_1 - \mathcal{P}_0^D$ elements	2.9132	1.086	13.8016	6.4693
$\mathcal{M}_2 - \mathcal{P}_1^D$ elements	2.6456	1.091	11.1457	6.6314
$\mathcal{M}_2 - \mathcal{M}_1$ elements	2.2615	1.078	10.2077	6.9639
CPE6MH elements	1.4308	0.622	4.4819	3.6324

Table 12.1 Comparison of maximum global deformation and the maximum local deformation (maximum principal stretch) in the final deformed configurations obtained by different types of elements. Both matrix and interphase are neo-Hookean materials.

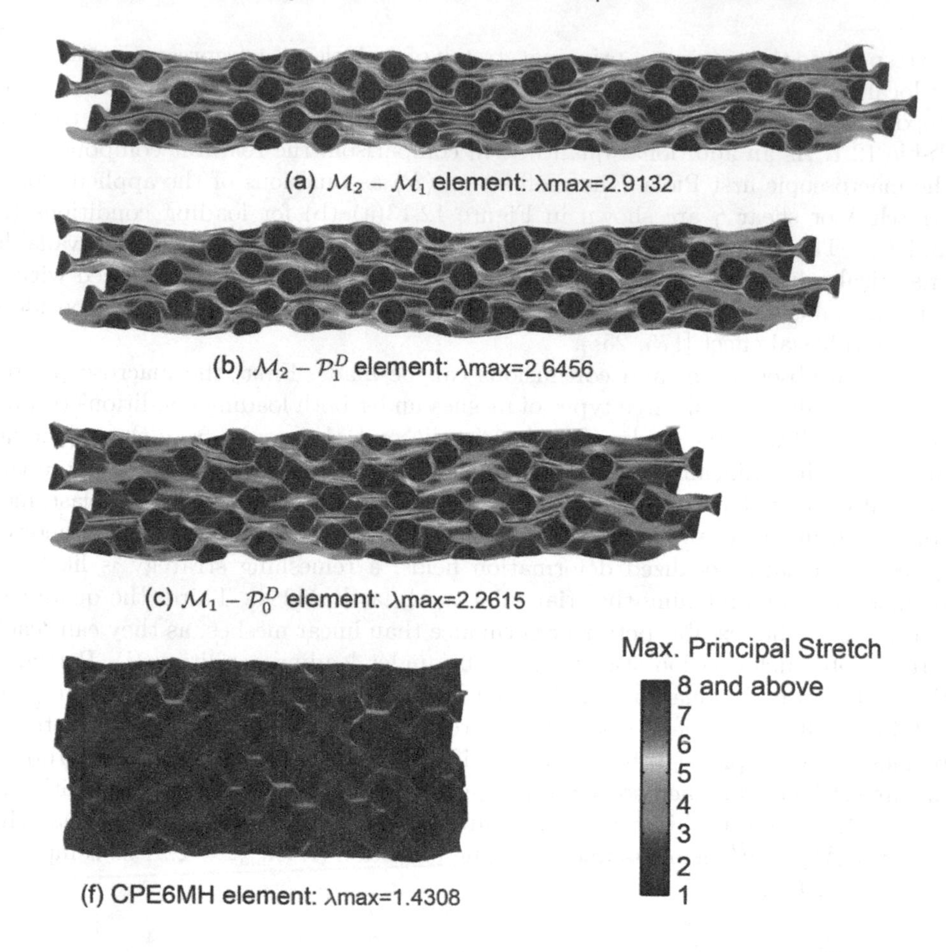

Figure 12.12 For loading condition (i), displaying pure shear with average deformation gradient $\langle \boldsymbol{F} \rangle = \lambda \boldsymbol{e}_1 \otimes \boldsymbol{e}_1 + \lambda^{-1} \boldsymbol{e}_2 \otimes \boldsymbol{e}_2$, $\lambda \geq 1$, the figure illustrates the final deformed configuration reached by (a) $\mathcal{M}_2 - \mathcal{M}_1$ elements, (b) $\mathcal{M}_2 - \mathcal{P}_1$ elements, (c) $\mathcal{M}_1 - \mathcal{P}_0$ elements, and (d) CPE6HM elements (solved in ABAQUS). This figure is adapted from [100]. See color insert.

following stored-energy function [259]

$$W(\boldsymbol{F}) = \begin{cases} \dfrac{3^{1-\alpha_1}}{2\alpha_1}\mu_1[I_1^{\alpha_1} - 3^{\alpha_1}] + \dfrac{3^{1-\alpha_2}}{2\alpha_2}\mu_2[I_1^{\alpha_2} - 3^{\alpha_2}], & \text{if } \det \boldsymbol{F} = 1, \\ +\infty, & \text{otherwise,} \end{cases}$$

with $I_1 = F_{ij}F_{ij}$, $\mu_1 = 0.032$ MPa, $\mu_2 = 0.3$ MPa, $\alpha_1 = 3.837$, $\alpha_2 = 0.559$. Thus, the initial shear modulus of the matrix is $\mu_m = \mu_1 + \mu_2 = 0.332$ MPa, and that of the interphase is $\mu_i = 3.32$ MPa. Again, both loading conditions (i) and (ii) are considered.

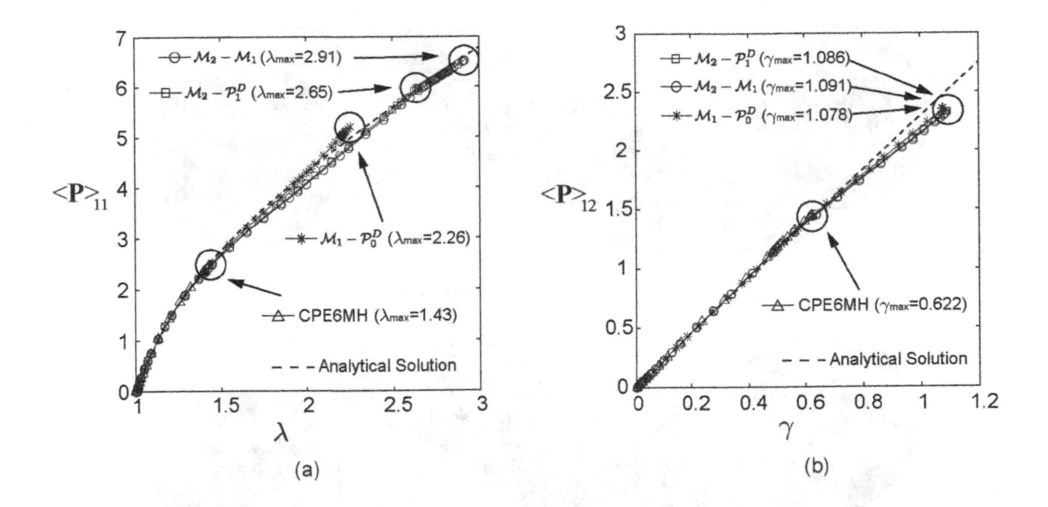

Figure 12.13 Plots of the macroscopic Piola–Kirchhoff stress function as the applied stretch/shear for different types of elements under (a) pure shear ($\lambda \geq 1$) and (b) simple shear.

Figure 12.14(a)–(d) shows the final deformed configurations of the unit cell obtained by $\mathcal{M}_2-\mathcal{M}_1$, $\mathcal{M}_2-\mathcal{P}_1^D$, $\mathcal{M}_1-\mathcal{P}_0^D$, and CPE6MH elements under loading condition (ii) at their respective maximum global deformation states. The color scale in each configuration corresponds to the maximum principal stretch of each element, with those having a value of 4 and above plotted in red. The maximum global and local stretches reached by each mesh are summarized in Table 12.2. Furthermore, the macroscopic responses predicted by each type of element under both loading conditions (i) and (ii) are plotted in Figure 12.15(a) and (b) as functions of applied global stretch λ or shear γ. Similar to the preceding subsection, we conclude that polygonal discretizations (both linear and quadratic elements)

	Max. global deformation		Max. local deformation	
	Pure shear	Simple shear	Pure shear	Simple shear
$\mathcal{M}_1-\mathcal{P}_0^D$ elements	1.911	1.482	4.256	4.545
$\mathcal{M}_2-\mathcal{P}_1^D$ elements	2.064	1.548	4.543	4.562
$\mathcal{M}_2-\mathcal{M}_1$ elements	1.910	1.551	4.257	4.799
CPE6MH elements	1.349	0.595	2.777	2.699

Table 12.2 Comparison of maximum global deformation and the maximum local deformation (maximum principal stretch) in the final deformed configurations obtained by different types of elements. Both matrix and interphase are considered to be typical silicone rubber.

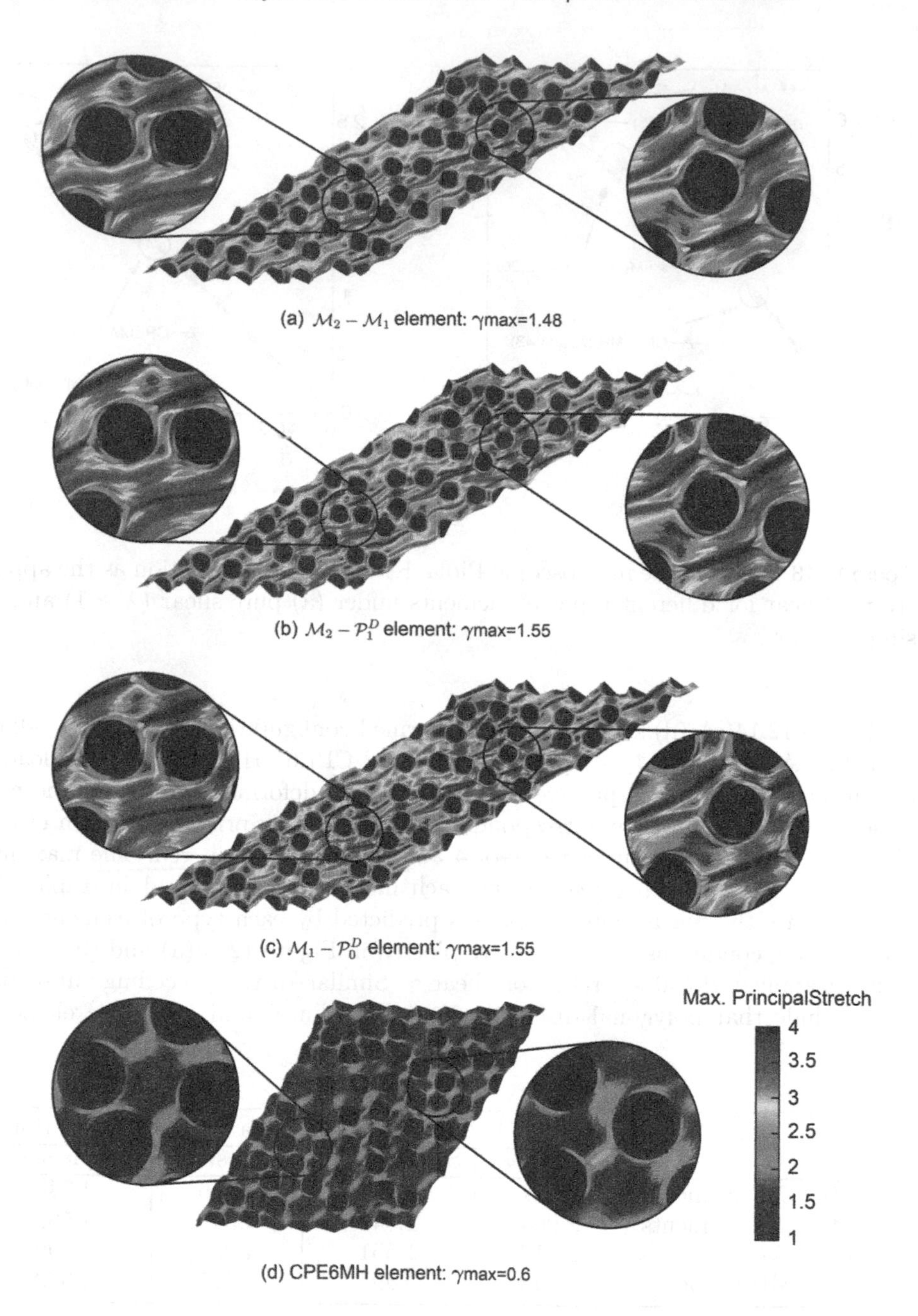

Figure 12.14 For loading condition (ii), displaying $\langle \boldsymbol{F} \rangle = \boldsymbol{I} + \gamma \boldsymbol{e}_1 \otimes \boldsymbol{e}_2$, $\gamma \geq 0$, the figure illustrates the final deformed configurations reached by (a) $\mathcal{M}_2 - \mathcal{M}_1$ elements, (b) $\mathcal{M}_2 - \mathcal{P}_1^D$ elements, (c) $\mathcal{M}_1 - \mathcal{P}_0^D$ elements, and (d) CPE6HM elements (solved in ABAQUS). Both matrix and interphase are considered to be typical silicone rubber. See color insert.

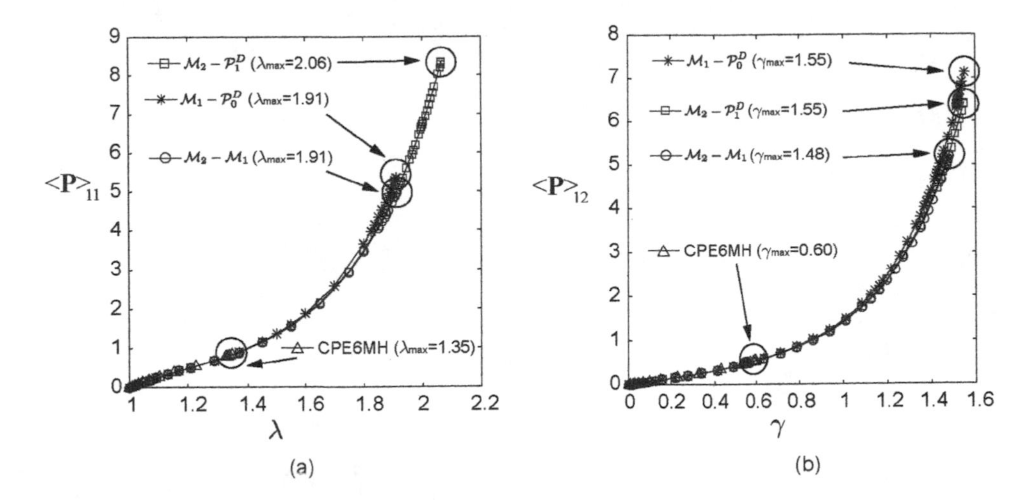

Figure 12.15 Plots of the macroscopic first Piola–Kirchhoff stress as functions of the applied stretch/shear for different types of elements under (a) pure shear ($\lambda \geq 1$) and (b) simple shear. Both matrix and interphase are considered to be typical silicone rubber.

are capable of reaching much larger stretches at both global and local levels than quadratic triangle elements under loading conditions (i) and (ii). However, unlike in the preceding subsection, the quadratic polygonal mesh with $\mathcal{M}_2 - \mathcal{P}_1^D$ elements reaches a larger global stretch than the one with $\mathcal{M}_2 - \mathcal{M}_1$ elements in terms of the maximum global stretch/shear reached. This difference in results indicates that the comparison of performance of the $\mathcal{M}_2 - \mathcal{P}_1^D$ element and $\mathcal{M}_2 - \mathcal{M}_1$ element may be problem-dependent.

CHAPTER 13

Maximum-Entropy Meshfree Coordinates in Computational Mechanics

Marino Arroyo

Universitat Politècnica de Catalunya, Barcelona, Spain

CONTENTS

CONVENTIONAL BARYCENTRIC COORDINATES are defined relative to vertices defining the boundary of a polytope. Here, we develop barycentric coordinates relative to a cloud of vertices sampling not only the boundary, but also the interior of a region in $\mathbb{R}^d$, with the objective of using these barycentric coordinates as basis functions to approximate partial differential equations or parametrize surfaces. We show that entropy maximization provides a rational way to define smooth barycentric coordinates, but for the resulting basis functions to be localized, and hence lead to sparse matrices in computational mechanics, entropy maximization needs to be biased by a suitable notion of locality. The basis functions that result from this approach are smooth, reproduce polynomials, are localized around their corresponding vertex, and their definition does not rely on an underlying mesh or grid but rather on less structured neighborhood relations between vertices. Thus, they can be viewed as meshfree basis functions [159, 245, 380].

The theory and practical evaluation behind these basis functions is reviewed next, and selected applications in computational mechanics are presented.

13.1 INTRODUCTION

A standard approach to numerically representing a function u over a domain $\Omega \subset \mathbb{R}^d$ is to expand it in terms of a finite set of basis functions, that is

$$u^h(\boldsymbol{x}) = \sum_{i=1}^{n} \phi_i(\boldsymbol{x}) u_i, \tag{13.1}$$

where $\phi_i : \bar{\Omega} \to \mathbb{R}$ is the i-th basis function and u_i its corresponding coefficient. Such a representation is the basis of Galerkin methods to solve partial differential equations (PDEs), see [204]. Here, rather than choosing *a priori* the set of basis functions, such as piecewise polynomials defined over a mesh, we initially leave their definition open, and then motivate and make explicit the specific choices leading to maximum-entropy meshfree basis functions.

To show convergence of numerical solutions obtained by a Galerkin method to the exact solution, the basis functions need to satisfy the so-called reproducibility conditions. These conditions ensure that polynomials up to degree p are exactly represented by the basis functions. For second-order PDEs such as the heat equation and the most common systems arising in fluid and solid mechanics, consistency conditions up to $p = 1$ are required at all points $\boldsymbol{x} \in \Omega$, namely

$$\sum_{i=1}^{n} \phi_i(\boldsymbol{x}) = 1, \tag{13.2a}$$

$$\sum_{i=1}^{n} \phi_i(\boldsymbol{x}) \boldsymbol{v}_i = \boldsymbol{x}, \tag{13.2b}$$

for some vector coefficients $\boldsymbol{v}_i \in \mathbb{R}^d$. If these conditions are met, then any affine function $w(\boldsymbol{x}) = \boldsymbol{a} \cdot \boldsymbol{x} + b$, where $\boldsymbol{a}$ is a vector in $\mathbb{R}^d$ and $\cdot$ denotes the scalar product, can be exactly represented in Ω as $w(\boldsymbol{x}) = \sum_{i=1}^{n} \phi_i(\boldsymbol{x})(\boldsymbol{a} \cdot \boldsymbol{v}_i + b)$.

Interpreting the coefficients $\boldsymbol{v}_i$ as the coordinates of points or vertices, we can naturally associate each basis function to a vertex, and view the set of vertices $V = \{\boldsymbol{v}_1, \boldsymbol{v}_2, \ldots, \boldsymbol{v}_n\}$ as points sampling the domain Ω. With this interpretation, we only need to require the basis functions to be non-negative,

$$\phi_i(\boldsymbol{x}) \geq 0 \quad \forall \boldsymbol{x} \in \Omega, \; i = 1, 2, \ldots, n \tag{13.3}$$

and recall (13.2) to realize that the functions $\phi_i(\boldsymbol{x})$, $i = 1, 2, \ldots, n$ define a set of generalized barycentric coordinates relative to V. A difference with respect to previous chapters is that now we allow for interior vertices to the domain, and for multiple vertices along a face of the domain.

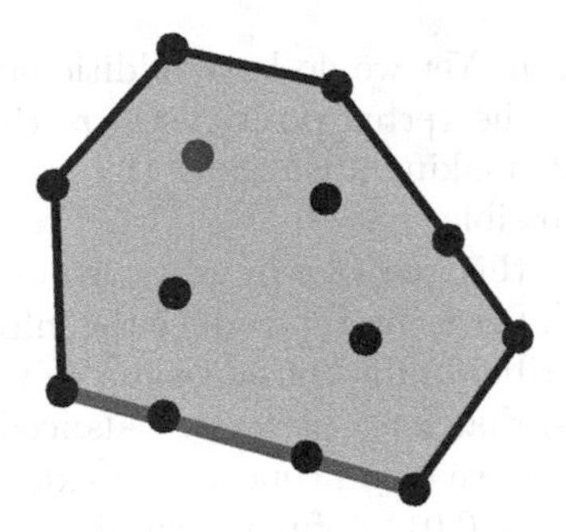

Figure 13.1 Set of vertices V in two dimensions, along with the convex hull P. See color insert.

Because generalized barycentric coordinates at $\boldsymbol{x}$ are coefficients of a convex combination of the vertices resulting in $\boldsymbol{x}$, see (13.2b), it immediately follows that such basis functions can only be defined in the convex hull of V, and therefore $\Omega \subset \text{conv } V$. Furthermore, exploiting elementary facts of convex geometry, it is possible to characterize the behavior of *convex approximation schemes* of V, that is, barycentric coordinates defined in the vertex set V, on the boundary of the polytope $P = \text{conv } V$ [15]. In particular, it can be shown that if F denotes a face of P in the sense of convex geometry (in $\mathbb{R}^3$ 0-dimensional faces are extreme points, 1-dimensional faces are edges, and 2-dimensional faces are proper faces) and $\boldsymbol{v}_i \notin F$, then the corresponding basis function ϕ_i vanishes on F [15].

Immediate consequences of this fact are that: (1) basis functions of interior vertices (see Figure 13.1) vanish on the boundary of P, (2) basis functions of extreme points $\boldsymbol{v}_i$ of P satisfy the Kronecker-delta property, $\phi_i(\boldsymbol{v}_j) = \delta_{ij}$, and (3) on a given face F (in Figure 13.1), only the basis functions of vertices lying on F are non-zero in F. As a result of (3), if we denote by I_F the set of indices of nodes lying on F, then (13.2a,13.2b) become $\sum_{i \in I_F} \phi_i(\boldsymbol{x}) = 1$ and $\sum_{i \in I_F} \phi_i(\boldsymbol{x}) \boldsymbol{v}_i = \boldsymbol{x}$ for all $\boldsymbol{x} \in F$, that is, the basis functions of nodes on a face maintain the full approximation properties on that face. Furthermore, (3) also implies that the basis functions on a face can be defined without reference to information in the higher-dimensional ambient space. These properties can be referred to as a weak Kronecker-delta property, which facilitate the imposition of boundary conditions in the numerical approximation of PDEs.

13.2 SELECTING BARYCENTRIC COORDINATES THROUGH ENTROPY MAXIMIZATION

On examining (13.2a) and (13.3), it is clear that for each point $\boldsymbol{x} \in P$, the barycentric coordinates $\{\phi_1(\boldsymbol{x}), \phi_2(\boldsymbol{x}), \ldots, \phi_n(\boldsymbol{x})\}$ can be viewed as a discrete probability distribution for n events. Defining the barycentric coordinates then becomes a problem of statistical inference. In the absence of additional information and if one tries to avoid any bias, Laplace's principle of insufficient reason would suggest selecting

$\phi_i(\boldsymbol{x}) = 1/n$ for $i = 1, \ldots, n$. Yet we do have additional information, since (13.2b) states that the average of the vertex positions is $\boldsymbol{x}$. How can we incorporate this piece of information, while making a choice of the probabilities (barycentric coordinates) as unbiased as possible?

One classical answer to this question in information theory is Jaynes's principle of maximum entropy [211]. Let us first introduce the information entropy associated to a finite probability distribution by considering a coin toss. In this case, there are two events (heads or tails) that for a perfectly balanced coin have equal probabilities, $p_1 = p_2 = 1/2$. Instead, we could conceive an extremely unbalanced coin with probabilities $p_1 = 0.99$, $p_2 = 0.01$. A fundamental question in information theory is how to quantify the uncertainty of a probability distribution, or in other words, how to quantify the amount of information gained by realizing a coin toss. It is obvious that the balanced coin leads to a much more uncertain outcome, whereas, for the unbalanced one, we will learn very little by throwing it since we know that the outcome will most likely be heads. It can be shown that a canonical measure of uncertainty or information is given by the Shannon entropy function,

$$H(p_1, p_2, \ldots, p_n) = -\sum_{i=1}^{n} p_i \log p_i, \tag{13.4}$$

where the function $x \log x$ is extended by continuity at zero, i.e., $0 \log 0 = 0$. See [223] for a detailed motivation of this measure of information and its properties. By evaluating Shannon entropy on the previous two probability distributions, we find that the first coin is about 12 times more uncertain than the second one. In fact, it can be shown that the function in (13.4) is strictly concave with a maximum at the uniform distribution $p_i = 1/n$, $i = 1, 2, \ldots, n$. This last fact shows that maximizing the entropy is consistent with Laplace's principle of insufficient reason.

Let us now introduce Jaynes's principle using another simple example, that of a dice with six faces, $A_1, A_2, \ldots, A_6$. Consider the random function that assigns to a given face its numerical value, $f(A_i) = i$. For a perfectly balanced dice, the average of f is $\sum_{i=1}^{6} i/6 = 7/2$. Suppose, however, that an accurate sample mean is computed for this quantity after a large number of trials, giving a result of 4. Clearly, this new piece of information is inconsistent with the uniform probabilities that we expect for a perfectly balanced dice. We would like to incorporate this information in the inference of the probabilities $p_1, p_2, \ldots, p_6$ associated to each face of the dice, but clearly it is not sufficient to uniquely determine these probabilities. Jaynes's recipe is to consider the most uncertain probability distribution, that is, the one that maximizes entropy, while being consistent with the known partial information [211]. Mathematically, this statement can be formulated as an optimization program with constraints

$$\max_{p_1,\ldots,p_6} \quad -\sum_{i=1}^{6} p_i \log p_i$$

$$\text{subject to} \quad p_i \geq 0, \quad \sum_{i=1}^{6} p_i = 1, \quad \sum_{i=1}^{6} i p_i = 4,$$

where the last constraint encodes the additional information about the system. The solution to this constrained optimization problem is $p_1 = 0.103$, $p_2 = 0.123$, $p_3 = 0.146$, $p_4 = 0.174$, $p_5 = 0.207$, and $p_6 = 0.247$, showing that the higher expected value for f biases the probabilities towards faces with a higher numerical value.

Going back to the problem of inferring the generalized barycentric coordinates, it is now clear that application of Jaynes's principle of maximum entropy leads to an optimization program at each point $\boldsymbol{x} \in P$ and is given by

$$\begin{aligned} &\max_{\phi_1,\ldots,\phi_n} \quad -\sum_{i=1}^{n} \phi_i \log \phi_i \\ &\text{subject to} \quad \phi_i \geq 0, \quad \sum_{i=1}^{n} \phi_i = 1, \quad \sum_{i=1}^{n} \phi_i \boldsymbol{v}_i = \boldsymbol{x}, \end{aligned} \tag{13.5}$$

where the dependence on $\boldsymbol{x}$ only enters through the last vectorial constraint. The solution to the program is the value of the barycentric coordinates selected by entropy maximization at $\boldsymbol{x}$, $\phi_i(\boldsymbol{x})$, $i = 1, 2, \ldots, n$. Because the entropy function is strictly concave, this program has a unique solution if and only if it is feasible, that is, whenever $\boldsymbol{x} \in \bar{P}$ [67]. This method, reviewed in Section 1.2.9, was introduced in [372] to generate basis functions for polygons.

Figure 13.2 shows the basis functions defined by entropy maximization for a set of vertices that not only defines a polygon, but also samples its interior and its faces. It can be observed that in agreement with the general results for convex approximation schemes mentioned earlier, the basis function associated to the extreme vertex $\boldsymbol{v}_1$ is 1 at $\boldsymbol{v}_1$, and therefore all other basis functions are zero at this point. Also in agreement with the general results, the basis function of the interior vertex $\boldsymbol{v}_3$ vanishes on the boundary of P, and the basis function ϕ_2 does not satisfy the Kronecker-delta property because $\boldsymbol{v}_2$ is on the boundary but is not an extreme point. Yet this set of basis functions presents a significant disadvantage when used to approximate functions in general, or solutions of PDEs in particular: the basis functions are completely nonlocal. The most dramatic case is that of basis functions of interior nodes, such as ϕ_3, which spread through the entire domain and are as uniform as allowed by the convex structure of the optimization program. This is not a desirable feature to approximate functions over P because with such a basis it is not possible to define a localized feature, say around vertex $\boldsymbol{v}_i$, and correlate this feature with the nodal value u_i in (13.1). Besides the lack of local resolution, such a global set of basis functions leads to a dense stiffness matrix when used in a Galerkin method to approximate partial differential equations [204].

This discussion suggests that in defining the generalized coordinates by mere entropy maximization, we have been "too unbiased." We have ignored the fundamental requirement of locality, by which the approximation in (13.1) at a given point $\boldsymbol{x}$ should more strongly depend on coefficients u_i associated to vertices close to $\boldsymbol{x}$. Local basis functions should concentrate their "mass" in the vicinity of their associate vertex, where their value should be closer to 1 than, for instance, ϕ_3 in Figure 13.2.

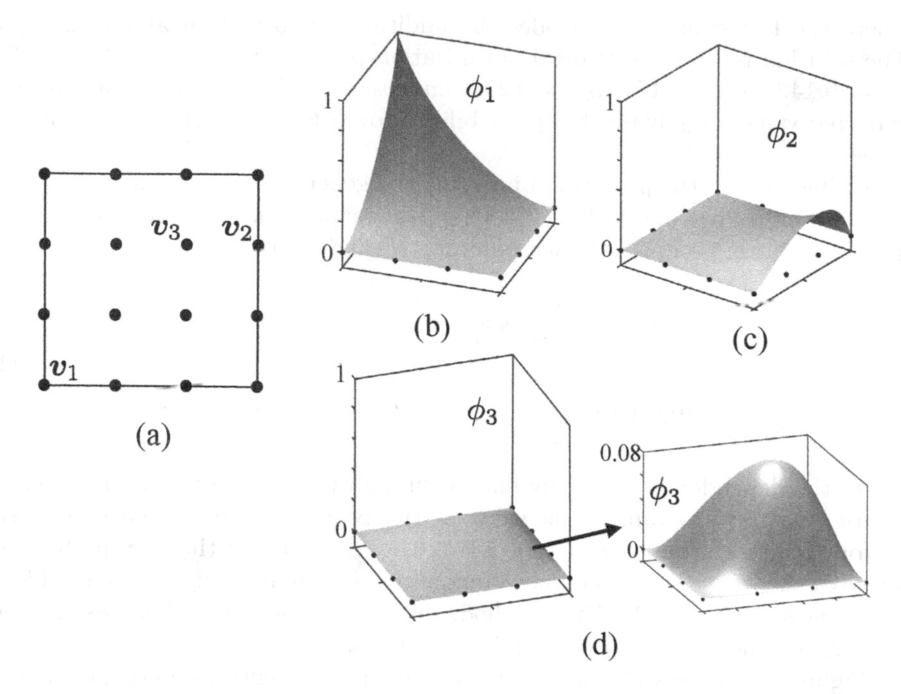

Figure 13.2 Basis functions selected by entropy maximization for the set of vertices shown in (a). Selected basis functions for vertices at extreme positions of P (b), along a face of P (c), and in the interior of P (d).

13.3 INTRODUCING LOCALITY: LOCAL MAXIMUM-ENTROPY APPROXIMANTS

Clearly, locality and entropy maximization are conflicting objectives. For a given point $\boldsymbol{x}$, locality implies that only a few basis functions (corresponding to vertices $\boldsymbol{x}_i$ that are closer to $\boldsymbol{x}$) have values significantly larger than 0, while others are close to 0. In contrast, maximizing entropy delivers $\phi_i(\boldsymbol{x})$ that are as uniform as possible. Interestingly, there is another family of generalized barycentric coordinates associated to V that can be characterized through an optimization program analogous to that in (13.5): the piecewise linear basis functions associated to the Delaunay triangulation of V. Indeed, if the vertices in V are in *general positions* (there are no $d+1$ cospherical vertices in V), then there is a unique Delaunay triangulation with vertices V and the solution to the constrained optimization problem

$$\begin{aligned} \min_{\phi_1,\ldots,\phi_n} \quad & \sum_{i=1}^{n} \phi_i \|\boldsymbol{x} - \boldsymbol{v}_i\|^2 \\ \text{subject to} \quad & \phi_i \geq 0, \quad \sum_{i=1}^{n} \phi_i = 1, \quad \sum_{i=1}^{n} \phi_i \boldsymbol{v}_i = \boldsymbol{x}, \end{aligned} \tag{13.6}$$

are the barycentric coordinates of the simplex in the d-dimensional Delaunay triangulation where $\boldsymbol{x}$ belongs [314]. Thus, at each point there are at most $d+1$ non-zero basis functions. See [15, 314] for a discussion about the situation in which the vertices are not in general positions. Arguably, these are the most local generalized barycentric coordinates associated to V.

Comparison of (13.5) and (13.6) suggests combining both objective functions to define the following optimization problem

$$\begin{aligned} \min_{\phi_1,\dots,\phi_n} \quad & \beta \sum_{i=1}^{n} \phi_i \|\boldsymbol{x} - \boldsymbol{v}_i\|^2 + \sum_{i=1}^{n} \phi_i \log \phi_i \\ \text{subject to} \quad & \phi_i \geq 0, \quad \sum_{i=1}^{n} \phi_i = 1, \quad \sum_{i=1}^{n} \phi_i \boldsymbol{v}_i = \boldsymbol{x}, \end{aligned} \tag{13.7}$$

which defines Pareto optima harmonizing the two conflicting objectives: information theory optimality and locality. Depending on β, or its non-dimensional counterpart $\gamma = \beta h^2$ where h is the typical spacing between vertices, the basis functions will more closely resemble nonlocal approximants such as those shown in Figure 13.2, or highly local piecewise affine barycentric coordinates supported on a Delaunay triangulation. In fact, the locality parameter β, controlling the aspect ratio of the basis functions, can be defined locally by considering the following objective function in (13.7)

$$\sum_{i=1}^{n} \beta_i \phi_i \|\boldsymbol{x} - \boldsymbol{v}_i\|^2 + \sum_{i=1}^{n} \phi_i \log \phi_i,$$

where β_i is the aspect ratio parameter associated to vertex i. We call the set of basis functions obtained through the optimization program (13.7), $\phi_i(\boldsymbol{x})$, $i = 1, 2, \dots, n$, *local maximum-entropy* (LME) approximants. Figure 13.3 illustrates the LME basis functions in one and two dimensions, although the formulation can be implemented in principle in any dimension. The smooth basis functions closely resemble other meshfree basis functions, such as those generated by the moving least-squares method [380]. The crucial difference, however, is that such alternative methods do not produce in general generalized barycentric coordinates because the basis functions can be negative. The fact that LME functions are non-negative, and thus convex approximants with a weak Kronecker-delta property, facilitates the imposition of essential boundary conditions in the numerical solution of PDEs.

Note that while (13.6) is a linear program with a unique solution only when vertices are in general positions, the one-parameter family of optimization programs in (13.7) is nonlinear and strictly convex as a result of the entropy function. Thus, since it is feasible for all $\boldsymbol{x} \in \bar{P}$, it has a unique solution. Thus, the entropy functional regularizes in some sense the linear program in (13.6). In fact, as shown in [15], entropy regularization leads to a unique well-defined set of generalized Delaunay barycentric coordinates in the limit $\beta \to +\infty$.

Another important consequence of the entropy functional in (13.7) is the smoothness of the resulting LME barycentric coordinates, as compared to those resulting from (13.6). While the Delaunay barycentric coordinates are only C^0 in P,

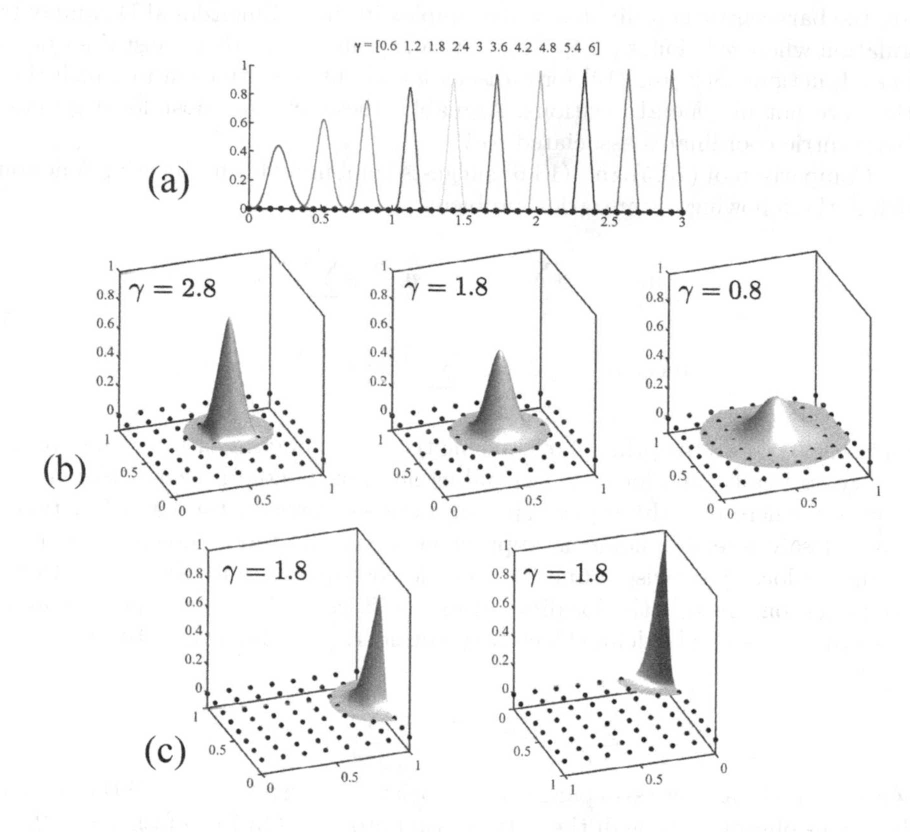

Figure 13.3 Local maximum-entropy (LME) basis functions. (a) Selected basis functions for a one-dimensional vertex set, with varying non-dimensional aspect ratio parameter $\gamma_i = \beta_i h^2$, see main text. (b) Basis functions for an interior node in a two-dimensional set of vertices and different aspect ratio parameters, and (c) representative basis functions of boundary nodes. See color insert.

since they have discontinuous derivatives along the faces of the Delaunay simplices, it can be shown using the implicit function theorem that the LME approximants are C^∞ in P with respect to $\boldsymbol{x}$ (also with respect to β_i).

The LME basis functions have been defined implicitly so far through a convex constrained optimization program with as many unknowns as vertices in V. Through standard duality methods, however, it is possible to obtain an almost explicit form of the basis functions. For this, let us first define the partition function

$$Z(\boldsymbol{x}, \boldsymbol{\lambda}) = \sum_{i=1}^{n} \exp\left[-\beta_i \|\boldsymbol{x} - \boldsymbol{v}_i\|^2 + \boldsymbol{\lambda} \cdot (\boldsymbol{x} - \boldsymbol{v}_i)\right]. \tag{13.8}$$

Then, the LME basis functions can be expressed at any point $\boldsymbol{x} \in P$ as

$$\phi_i(\boldsymbol{x}) = \frac{1}{Z\left(\boldsymbol{x}, \boldsymbol{\lambda}^*(\boldsymbol{x})\right)} \exp\left[-\beta_i\|\boldsymbol{x} - \boldsymbol{v}_i\|^2 + \boldsymbol{\lambda}^*(\boldsymbol{x}) \cdot (\boldsymbol{x} - \boldsymbol{v}_i)\right]. \tag{13.9}$$

This expression is not fully explicit because to compute $\boldsymbol{\lambda}^*(\boldsymbol{x})$ an unconstrained optimization must be solved: the minimization of the strictly convex function $\log Z(\boldsymbol{x}, \boldsymbol{\lambda})$. It can be observed that the functions can be viewed as the product of a Gaussian function around vertex $\boldsymbol{v}_i$, a normalizing factor (Z) so that all functions add up to one, and an exponential factor that depends on the Lagrange multiplier $\boldsymbol{\lambda}^*$ enforcing the linear reproducing condition. Examination of (13.9) shows that, strictly speaking, these basis functions have global support. However, due to the fast decay of the Gaussian function, for all practical purposes they can be considered as compactly supported. Furthermore, the sum in calculation of Z in (13.8) effectively involves only a handful of terms corresponding to vertices in the neighborhood of $\boldsymbol{x}$, which makes the evaluation of the basis functions efficient.

We emphasize the dependence of $\boldsymbol{\lambda}^*$ on position to highlight the fact that at each evaluation point, an unconstrained nonlinear optimization problem must be solved. This problem, however, is only d-dimensional, has a unique solution, and can be efficiently solved numerically using Newton's method. See [15, 279, 335] for mathematical and computational details, as well as for the explicit expressions to compute first and second spatial derivatives of the basis functions, and see [177] for the calculation of the gradients of the basis functions on the boundary of P.

We end this section by presenting an alternative method to introduce locality into the maximum-entropy framework, which in addition provides a means to define strictly compactly supported basis functions. It is very easy to define a set of non-negative smooth basis functions $\psi_i(\boldsymbol{x})$ relative to V, localized around the corresponding vertex, and satisfying only the zeroth-order reproducing condition, $\sum_{i=1}^{n} \psi_i(\boldsymbol{x}) = 1$. Indeed, consider for instance a scalar, non-negative, and smooth window function $w : [0, +\infty) \to \mathbb{R}$, which decays when the argument is large compared to 1 and satisfies that all odd derivatives up to k at 0 vanish (to guarantee that the even extension of w to $\mathbb{R}$ is C^k). Then, for a large enough ρ the functions

$$\psi_i(\boldsymbol{x}) = \frac{w\left(\|\boldsymbol{x} - \boldsymbol{v}_i\|/\rho\right)}{\sum_{j=1}^{n} w\left(\|\boldsymbol{x} - \boldsymbol{v}_j\|/\rho\right)}$$

are clearly non-negative, localized around their corresponding vertex $\boldsymbol{v}_i$ with typical width ρ, smooth, and add up to one. Furthermore if w is compactly supported, then the functions $\psi_i(\boldsymbol{x})$ are also compactly supported. Thus, at each point, the numbers $\psi_1(\boldsymbol{x}), \psi_2(\boldsymbol{x}), \ldots, \psi_n(\boldsymbol{x})$ define a discrete probability distribution, which contains information about locality ($\psi_i(\boldsymbol{x})$ is larger if $\boldsymbol{x}$ is closer to $\boldsymbol{v}_i$) but does not satisfy the first-order reproducing constraint (13.2b). This constraint can be recovered using the concept of *relative entropy*, also called Kullback–Leibler distance, which measures the amount of information required to obtain a discrete probability distribution from a prior distribution. Viewing $\psi_i(\boldsymbol{x}),\ i = 1, \ldots, n$ as a prior distribution, the relative entropy can be written as

$$D(\boldsymbol{\phi}|\boldsymbol{\psi}(\boldsymbol{x})) = \sum_{i=1}^{n} \phi_i \log \frac{\phi_i}{\psi_i(\boldsymbol{x})}.$$

Minimization of this function with respect to ϕ_i, $i = 1, \dots, n$, subject to (13.2) and (13.3), will produce the closest set of barycentric coordinates to the prior from an information theoretical viewpoint, i.e., introducing the least extra information [16, 202, 380]. A direct calculation using duality methods shows that the resulting basis function ϕ_i can be written as the product of ψ_i and another factor enforcing the linear reproducing condition, and thus the prior and the relative maximum-entropy functions have the same support. In particular, if $w(r) = e^{-r^2}$ and $\rho = 1/\sqrt{\beta}$, then these functions coincide with the LME basis functions.

13.4 FURTHER EXTENSIONS

We summarize next extensions to the meshfree and convex basis functions generated by the LME optimization program. An obvious question is whether these approximants can be extended to higher-order, that is, if in addition to exactly reproducing affine functions due to (13.2), they can be designed to exactly reproduce higher-order polynomials. This would lead to a higher rate of convergence when approximating PDEs. Focusing in 1D for simplicity, a naive extension would be to maximize the LME objective function subject to

$$\sum_{i=1}^{n} \phi_i = 1, \qquad \sum_{i=1}^{n} \phi_i v_i = x, \qquad \sum_{i=1}^{n} \phi_i v_i^2 = x^2, \tag{13.10}$$

which are a set of conditions met, for instance, by quadratic finite element spaces. However, as identified in [15], conditions in (13.10) together with the non-negativity constraint $\phi_i \geq 0$ are infeasible for almost all points in P. This may seem contradictory at first sight, since there are examples of non-negative approximants that can reproduce quadratic and higher-order functions, notably B-splines. One option is to give up non-negativity and consider a notion of entropy for signed probability distributions, which however destroys the convex structure of the approximation scheme, does not produce generalized barycentric coordinates, and can lead to wiggly basis functions [62, 373]. Insisting on non-negativity, the apparent contradiction referred to above is resolved by noting that for the second-order consistency condition to hold, the coefficients that multiply the generalized barycentric coordinates do not need to be v_i^2. By designing appropriate coefficients, it is possible to restore feasibility, and thus define second-order maximum entropy approximants in multiple dimensions [111, 336], see Figure 13.4(a). The construction of feasible constraints for higher-order reproducing conditions, however, is an open question.

When applied to the numerical solution of PDEs with Galerkin methods, one aspect complicating the implementation and diminishing the efficiency of LME approximations is the "uncontrolled" support of the basis functions. This is important

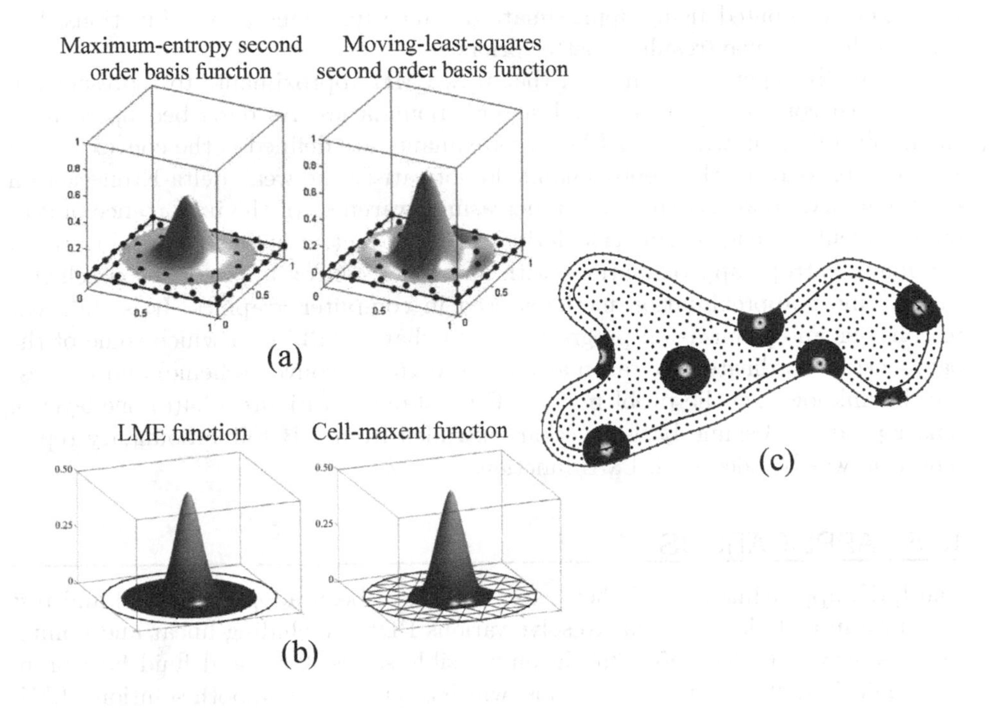

Figure 13.4 Extensions of maximum-entropy approximations. (a) Representative second-order maximum entropy basis function, compared to a moving-least-squares function, which is wiggly and negative at some points. (b) Comparison of a standard LME basis function and a cell-based maximum entropy basis function supported on a triangulation, which results in a structured adjacency relation and more efficient calculations to approximate PDEs for a comparable accuracy. (c) Combination of LME approximants with a boundary representation based on B-splines. See color insert.

because any two basis functions with overlapping support will generate an entry in the stiffness matrix resulting from the Galerkin method. Because the LME basis functions have completely unstructured support, the underlying adjacency structure setting the matrix structure is quite dense, but with many very small entries [305]. To alleviate this, a modification of the LME basis functions was proposed, and called cell-based maximum entropy approximations, in which prior functions ($\psi_i(\boldsymbol{x})$ in the previous section) were designed to be supported on the k-ring of simplices around $\boldsymbol{v}_i$ of a triangulation, see Figure 13.4(b). In this way, the adjacency structure of the stiffness matrix is given directly by the topology of the triangulation and is much sparser, leading to more efficient calculations without nearly any degradation in accuracy [281]. In this reference, the smooth and compactly supported prior functions were computed using approximate distance functions and R-functions, but other techniques may result in better priors.

We finally report on a method that uses LME approximants to represent domains with complex geometry and smooth boundaries. As described up to now, the natural domain where the LME approximants are defined is the convex hull of P. For subsets of P, the approximants do not satisfy the weak delta-Kronecker on the boundary. However, there is an increasing awareness of the importance of geometric fidelity in engineering calculations [205]. In fact, it is quite natural to blend maximum-entropy approximations with any other convex approximation scheme, such as most approximation methods used in computer graphics. It is sufficient to consider an optimization program such as that in (13.7), in which some of the barycentric coordinates are taken as data (the known convex scheme) and the rest are the unknowns of the problem [334]. For instance, in Figure 13.4(c) one layer of tensor product B-spline basis functions extruded from a B-spline boundary representation was blended with LME functions.

13.5 APPLICATIONS

The LME approximants and their extensions have been adopted as trial and test functions in Galerkin methods to solve various PDEs, including linear and nonlinear elasticity [15, 334, 335, 336], incompressible solids [296], and fluid flow problems [243, 292, 306]. In these problems, which often exhibit smooth solutions, LME approximations provide highly accurate numerical approximations with relatively coarse sets of points, as compared to standard C^0 finite elements. Furthermore, the meshfree basis functions can accommodate very large grid distortions in Lagrangian large deformation simulations [243, 306]. However, some applications truly benefit from the smoothness of the LME approximants for general unstructured sets of points. We next outline two examples: the numerical approximation of high-order PDEs and the approximation of manifolds defined by scattered sets of points.

13.5.1 High-order partial differential equations

Many physical phenomena of interest in science and engineering are modeled using higher-order PDEs. For instance, the Kirchhoff–Love equations of thin shells are fourth-order PDEs. Similarly, many problems involving evolving discontinuities can be modeled using phase-field models, where the discontinuity is smeared out and tracked by a field governed by a PDE. In many cases, such a PDE is of fourth order, including models for crack propagation [64], crystal growth [396], or the mechanics of fluid membranes [128]. The numerical treatment of such problems with C^0 finite elements is cumbersome and involves independently interpolating the field and its gradient [129]. In contrast, a Galerkin approach is straightforward if smooth basis functions are used. LMEs are not only smooth, but also easily amenable to adaptive methods and have monotonicity properties that help them accurately describe the sharp variations characteristic of phase-field solutions. See Figure 13.5 for an illustration of LME approximants applied to a phase-field model

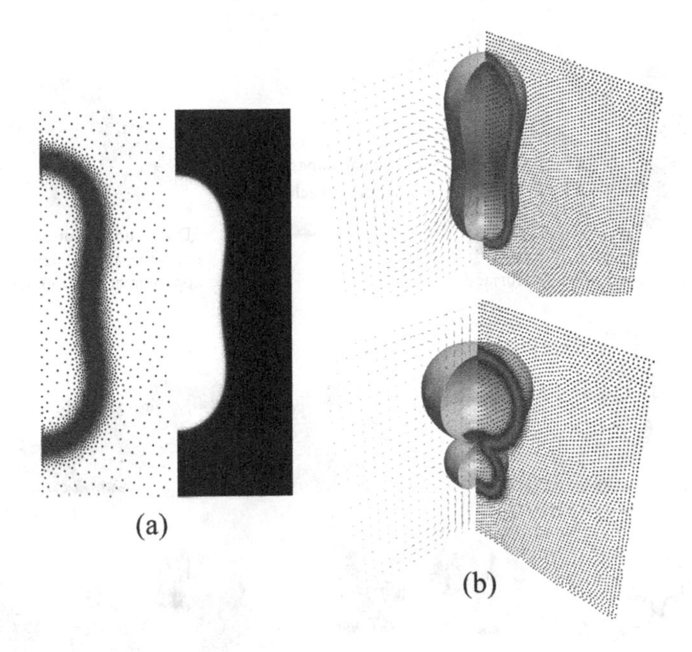

Figure 13.5 Application of the LME approximants to high-order PDEs. (a) In phase-field models of biomembranes, the unknown field defining the shape of a moving interface is governed by a fourth-order PDE. Therefore, a Galerkin method requires smooth basis functions (with square integrable second derivatives). The phase field develops sharp gradients, calling for an adaptive solution and an approximation scheme devoid of Gibbs phenomenon. All these conditions are met by LME approximations. (b) shows two snapshots in a dynamical simulation tracking the shape of a deforming vesicle made out of a fluid membrane with curvature elasticity, together with the hydrodynamics of the surrounding fluid. The resolution of the set of vertices follows the sharp variations of the phase-field. See color insert.

of vesicle dynamics [306, 337]. LME approximants have also been applied to phase-field models of fracture [244], and to other high-order PDEs such as those arising in the mechanics of thin shells [280] or in the coupled electromechanics of flexoelectric materials [1, 2].

13.5.2 Manifold approximation

Finally, LME approximations have been used to smoothly parametrize manifolds sampled by clouds of points in an automated way and without the need for a mesh, using an atlas of partially overlapping charts [280]. See [263] for recent work along these lines, with a review of similar approaches in computer graphics. Figure 13.6(a) illustrates the procedure, in which the point-set on a d-dimensional

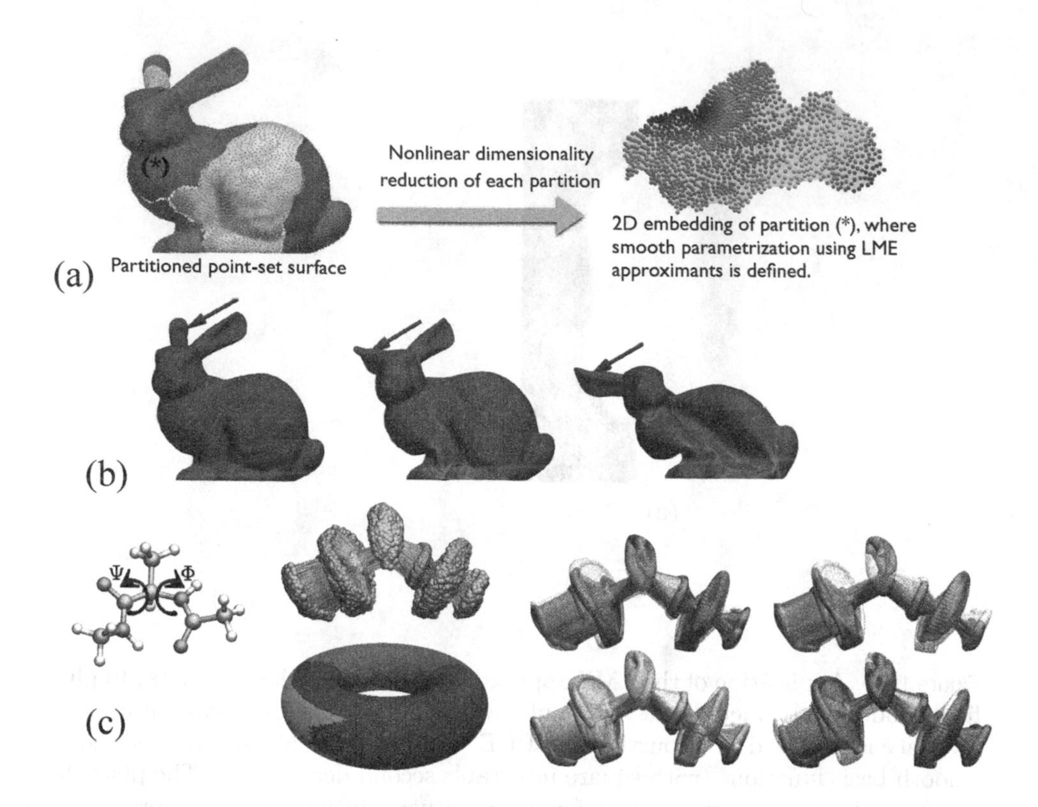

Figure 13.6 Approximation of manifolds using an atlas of smooth LME parametrizations. (a) Methodology to define local parametrizations mapping partitions of a surface defined by a set of points. (b) Application of this method to solve the high-order PDE for a nonlinear Kirchhoff–Love shell. (c) Application of the method to describe the molecular conformations of alanine dipeptide, a small molecule. An ensemble of conformations of this molecule samples an underlying configurational manifold homeomorphic to a torus. With the method presented in (a), this manifold is partitioned into four pieces, each of which is parametrized with LME approximants. This method allows us to define smooth collective variables for this system, required to enhance sampling in molecular dynamics simulations. See color insert.

manifold is systematically partitioned until each partition is homeomorphic to an open set in $\mathbb{R}^d$. Then, each partition is embedded in $\mathbb{R}^d$ using nonlinear dimensionality reduction methods, and this embedding is used as a parametric domain for a local parametrization of the partition. Parametrizations of adjacent partitions are "glued" with partition-of-unity functions. This natural but powerful procedure has been used to model thin shells of complex geometry, see Figure 13.6(b); to parametrize the gait of microscopic swimming cells [14]; or to define collective variables for molecular systems based on parametrizing the so-called intrinsic manifold underlying molecular flexibility [186], see Figure 13.6(b).

13.6 OUTLOOK

In summary, the principle of maximum entropy is a conceptually appealing and computationally practical approach to defining generalized barycentric coordinates for clouds of points. The resulting barycentric coordinates can be viewed as meshfree basis functions and used in computational mechanics, exploiting their smoothness and the relative ease of implementing adaptive strategies. The fact that these basis functions are barycentric coordinates makes it easier to impose boundary conditions, but also provides a natural framework to couple maximum entropy basis functions with multivariate spline techniques. We have demonstrated this by blending maximum entropy approximations to B-spline boundary representations, but another attractive idea is using such blending to resolve the issues associated with irregular vertices in spline patches [263]. Open issues include the systematic formulation of feasible higher-order consistency conditions in multiple dimensions (see the discussion following (13.10)), and the accurate and efficient numerical quadrature of the integrals appearing in Galerkin methods, which involve products of derivatives of the non-polynomial LME basis functions. Finally, the applicability of the LME approach in higher (greater than three) space dimensions has not yet been exploited in applications.

CHAPTER 14

BEM-Based FEM

Steffen Weißer

Saarland University, Saarbrücken, Germany

CONTENTS

THE Boundary Element Method (BEM)-based Finite Element Method (FEM) is an approach to approximate solutions of boundary-value problems over polygonal and polyhedral meshes. The approximation spaces are defined implicitly over the polygonal or polyhedral elements. Harmonic coordinates are recovered for the lowest order approximation in the special case of the diffusion problem. Thus, this method can be viewed as a generalization of harmonic coordinates to higher-order approximations. The definition and treatment of basis functions as well as the finite element formulation are discussed in detail. Furthermore, the BEM-based FEM is applied in the context of an adaptive FEM strategy yielding locally refined polygonal meshes. Theoretical evidence and numerical experiments are presented that establish optimal rates of convergence for uniform and adaptive mesh refinement strategies.

14.1 INTRODUCTION

In this chapter, we study a continuous-Galerkin finite element formulation on polygonal and polyhedral meshes, which is applied to boundary-value problems. The approach was proposed in [107] and established in a sequence of papers [195, 196, 329, 330, 420, 421, 423]. First, the general situation is described in the introduction, and afterwards the Poisson problem is considered as a simple model problem to explain the approach.

Let $\Omega \subset \mathbb{R}^d$, $d = 2, 3$ be a bounded polygonal or polyhedral domain with Lipschitz boundary $\partial\Omega = \Gamma_D \cup \Gamma_N$, which is split into a Dirichlet and a Neumann part that are denoted by Γ_D and Γ_N, respectively. We seek to find the function $u : \Omega \to \mathbb{R}$ that solves the following second-order convection-diffusion-reaction boundary-value problem:

$$\begin{aligned} Lu = -\operatorname{div}(\boldsymbol{A}\nabla u) + \boldsymbol{b}\cdot\nabla u + cu &= f \quad &&\text{in } \Omega,\\ u &= g_D &&\text{on } \Gamma_D,\\ \frac{\partial u}{\partial \boldsymbol{n}} &= g_N &&\text{on } \Gamma_D, \end{aligned}$$

where $\boldsymbol{n}$ denotes the outward unit normal vector to Ω. $\boldsymbol{A} \in \mathbb{R}^{d\times d}$, $\boldsymbol{b} \in \mathbb{R}^d$, $c \in \mathbb{R}$ are some coefficients, f is a source term, and g_D, g_N are boundary data. We assume that the operator L is elliptic and the boundary-value problem has a unique solution. This is guaranteed by the usual assumptions on the coefficients and the data.

To obtain an approximation of the unknown solution u, the domain Ω is discretized by non-overlapping polygonal or polyhedral elements and a finite element formulation is applied. The idea of the BEM-based FEM is to use a Trefftz-like approximation space. Its basis functions are defined to fulfil certain local boundary-value problems related to the differential operator L on each element P. In these local problems, the coefficients of L are chosen to be constant. The global continuity of the basis functions is ensured by prescribing the boundary data on ∂P. Due to the special choice of basis functions, it is possible to restate the finite element formulation such that only integrals over ∂P appear in the finite element matrix and we do not have to evaluate the basis functions explicitly in the interior of P. However, this procedure involves boundary integral operators, which are treated by means of the BEM in the realization.

All these steps are treated more precisely in the following sections. To simplify the presentation, we restrict ourselves to $L = -\Delta$ and zero Dirichlet data on Γ_D with $|\Gamma_D| > 0$. Thus, we consider the problem

$$-\Delta u = f \text{ in } \Omega, \qquad u = 0 \text{ on } \Gamma_D, \qquad \frac{\partial u}{\partial \boldsymbol{n}} = g_N \text{ on } \Gamma_N. \tag{14.1}$$

The well-known variational formulation reads: find $u \in V$ such that

$$b(u,v) = (f,v)_{L_2(\Omega)} + (g_N, v)_{L_2(\Gamma_N)} \quad \forall v \in V, \tag{14.2}$$

where

$$b(u,v) = \int_\Omega \nabla u \cdot \nabla v \, d\boldsymbol{x} \quad \text{and} \quad V = \{v \in H^1(\Omega) : v = 0 \text{ on } \Gamma_D\}.$$

For $f \in L_2(\Omega)$ and $g_N \in L_2(\Gamma_N)$, problem (14.2) admits a unique solution according to the Lax–Milgram Lemma, since the bilinear form $b(\cdot,\cdot)$ is bounded and coercive on V. To obtain a discrete finite element solution we replace V in (14.2) by a finite-dimensional conforming subspace $V_h \subset V$ that is spanned by some basis functions, i.e., $V_h = \operatorname{span}\{\varphi_1, \ldots, \varphi_m\}$. The ansatz $u_h = \sum_{i=1}^{m} u_i \varphi_i$ yields the finite element system of linear equations

$$\boldsymbol{K}\boldsymbol{u} = \boldsymbol{f} \quad \text{for} \quad \boldsymbol{u} = (u_1, \ldots, u_m) \in \mathbb{R}^m, \tag{14.3}$$

where

$$\boldsymbol{K} = \left(b(\varphi_j, \varphi_i)\right)_{i,j=1}^{m} \in \mathbb{R}^{m\times m}, \quad \boldsymbol{f} = \left((f, \varphi_i)_{L_2(\Omega)} + (g_N, \varphi_i)_{L_2(\Gamma_N)}\right)_{i=1}^{m} \in \mathbb{R}^m.$$

The basis functions φ_i are defined over the polygonal or polyhedral mesh $\mathcal{T}_h$ in such a way that they have local support. Consequently, the finite element stiffness matrix $\boldsymbol{K}$ is sparse, and it is symmetric as well as positive definite because of the properties of $b(\cdot,\cdot)$. Therefore, efficient numerical solvers such as those based on the conjugate gradient method are available to solve (14.3).

In the first part of this chapter we consider the two-dimensional case $d = 2$. We already mentioned that the basis functions φ_i are defined locally on each polygonal element P with the help of the differential operator $L = -\Delta$ in the BEM-based FEM. For the first-order approximation space, we prescribe linear data on ∂P and use $-\Delta\varphi_i = 0$ in P. As a consequence, we recover harmonic coordinates that were first proposed in [215] for applications in computer graphics, see Section 1.2.8 as well. These harmonic coordinates belong to the class of generalized barycentric coordinates, which also find applications in polygonal finite element methods, see Chapter 11. The general construction of basis functions for high-order approximation spaces within the BEM-based FEM is discussed in Section 14.2. This approach generalizes harmonic coordinates to higher order basis functions and shows similarities to the virtual element method (VEM). The VEM was first introduced in [27] and is discussed in Chapter 15. In Section 14.2, we additionally discuss the treatment of the locally implicit basis functions and the computation of the finite element matrix $\boldsymbol{K}$. This involves the application of BEM locally, for which we give a short introduction. One of the promising application areas of polygonal meshes is in adaptive mesh refinement, where the flexibility of using different element shapes in the meshes can be exploited. In Section 14.3, an adaptive FEM strategy is reviewed, one that makes use of the BEM-based FEM and a residual-based *a posteriori* error estimator. Numerical experiments are presented that demonstrate the application of polygonal meshes to solve two-dimensional boundary-value problems. Finally, in Section 14.4, we present an outlook to further developments in 2D and 3D.

14.2 HIGH-ORDER BEM-BASED FEM IN 2D

In this section, we address the finite element formulation and give theoretical and numerical rates of convergence for the BEM-based FEM on sequences of uniformly refined meshes. First, the domain $\Omega \subset \mathbb{R}^2$ is decomposed into a finite number of non-overlapping polygonal elements P such that $\bar{\Omega} = \bigcup\{\bar{P} : P \in \mathcal{T}_h\}$. Here, $\mathcal{T}_h$ is the set of all these polygonal elements, and it is called the polygonal mesh. We assume that $\mathcal{T}_h$ is regular, i.e., all polygonal elements $P \in \mathcal{T}_h$ are star-shaped with respect to a circle such that the aspect ratio h_P/ρ_P of their diameter h_P and the radius ρ_P of the circle is uniformly bounded for all elements, and that the lengths $h_{[\boldsymbol{v}_i,\boldsymbol{v}_{i+1}]}$ of the edges of P can be uniformly bounded from below such that $c_{\mathcal{T}} h_P \le h_{[\boldsymbol{v}_i,\boldsymbol{v}_{i+1}]}$ for a fixed constant $c_{\mathcal{T}}$.

14.2.1 Construction of basis functions

The global approximation space $V_h = \operatorname{span}\{\varphi_1, \ldots, \varphi_m\}$ is also denoted by V_h^k to specify its approximation order $k \in \mathbb{N}$. It is constructed by prescribing its basis functions φ_i over the single elements $P \in \mathcal{T}_h$ such that

$$V_h^k = \{v \in V : v|_P \in V_h^k(P)\, \forall P \in \mathcal{T}_h\},$$

and we only have to give a local basis for $V_h^k(P)$.

Let $\mathcal{P}^k(\omega)$ be the space of polynomials of degree less than or equal to k over ω, where ω is either an element P or an edge $[\boldsymbol{v}_i, \boldsymbol{v}_{i+1}]$. Before we address $V_h^k(P)$, we introduce an auxiliary space over ∂P. Let

$$\mathcal{P}_{\mathrm{pw}}^k(\partial P) = \{v \in C(\partial P) : v|_{[\boldsymbol{v}_i,\boldsymbol{v}_{i+1}]} \in \mathcal{P}^k([\boldsymbol{v}_i, \boldsymbol{v}_{i+1}]),\ i = 1, \ldots, n\}$$

be the space of piecewise functions on ∂P that are polynomials of degree smaller or equal to k over each edge $[\boldsymbol{v}_i, \boldsymbol{v}_{i+1}]$ and that are continuous at the vertices $\boldsymbol{v}_i$. Furthermore, let $\{\lambda_i : i = 1, \ldots, kn\}$ be a basis set of $\mathcal{P}_{\mathrm{pw}}^k(\partial P)$, where λ_i, $i = 1, \ldots, n$ is linear on each edge $[\boldsymbol{v}_j, \boldsymbol{v}_{j+1}]$ and fulfils $\lambda_i(\boldsymbol{v}_j) = \delta_{ij}$. For $i > n$, λ_i are chosen as polynomial bubble functions that vanish at the vertices and have local support on one edge. For example, λ_i, $i = n+1, \ldots, 2n$ is a quadratic polynomial over $[\boldsymbol{v}_i, \boldsymbol{v}_{i+1}]$, λ_i, $i = 2n+1, \ldots, 3n$ is a cubic polynomial over $[\boldsymbol{v}_i, \boldsymbol{v}_{i+1}]$, and so on.

Next, the space $V_h^k(P)$ is given by prescribing its basis functions. They satisfy certain boundary-value problems on P, as mentioned in the introduction. Since $L = -\Delta$ for the model problem, we define ϕ_i, $i = 1, \ldots, kn$ as the unique solution of

$$-\Delta \phi_i = 0 \quad \text{in } P \quad \text{and} \quad \phi_i = \lambda_i \quad \text{on } \partial P. \tag{14.4}$$

For the lowest-order case $k = 1$, we recover harmonic coordinates, compare Section 1.2.8. For $k > 1$, we have additional harmonic basis functions with polynomial data on ∂P. Furthermore, we introduce element bubble functions $\phi_{i,j}$ for $k > 1$ with $i = 0, \ldots, k-2$ and $j = 0, \ldots, i$ as the unique solutions of

$$-\Delta \phi_{i,j} = p_{i,j} \quad \text{in } P \quad \text{and} \quad \phi_{i,j} = 0 \quad \text{on } \partial P, \tag{14.5}$$

where $p_{i,j}(\boldsymbol{x}) = (x-\bar{x})^{i-j}(y-\bar{y})^j$ and $\bar{\boldsymbol{x}} = (\bar{x}, \bar{y})$ is the center of mass of P. Later on, it is convenient to reformulate the local boundary-value problems for the element bubble functions $\phi_{i,j}$ such that they reduce to Laplace problems. Afterwards, $\phi_{i,j}$ can be expressed in terms of the harmonic basis functions ϕ_i. This yields the form

$$\phi_{i,j} = q_{i,j} - \sum_{\ell=1}^{n} \alpha_\ell \phi_\ell, \tag{14.6}$$

where $q_{i,j} \in \mathcal{P}^k(P)$ is such that $-\Delta q_{i,j} = p_{i,j}$. The polynomial $q_{i,j}$ is always constructible and it is not unique. The coefficients α_ℓ are chosen in accordance with $q_{i,j}|_{\partial P} - \sum_{\ell=1}^{n} \alpha_\ell \lambda_i = 0$.

Since the basis functions are defined as solutions of boundary-value problems, the theory of partial differential equations provides regularity results. It is known that the solution of (14.4) and (14.5) belongs to $C^2(P) \cap C(\bar{P})$ for convex elements P. In the case of non-convex elements, the classical notion of smoothness is lost and the differential equations have to be understood in the weak sense. Nevertheless, the basis functions still belong to $H^1(P) \cap C(\bar{P})$, which is sufficient to ensure conformity.

Finally, the space $V_h^k(P)$ is defined as the span of the functions ϕ_i and $\phi_{i,j}$. Since $\{p_{i,j} : i = 0, \ldots, k-2 \text{ and } j = 0, \ldots, i\}$ is a basis set of $\mathcal{P}^{k-2}(P)$, we obtain

$$V_h^k(P) = \{v \in H^1(P) : -\Delta v \in \mathcal{P}^{k-2}(P) \text{ and } v|_{\partial P} \in \mathcal{P}_{\mathrm{pw}}^k(\partial P)\}. \tag{14.7}$$

Furthermore, we observe that $\mathcal{P}^k(P) \subset V_h^k(P)$. This is a consequence of the unique solvability of the Poisson problem with Dirichlet boundary data. For $p \in \mathcal{P}^k(P)$, it is evidently $p|_{\partial P} \in \mathcal{P}_{\mathrm{pw}}^k(\partial P)$ and $-\Delta p \in \mathcal{P}^{k-2}(P)$ and thus $p \in V_h^k(P)$.

The VEM in Chapter 15 also uses the local space defined in (14.7). Therefore, the BEM-based FEM and the VEM seek the approximation of the solution of the boundary-value problem for the Poisson equation (14.1) in the same discrete space. The VEM reduces all computations to carefully chosen degrees of freedom and to local projections into polynomial spaces. The BEM-based FEM in contrast makes use of the explicit knowledge of the basis functions and thus enables the evaluation of the approximation inside the elements. Both methods rely on clever reformulations to avoid volume integration. Since the BEM-based FEM applies Trefftz-like basis functions, which are related to the differential equation of the global problem, the discrete space for the BEM-based FEM and the VEM differ as soon as more general boundary-value problems are considered.

14.2.2 Finite element method

The conforming approximation space $V_h^k \subset H^1(\Omega)$ can be utilized in an FE computation. The variational formulation of the model problem is already given in (14.2) and its approximation yields the system of linear equations (14.3). The approximation properties of the Galerkin formulation have been studied in [329, 421]. The space V_h^k yields the same error estimates that are known for classical approximation spaces over triangulations, but on much more general meshes.

Theorem 14.1. *Let $\mathcal{T}_h$ be a regular polygonal mesh, $u \in V$ be the solution of* (14.2), *and $u_h \in V_h^k$ be the Galerkin approximation obtained by* (14.3). *Then,*

$$\|u - u_h\|_{H^\ell(\Omega)} \le c\, h^{k+1-\ell} |u|_{H^{k+1}(\Omega)} \quad \text{for } u \in H^{k+1}(\Omega) \text{ and } \ell = 0, 1,$$

where the constant c only depends on the mesh regularity and for $\ell = 0$ some additional regularity of the dual problem is assumed.

It remains to discuss the setup of (14.3) and how to compute the involved integrals. The formation of the matrix $\boldsymbol{K}$ and the right-hand side $\boldsymbol{f}$ is done as in FEM by agglomerating local (element-wise) stiffness matrices. Therefore, we write

$$b(\varphi_j, \varphi_i) = \sum_{P \in \mathcal{T}_h} \int_P \nabla\varphi_i \cdot \nabla\varphi_j \, d\boldsymbol{x}. \tag{14.8}$$

The difficulty arises in the application of the bilinear form to the implicitly defined basis functions and to evaluate the integral $(f, \varphi_i)_{L_2(P)}$ on the right hand side. The computation of the boundary integral $(g_N, \varphi_i)_{L_2(\Gamma_N)}$ is straightforward, since the basis functions φ_i are known explicitly along the edges. Thus, these integrals are treated over each edge in Γ_N by numerical quadrature.

To evaluate the integrals in (14.8), we distinguish three cases and apply Green's first identity locally on each element $P \in \mathcal{T}_h$. The details are left to the reader. If $\varphi_i|_P$ and $\varphi_j|_P$ correspond to harmonic basis functions ϕ_i and ϕ_j, we obtain

$$\int_P \nabla\phi_i \cdot \nabla\phi_j \, d\boldsymbol{x} = \int_{\partial P} \frac{\partial \phi_i}{\partial \boldsymbol{n}_P} \phi_j \, ds_{\boldsymbol{x}}. \tag{14.9}$$

If $\varphi_i|_P$ and $\varphi_j|_P$ correspond to element bubble functions $\phi_{i,j}$ and $\phi_{i',j'}$, then we can use (14.6) to obtain

$$\int_P \nabla\phi_{i,j} \cdot \nabla\phi_{i',j'} \, d\boldsymbol{x} = \int_{\partial P} \frac{\partial q_{i,j}}{\partial \boldsymbol{n}_P} q_{i',j'} \, ds_{\boldsymbol{x}} - \sum_{\ell=1}^{n} \alpha_\ell \int_{\partial P} \frac{\partial \phi_\ell}{\partial \boldsymbol{n}_P} q_{i',j'} \, ds_{\boldsymbol{x}} + \int_P p_{i,j} q_{i',j'} \, d\boldsymbol{x}.$$

And if $\varphi_i|_P$ corresponds to a harmonic basis function and $\varphi_j|_P$ to an element bubble function, we obtain $\int_P \nabla\varphi_i \cdot \nabla\varphi_j \, d\boldsymbol{x} = 0$. Therefore, the system of linear equations decouples in the FE computation. Furthermore, besides the unknown term $\partial\phi_i/\partial\boldsymbol{n}_p$ all others are (piecewise) polynomials such that the first and third integral in the last formula can be analytically computed. The approximation of $\partial\phi_i/\partial\boldsymbol{n}_p$ is addressed in the next subsection.

For the computation of $(f, \phi_i)_{L_2(\Omega)}$, we proceed as in (14.8). Then, we approximate the local integrals $\int_P f\varphi_i \, d\boldsymbol{x}$ by a numerical quadrature over polygonal elements. This is realized by subdividing the polygon into triangles and applying standard quadrature rules over them, cf. [378]. Alternatively, we might also apply quadrature rules over polygons directly, see for example [102]. If $\varphi_i|_P$ corresponds to a harmonic basis function ϕ_i, we make use of the representation formula

$$\phi_i(\boldsymbol{x}) = -\frac{1}{2\pi} \int_{\partial P} \ln|\boldsymbol{x} - \boldsymbol{y}| \frac{\partial \phi_i(\boldsymbol{y})}{\partial \boldsymbol{n}_P} \, ds_{\boldsymbol{y}} - \frac{1}{2\pi} \int_{\partial P} \frac{\boldsymbol{n}_P \cdot (\boldsymbol{x} - \boldsymbol{y})}{|\boldsymbol{x} - \boldsymbol{y}|^2} \phi_i(\boldsymbol{y}) \, ds_{\boldsymbol{y}} \tag{14.10}$$

for $\boldsymbol{x} \in P$, see below. If $\varphi_i|_P$ corresponds to an element bubble function, we apply (14.6). We stress again that the knowledge of $\partial\phi_i/\partial\boldsymbol{n}_p$ is important.

14.2.3 Introduction to boundary element methods

The representation formula (14.10) is crucial for boundary element methods, which are utilized to approximate boundary integral formulations. To be mathematically precise, see [367], we denote by $\gamma_0 : H^1(P) \to H^{1/2}(\partial P)$ the usual trace operator that satisfies $\gamma_0 v = v|_{\partial P}$ for continuous functions. Furthermore, let $\gamma_1 v \in H^{-1/2}(\partial P)$ be the conormal derivative, also called Neumann trace, of a function $v \in H^1(\Omega)$ with Δv in the dual of $H^1(\Omega)$. The Neumann trace $\gamma_1 v$ is defined by Green's identity

$$\int_P \nabla v \cdot \nabla w \, d\boldsymbol{x} = \int_{\partial P} \gamma_1 v \, \gamma_0 w \, ds_{\boldsymbol{x}} - \int_P w \Delta v \, d\boldsymbol{x}$$

for $w \in H^1(P)$ and coincides with $\partial v / \partial \boldsymbol{n}_P$ for sufficient regular functions. Thus, we write (14.10) in terms of boundary integral operators for a harmonic function v as $v = \tilde{\mathcal{V}}(\gamma_1 v) - \mathcal{W}(\gamma_0 v)$ in P. Applying the trace operators γ_0 and γ_1 to the representation formula yields the Calderon projector

$$\begin{pmatrix} \gamma_0 v \\ \gamma_1 v \end{pmatrix} = \begin{pmatrix} \frac{1}{2}\mathcal{I} - \mathcal{K} & \mathcal{V} \\ \mathcal{D} & \frac{1}{2}\mathcal{I} + \mathcal{K}' \end{pmatrix} \begin{pmatrix} \gamma_0 v \\ \gamma_1 v \end{pmatrix}, \tag{14.11}$$

where $\mathcal{I}$ is the identity operator. The single layer, double layer, and adjoint double layer potential as well as the hypersingular integral operator have the mapping properties

$$\mathcal{V} : H^{-1/2}(\partial P) \to H^{1/2}(\partial P), \qquad \mathcal{K} : H^{1/2}(\partial P) \to H^{1/2}(\partial P),$$
$$\mathcal{K}' : H^{-1/2}(\partial P) \to H^{-1/2}(\partial P), \qquad \mathcal{D} : H^{1/2}(\partial P) \to H^{-1/2}(\partial P).$$

For a detailed study of these operators, see [367]. The first equation in (14.11) gives the boundary integral equation $\mathcal{V}(\gamma_1 v) = (\frac{1}{2}\mathcal{I} + \mathcal{K})(\gamma_0 v)$. Consequently, if $\gamma_0 v$ is known, the Neumann trace $t = \gamma_1 v$ is given as the solution of the Galerkin formulation

$$(\mathcal{V}t, \tau)_{L_2(\partial P)} = \left((\tfrac{1}{2}\mathcal{I} + \mathcal{K})\gamma_0 v, \tau\right)_{L_2(\partial P)} \quad \forall \tau \in H^{-1/2}(\partial P).$$

This formulation admits a unique solution, since $\mathcal{V}$ is known to be invertible for $h_P < 1$. To obtain an approximation of $t \in H^{-1/2}(\partial P)$, we apply the BEM. We discretize $H^{1/2}(\partial P)$ by $\mathcal{P}^k_{\mathrm{pw}}(\partial P)$ and $H^{-1/2}(\partial P)$ by $\mathcal{P}^{k-1}_{\mathrm{pw,d}}(\partial P)$, where

$$\mathcal{P}^{k-1}_{\mathrm{pw,d}}(\partial P) = \{v \in L_2(\partial P) : v|_{[\boldsymbol{v}_i, \boldsymbol{v}_{i+1}]} \in \mathcal{P}^{k-1}([\boldsymbol{v}_i, \boldsymbol{v}_{i+1}]),\ i = 1, \ldots, n\}$$

is the space of piecewise polynomials of order smaller or equal to $k-1$ that allow discontinuities at the vertices of P. Let $\{\tau_i : i = 1, \ldots, kn\}$ be a basis set of $\mathcal{P}^{k-1}_{\mathrm{pw,d}}(\partial P)$ that is constructed analogously to $\{\lambda_i : i = 1, \ldots, kn\}$. For polynomial data $\gamma_0 v = \sum_{i=1}^{kn} v_i \lambda_i \in \mathcal{P}^k_{\mathrm{pw}}(\partial P)$, the ansatz $t \approx t_h = \sum_{i=1}^{kn} t_i \tau_i \in \mathcal{P}^{k-1}_{\mathrm{pw,d}}(\partial P)$ yields the system of linear equations

$$\boldsymbol{\mathcal{V}} \boldsymbol{t} = \left(\tfrac{1}{2}\boldsymbol{\mathcal{M}} + \boldsymbol{\mathcal{K}}\right) \boldsymbol{v}, \tag{14.12}$$

with $\boldsymbol{t} = (t_1, \ldots, t_{kn})$ and $\boldsymbol{v} = (v_1, \ldots, v_{kn})$, where

$$\boldsymbol{\mathcal{V}} = \left((\mathcal{V}\tau_j, \tau_i)_{L_2(\partial P)}\right)_{i,j=1}^{kn}, \qquad \boldsymbol{\mathcal{M}} = \left((\lambda_j, \tau_i)_{L_2(\partial P)}\right)_{i,j=1}^{kn},$$
$$\boldsymbol{\mathcal{K}} = \left((\mathcal{K}\lambda_j, \tau_i)_{L_2(\partial P)}\right)_{i,j=1}^{kn}, \qquad \boldsymbol{\mathcal{D}} = \left((\mathcal{D}\lambda_j, \lambda_i)_{L_2(\partial P)}\right)_{i,j=1}^{kn}.$$

If we apply analogously a Galerkin formulation to the second equation in (14.11) and replace $\gamma_1 v$ with t_h on the right-hand side, then we obtain the matrix

$$\boldsymbol{\mathcal{S}} = \boldsymbol{\mathcal{D}} + \left(\tfrac{1}{2}\boldsymbol{\mathcal{M}}^\top + \boldsymbol{\mathcal{K}}^\top\right)\boldsymbol{\mathcal{V}}^{-1}\left(\tfrac{1}{2}\boldsymbol{\mathcal{M}} + \boldsymbol{\mathcal{K}}\right).$$

Because of $\gamma_0\phi_i = \lambda_i$, the matrix entries are good approximations for

$$\boldsymbol{\mathcal{S}} \approx \left((\gamma_1\phi_j, \gamma_0\phi_i)_{L_2(\partial P)}\right)_{i,j=1}^{kn}. \tag{14.13}$$

The entries of the matrices $\boldsymbol{\mathcal{V}}$, $\boldsymbol{\mathcal{K}}$, and $\boldsymbol{\mathcal{D}}$ are double integrals over ∂P. The inner integral, which corresponds to the action of the boundary integral operator on a basis function, can be evaluated in closed form. The outer integral, which corresponds to the L_2-product, is approximated by numerical quadrature.

At this point, we stress that in the previous description the boundary element method is directly applied on the naturally given discretization of the polygonal element P. For example, if we have a pentagon and we are interested in $k = 1$, the Dirichlet trace is described exactly by five degrees of freedom on ∂P and the Neumann trace is approximated by one constant per edge. In general, the BEM makes use of a finer discretization of ∂P, but for our purpose the coarsest and naturally given discretization is sufficient.

Finally, we draw the connection to the global BEM-based FEM. If we compare (14.13) with (14.8) and (14.9), we see that $\boldsymbol{\mathcal{S}}$ serves as a good approximation of the local stiffness matrices in the finite element formulation. Furthermore, the approximation t_h of $\gamma_1 v$ for $v = \phi_i$ obtained by (14.12) is used in the representation formula (14.10) to evaluate the basis functions ϕ_i in the interior of the polygonal element P. All boundary element matrices are set up once per element in a BEM-based FEM simulation and are used for all basis functions. Since the number of vertices per element is bounded and we do not discretize the element boundaries further, the local BEM matrices are rather small and the inversion of $\boldsymbol{\mathcal{V}}$ can be done efficiently with a LAPACK routine.

14.2.4 Numerical examples

The theoretical results of Theorem 14.1 are illustrated on a model problem. The BEM-based FEM is applied on a sequence of uniformly refined polygonal meshes. In each step of the refinement, the boundary-value problem

$$-\Delta u = f \quad \text{in } \Omega = (0,1)^2, \qquad u = 0 \quad \text{on } \Gamma$$

is solved, where f is chosen such that $u(\boldsymbol{x}) = \sin(\pi x)\,\sin(\pi y)$ is the unique solution. The initial mesh and some refinements are shown in Figure 14.1. The successively refined meshes are obtained by dividing each polygonal element as described in [420], also see the next section. The Galerkin error $\|u - u_h\|_{H^\ell(\Omega)}$ is computed for the H^1-norm ($\ell = 1$) and the L_2-norm ($\ell = 0$). In Figure 14.2, the relative errors are plotted with respect to the mesh size $h = \max\{h_P : P \in \mathcal{T}_h\}$ on a logarithmic scale. The slopes of the curves reflect the theoretical rates of convergence for the approximation orders $k = 1, 2, 3$.

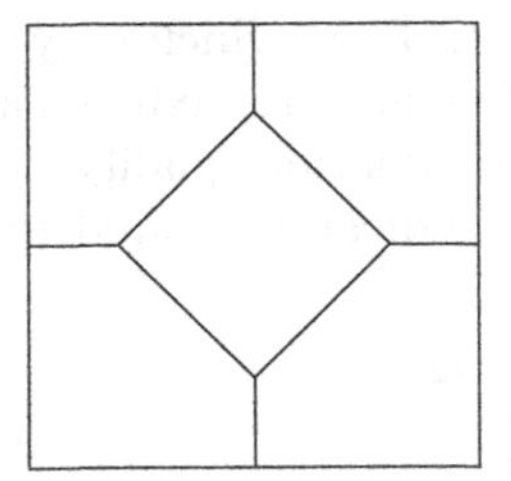
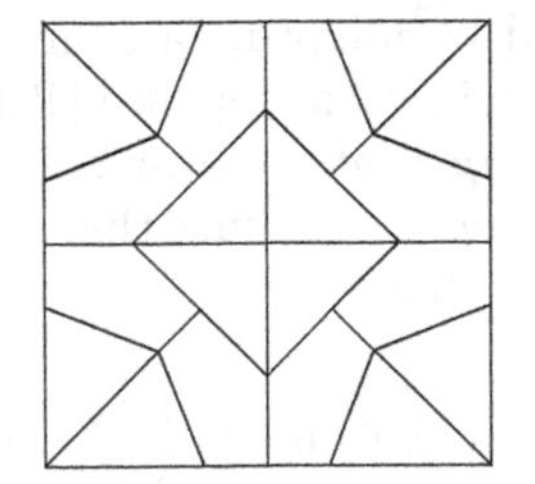
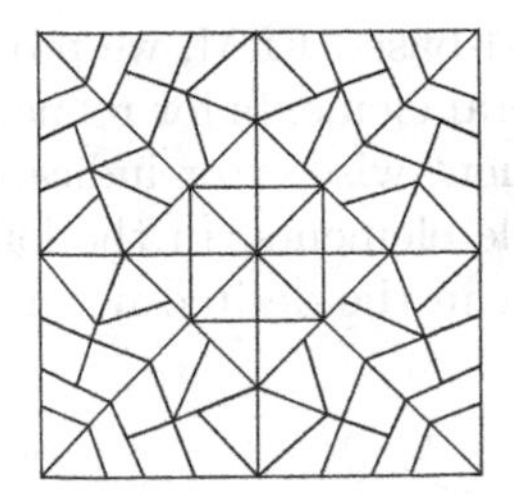

Figure 14.1 Initial mesh (left), refined mesh after two steps (middle), refined mesh after four steps (right).*

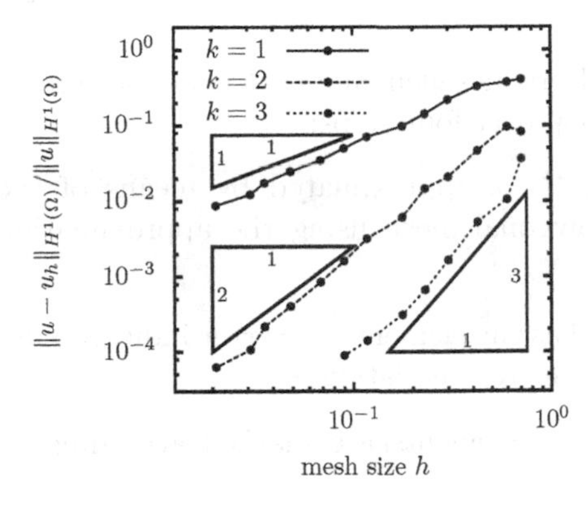

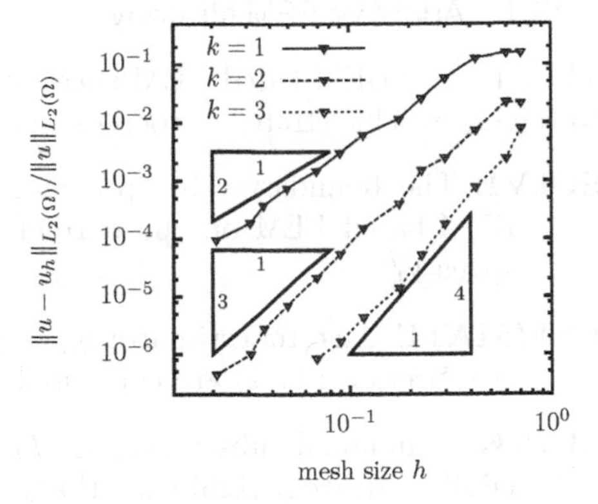

Figure 14.2 Relative error in H^1-norm (left) and L_2-norm (right) with respect to the mesh size h.†

14.3 ADAPTIVE BEM-BASED FEM IN 2D

In the uniform refinement strategy, each element of the mesh is split into two new elements to obtain the next finer mesh. This splitting process is performed as described below. To obtain the optimal rates of convergence according to Theorem 14.1, the solution has to be smooth. This assumption is often violated in practical applications, where singularities can arise due to discontinuous material parameters or reentrant corners in the geometry. Consequently, the convergence slows down for uniform refinement. Therefore, the meshes have to be adapted to the problem in order to recover optimal rates of convergence. For the adaptive

†Reprinted from Numerische Mathematik, Residual error estimate for BEM-based FEM on polygonal meshes, Vol. 118, 2011, pages 765–788, S. Weißer, © Springer-Verlag 2011, with permission of Springer.

*Reprinted from Computers & Mathematics with Applications, Vol. 67(7), S. Weißer, Arbitrary order Trefftz-like basis functions on polygonal meshes and realization in BEM-based FEM, pages 1390–1406, Copyright (2014), with permission from Elsevier.

BEM-based FEM, we proceed by adopting a common strategy, which only selects several elements for refinement in each step, see [403]. Essential ingredients are the element-wise error indicators η_P that monitor the approximation quality over the single elements. In the following we assume that these indicators bound the true error in the desired norm such that

$$\|u - u_h\| \le c\,\eta_R = c\left(\sum_{P\in\mathcal{T}_h} \eta_P^2\right)^{1/2}. \tag{14.14}$$

In Section 14.3.2, we present a residual-based error estimator, which bounds the error in the energy norm.

14.3.1 Adaptive FEM strategy

The adaptive BEM-based FEM successively refines elements that contribute most to the error. This strategy proceeds in a loop over four steps:

SOLVE The boundary-value problem (14.1) is approximated by means of the BEM-based FEM on the current polygonal mesh using the approximation space V_h^k.

ESTIMATE An error estimator η_R as well as the element-wise error indicators η_P (see Section 14.3.2) are computed on the discretization $\mathcal{T}_h$.

MARK A minimal subset $\mathcal{M}_h \subset \mathcal{T}_h$ of all elements are marked according to Dörflers strategy [125] such that

$$\left(\sum_{P\in\mathcal{M}_h} \eta_P^2\right)^{1/2} \ge (1-\theta)\,\eta_R,$$

where $0 \le \theta < 1$ is a user-defined parameter. To obtain a minimal set $\mathcal{M}_h$, it is possible to sort the elements according to their indicators η_P and mark those with the largest indicators. Instead of that, we implemented the marking algorithm given in [125] to achieve linear complexity. Furthermore, we choose $\theta = 0.5$ in the numerical experiments.

REFINE Each marked element is refined and we consequently obtain a new mesh for the next cycle in the loop. For the refinement of an element P, we bisect P through its barycenter $\bar{\boldsymbol{x}}$ orthogonal to its characteristic direction, see Figure 14.3, that is given by the eigenvector that corresponds to the largest eigenvalue of the matrix

$$M_{\text{Cov}} = \int_P (\boldsymbol{x}-\bar{\boldsymbol{x}})(\boldsymbol{x}-\bar{\boldsymbol{x}})^\top d\boldsymbol{x}, \qquad \bar{\boldsymbol{x}} = \frac{1}{|P|}\int_P \boldsymbol{x}\,d\boldsymbol{x}.$$

For more details, see [420]. Furthermore, we check the regularity of the mesh and refine additional elements if $c_{\mathcal{T}} h_P \le h_{[\boldsymbol{v}_i,\boldsymbol{v}_{i+1}]}$ for any $i = 1,\ldots,n$ is violated with a user-defined parameter $c_{\mathcal{T}}$.

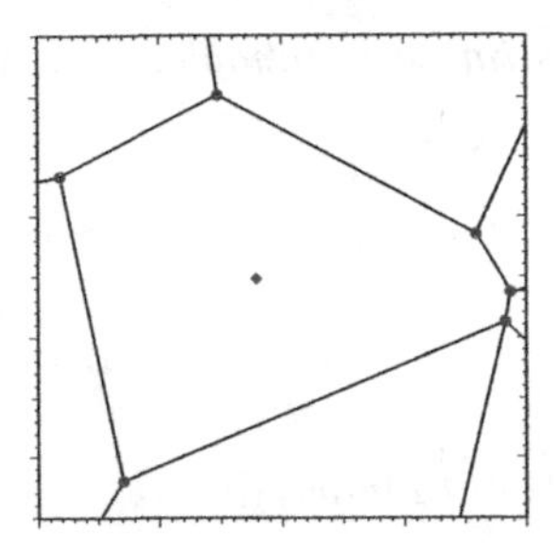
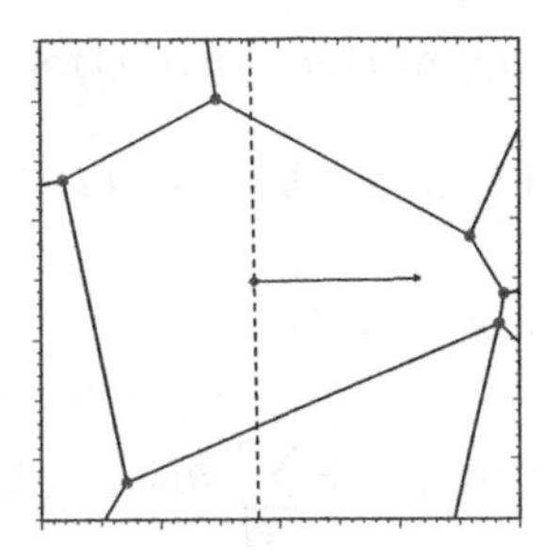
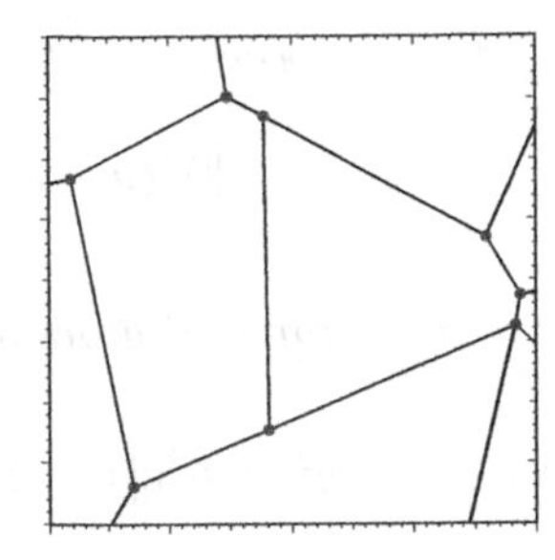

Figure 14.3 Refinement of an element: element with center $\bar{\boldsymbol{x}}$ (left), element with eigenvector (middle), two new elements (right).†

The adaptive mesh refinement process is kept very local. Only the marked and degenerated elements are bisected during the refinement. It is not necessary to resolve hanging nodes and keep the mesh admissible, as for example in the *red-blue-green* refinement procedure for triangular meshes, see [403]. This advantage is due to the polygonal meshes with very flexible elements.

14.3.2 Residual-based error estimate for polygonal meshes

The residual-based *a posteriori* error estimate bounds the difference of the exact solution and the Galerkin approximation in the energy norm associated to the bilinear form. Among others, the estimate contains the jumps of the conormal derivatives over the element edges. Let $P \in \mathcal{T}_h$ be a polygonal element with edge $[\boldsymbol{v}_i, \boldsymbol{v}_{i+1}]$ and $P' \in \mathcal{T}_h$ be the neighboring element that contains this edge. Then, the jump is given on $(\boldsymbol{v}_i, \boldsymbol{v}_{i+1})$ by

$$[\![u_h]\!] = \frac{\partial u_h|_P}{\partial \boldsymbol{n}_P} + \frac{\partial u_h|_{P'}}{\partial \boldsymbol{n}_{P'}}.$$

We define the so-called edge residual by

$$R_{P,i} = \begin{cases} 0 & \text{for } (\boldsymbol{v}_i, \boldsymbol{v}_{i+1}) \subset \Gamma_D, \\ g_N - \frac{\partial u_h}{\partial \boldsymbol{n}_P} & \text{for } (\boldsymbol{v}_i, \boldsymbol{v}_{i+1}) \subset \Gamma_N, \\ -\frac{1}{2}[\![u_h]\!] & \text{for } (\boldsymbol{v}_i, \boldsymbol{v}_{i+1}) \subset \Omega. \end{cases}$$

All ingredients are now available to state the residual-based error estimate that fulfils (14.14) for the energy norm $\|\nabla \cdot\|_{L_2(\Omega)}$.

Theorem 14.2 (Reliability). *Let $\mathcal{T}_h$ be a regular mesh. Furthermore, let $u \in V$ and $u_h \in V_h^k$ be the solutions of* (14.2) *and the Galerkin approximation obtained*

†Reprinted from Numerische Mathematik, Residual error estimate for BEM-based FEM on polygonal meshes, Vol. 118, 2011, pages 765–788, S. Weißer, © Springer-Verlag 2011, with permission of Springer.

by (14.3)*, respectively. Then the residual-based error estimate is reliable, i.e.,*

$$\|\nabla(u - u_h)\|_{L_2(\Omega)} \leq c\,\eta_R \quad \textit{with} \quad \eta_R^2 = \sum_{P \in \mathcal{T}_h} \eta_P^2,$$

where the error indicators are defined by

$$\eta_P^2 = h_P^2 \|f + \Delta u_h\|_{L_2(P)}^2 + \sum_{i=1}^{n} h_{[\boldsymbol{v}_i,\boldsymbol{v}_{i+1}]} \|R_{P,i}\|_{L_2([\boldsymbol{v}_i,\boldsymbol{v}_{i+1}])}^2.$$

The constant $c > 0$ only depends on the regularity parameters of the mesh and the approximation order k.

The error in the energy norm decreases at least as fast as the error estimator η_R according to the previous theorem. In order to have a good estimator, which can be used efficiently to estimate the true error, it is important that the error does not decrease faster than η_R. This behavior is shown in the next theorem, where the error indicators η_P are bounded in terms of the error plus some data oscillations. For the lower bound, the estimate involves the neighborhood ω_P of the element P. This patch of elements is given by $\bar{\omega}_P = \bigcup\{\bar{P}' : P' \in \mathcal{T}_h, \bar{P} \cap \bar{P}' \neq \emptyset\}$.

Theorem 14.3 (Efficiency)**.** *Under the assumptions of Theorem 14.2, the residual-based error indicator is efficient, i.e.,*

$$\eta_P \leq c \Bigg(\|\nabla(u - u_h)\|_{L_2(\omega_P)}^2 + h_P^2 \|f - \widetilde{f}\|_{L_2(\omega_P)}^2 + \sum_{\substack{i=1,\ldots,n: \\ [\boldsymbol{v}_i,\boldsymbol{v}_{i+1}] \not\subset \Gamma_D}} h_{[\boldsymbol{v}_i,\boldsymbol{v}_{i+1}]} \|g_N - \widetilde{g}_N\|_{L_2([\boldsymbol{v}_i,\boldsymbol{v}_{i+1}])}^2 \Bigg)$$

where $\widetilde{f}$ and $\widetilde{g}_N$ are piecewise polynomial approximations of the data f and g_N, respectively. The constant $c > 0$ only depends on the regularity parameters of the mesh and the approximation order k.

The terms $\|f - \widetilde{f}\|_{L_2(\omega_P)}$ and $\|g_N - \widetilde{g}_N\|_{L_2([\boldsymbol{v}_i,\boldsymbol{v}_{i+1}])}$ are often called data oscillations. They are usually of higher order.

14.3.3 Numerical examples

Let $\Omega = \big((-1,1) \times (-1,1)\big) \setminus \big([0,1] \times [0,-1]\big)$ and $\Gamma_D = \partial\Omega$. Using polar coordinates (r,φ) for $x = (r\cos\varphi, r\sin\varphi)$, the boundary data g_D is chosen in such a way that

$$u(r\cos\varphi, r\sin\varphi) = r^{2/3} \sin\left(\frac{2\varphi}{3}\right)$$

is the solution of the boundary-value problem

$$-\Delta u = 0 \quad \text{in } \Omega, \qquad u = g_D \quad \text{on } \Gamma.$$

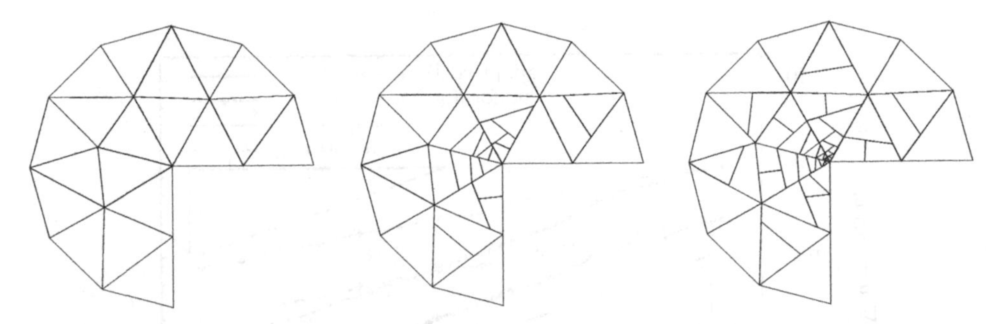

Figure 14.4 Initial mesh (left), adaptive refined mesh after five steps (middle), adaptive refined mesh after ten steps (right).†

The inhomogeneous Dirichlet data is treated by means of a discrete extension as usual in finite element methods. The boundary-value problem is discretized using the first order approximation space V_h^1 on uniformly and adaptively refined meshes. Because of the reentrant corner in the geometry, the solution does not meet the regularity assumptions of Theorem 14.1. The derivatives of u have a singularity at the origin of the coordinate system, and consequently, uniform refinement yields suboptimal rates of convergence. However, we still expect optimal rates of convergence for the adaptive BEM-based FEM computation.

On a sequence of uniform refined meshes, the following hold:

$$\text{DOFs} = \mathcal{O}(h^{-2}),$$

where DOFs denotes the number of degrees of freedom. Since the mesh size h is not strictly monotonically decreasing for adaptive refinement strategies, we study the convergence with respect to the number of degrees of freedom. Figure 14.4 shows the initial mesh as well as the adaptive meshes after five and ten refinement steps. The adaptive algorithm detects the singularity and tunes the mesh towards the origin of the coordinate system. The refinement is kept very local since hanging nodes do not have to be resolved by additional refinements. The convergence plot in Figure 14.5 confirms the predicted behavior. Whereas the convergence for uniform refinement slows down, the adaptive algorithm recovers linear convergence, which corresponds to a slope of $-1/2$. Furthermore, the residual-based error estimator η_R bounds the true error and represents the reduction of the error very well, see Figure 14.5.

In order to further stress the use and the flexibility of polygonal meshes in adaptive computations, we analyze the first two refinement steps. For this reason, the error distribution is visualized in Figure 14.6 for the first three meshes. Each element P is colored according to the value $\|\nabla(u-u_h)\|_{L_2(P)}$. The adaptive algorithm apparently marks and refines the elements with the largest error contribution. The introduced nodes on straight edges (hanging nodes for classical meshes) are not

†Reprinted from Numerische Mathematik, Residual error estimate for BEM-based FEM on polygonal meshes, Vol. 118, 2011, pages 765–788, S. Weißer, © Springer-Verlag 2011, with permission of Springer.

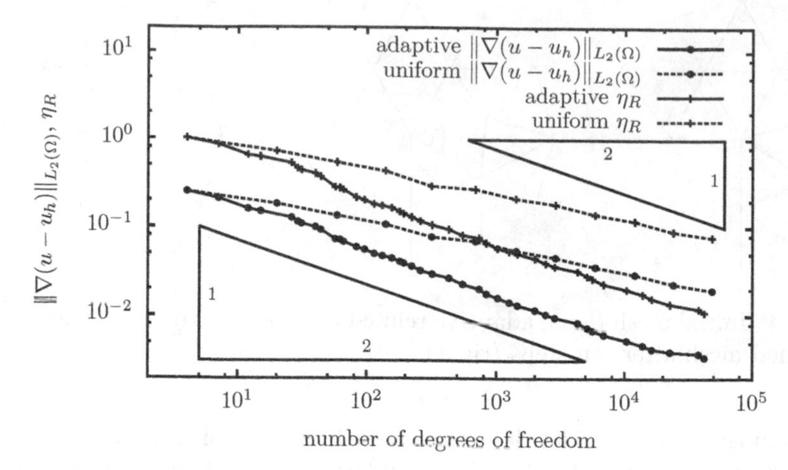

Figure 14.5 Convergence for singular solution on an arc, cf. Figure 14.4, on a logarithmic scale.†

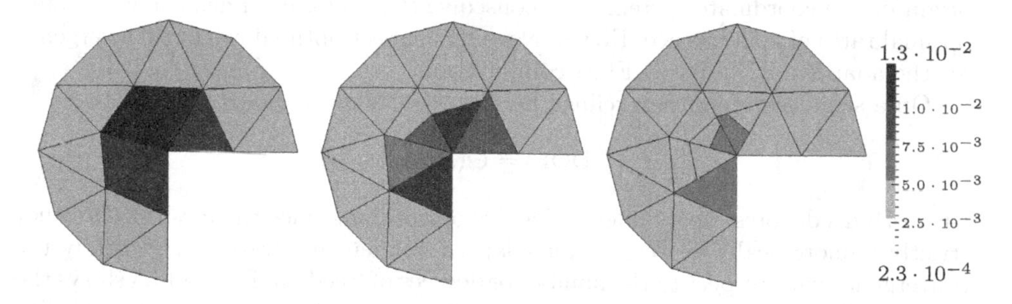

Figure 14.6 Error distribution $\|\nabla(u-u_h)\|^2_{L_2(P)}$ for the first three meshes.†

resolved. Each of these nodes corresponds to a degree of freedom in the FE computation and thus, improves the approximation within the neighboring elements. For example, the upper right triangle close to the reentrant corner in Figure 14.6 is not refined. But, the error reduces due to the additional nodes on the left edge, namely, the triangle became a pentagon in the right most mesh.

14.4 DEVELOPMENTS AND OUTLOOK

The idea of the BEM-based FEM has been applied to time-dependent problems [422] and in the discretization of $H(\text{div})$-conforming approximation spaces

†Reprinted from Numerische Mathematik, Residual error estimate for BEM-based FEM on polygonal meshes, Vol. 118, 2011, pages 765–788, S. Weißer, © Springer-Verlag 2011, with permission of Springer.

over polygonal meshes [138]. The application to three-dimensional problems on polyhedral meshes is discussed in the original work [107] and a generalization is given in [330]. Additionally, a finite element tearing and interconnecting (FETI) type solver has been developed to handle the large system of linear equations [198]. In this section, we present the essentials for 3D problems and provide an application to a convection-dominated boundary-value problem.

14.4.1 Hierarchical construction for 3D problems

In several publications on the BEM-based FEM, the approach is applied to three-dimensional problems. For simplicity we restrict ourselves to the first order method ($k = 1$) in the following. The construction of basis functions is straightforward, as soon as it is assumed that the faces of the polyhedral elements are triangles. In the original approach, the basis functions are defined analogously to the two-dimensional case. For each vertex $\boldsymbol{v}_i$, we have a function ϕ_i such that $\phi_i(\boldsymbol{v}_j) = \delta_{ij}$, it is linear on the triangular faces, and ϕ_i is harmonic inside the polyhedral element. Furthermore, a three-dimensional version of the boundary element method is available. Here, the boundary integral operators only differ in their kernel function and, of course, in the integration, which is done over two-dimensional surfaces of the polyhedra instead of one-dimensional boundaries of the polygons. In the treatment of the resulting boundary integral equation, in order to approximate the Neumann trace, the Sobolev spaces $H^{1/2}(\partial P)$ and $H^{-1/2}(\partial P)$ are discretized by continuous and piecewise linear polynomial functions over the surface triangles and by piecewise constant functions, respectively. All considerations on the assembling of the finite element matrix in 2D transfer to the 3D case.

What can be done in the case of general polyhedral elements with polygonal faces? A simple and practical idea is to triangulate the surface ∂P of the polyhedral element P first and then apply the strategy described above. However, this introduces additional (artificial) vertices to the polyhedral element, which lead to additional degrees of freedom in the finite element computation. An alternative approach has been proposed in [330] that is applicable on general polygonal faces.

In order to motivate the construction in 3D, we analyze once more the first-order basis functions ϕ_i for $i = 1, \ldots, n$ in 2D, or equivalently the harmonic coordinates from Section 1.2.8. These functions fulfil $\phi_i(\boldsymbol{v}_j) = \delta_{ij}$, are linear on the one-dimensional edges, and are harmonic inside the two-dimensional polygonal element. We observe that the second derivative of a linear function vanishes. Consequently, if we parametrize the edges linearly and treat ϕ_i as a function of one variable along the edge, we see that $\phi_i'' = 0$. Therefore, the linear function ϕ_i is harmonic in 1D on the edge and can be interpreted as the solution of a Laplace problem with boundary data that is correspondingly 0 and 1. From this viewpoint, it is clear how to proceed for the construction in 3D. Let Δ_1 and Δ_2 denote the one- and two-dimensional Laplace operators in the linear parameter spaces of an edge and face, respectively, and let Δ_3 be the usual Laplace operator in three dimensions.

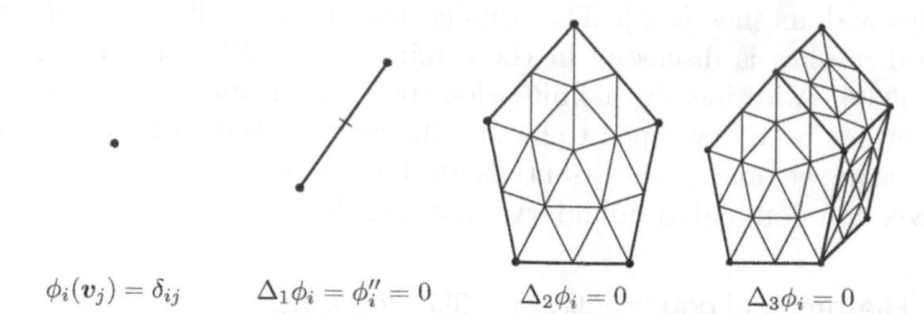

Figure 14.7 Construction of 3D basis functions for pentagonal prism P as well as auxiliary triangulation of the surface ∂P.

Then, the basis functions are defined by

$$\begin{aligned} \phi_i(\boldsymbol{v}_j) &= \delta_{ij} && j = 1, \dots, n, \\ -\Delta_1\phi_i &= 0 && \text{on each edge,} \\ -\Delta_2\phi_i &= 0 && \text{on each face,} \\ -\Delta_3\phi_i &= 0 && \text{in } P. \end{aligned}$$

This procedure is visualized in Figure 14.7, where the auxiliary triangulation of ∂P can be neglected for the definition. We prescribe the values at the vertices. Afterwards, we solve 1D Laplace problems on the edges with the given data at the vertices as boundary-values, which is equivalent to connecting the values by a linear function. Next, 2D Laplace problems are solved on each polygonal face with the previously defined data on the edges as boundary-values. And finally, the data on the faces is used as Dirichlet data for 3D Laplace problems in the element. Thus, we have a hierarchical construction for the basis functions starting from the definition at the vertices and going over the edges to the faces and finally to the elements. In the case of a polyhedral element P with triangular faces, the two strategies result in the same basis functions. If we have polygonal faces, however, the hierarchical strategy has fewer basis functions compared to the original one in general.

In the realization of the approach, we have to deal with an implicit definition of basis function on faces as well as on elements. To handle these two- and three-dimensional boundary-value problems, it has been proposed to introduce an auxiliary triangulation of the surface of P in [330]. For this reason the vertices of each face are connected with the center of mass of this face. The resulting triangulation can be refined successively such that several levels of nested triangular meshes are obtained. We denote the triangular mesh of the surface ∂P by $\mathcal{T}_h(\partial P)$. In Figure 14.7, $\mathcal{T}_h(\partial P)$ is shown for one successive refinement. Afterwards, the Sobolev spaces $H^{1/2}(\partial P)$ and $H^{-1/2}(\partial P)$ are discretized as usual on the surface triangulation $\mathcal{T}_h(\partial P)$, see above. The piecewise linear and continuous functions in the discrete space over $\mathcal{T}_h(\partial P)$ from the 3D BEM can be used for a 2D finite element method on each face to approximate the basis functions ϕ_i there. Consequently, standard 2D FEM tools are used to handle the boundary-value problems on the faces and standard 3D BEM tools are applied for the problems on the element. This approach yields optimal rates of convergence for polyhedral meshes in 3D, see [330].

14.4.2 Convection-adapted basis functions in 3D

In the following, we depart from the model Laplace equation and discuss the convection-diffusion equation

$$Lu = -\varepsilon \Delta u + \boldsymbol{b} \cdot \nabla u = 0 \tag{14.15}$$

on a bounded polyhedral domain $\Omega \subset \mathbb{R}^3$ with constant $\varepsilon > 0$, $\boldsymbol{b} \in \mathbb{R}^3$, $\boldsymbol{b} \neq 0$, and some boundary conditions. As already mentioned in the introduction, the basis functions ϕ_i are now constructed for the differential operator L. The lowest order basis functions fulfil $L\phi_i = 0$ instead of $-\Delta\phi_i = 0$ in each polyhedral element. While piecewise linear functions are recovered on triangular and tetrahedral meshes for the Laplace operator, we now build-in some non-linear features of the differential operator L into the functions ϕ_i.

The original version of the BEM-based FEM is applied in [197] to (14.15) on a fixed tetrahedral mesh in 3D. Thus, the basis functions ϕ_i are linear on the surface triangles of the tetrahedral elements and fulfil the differential equation (14.15) inside the elements. The case $\varepsilon \to 0$ is numerically analyzed. Standard methods without additional stabilization run into oscillations for small ε, since the equation degenerates to a first-order problem. The original BEM-based FEM formulation without explicit stabilization, however, shows less oscillations due to the built-in features. The hierarchical construction described in the previous subsection even improves this effect. In this case the basis functions ϕ_i fulfil certain projected convection-diffusion equations on the edges and faces of the tetrahedra, see Figure 14.7 and replace $-\Delta$ by L. The results presented in [199] are even promising when compared to a streamline upwind/Petrov-Galerkin (SUPG) finite element method, which has explicit stabilization to treat the convection-dominated case.

CHAPTER 15

Virtual Element Methods for Elliptic Problems on Polygonal Meshes

Andrea Cangiani, Oliver J. Sutton

University of Leicester, Leicester, UK

Vitaliy Gyrya, Gianmarco Manzini

Los Alamos National Laboratory, Los Alamos, USA

CONTENTS

THIS CHAPTER establishes a connection between the harmonic generalized barycentric coordinates (GBCs) and the lowest order virtual element method, resulting from the fact that the discrete function space is the same for both methods. This connection allows us to look at the high order virtual element spaces as a further generalization of the harmonic GBC in both two and three spatial dimensions. We also discuss how the virtual element methodology can be used to compute approximate solutions to PDEs without requiring any evaluation of functions in the local discrete spaces, which are implicitly defined through local boundary-value problems.

15.1 INTRODUCTION

The *virtual element method* (VEM) was originally introduced in [27] as a variational reformulation of the arbitrary order mimetic finite difference method [31, 71]; see also [32] and the review paper [253]. The VEM is a generalization of the finite element method (FEM), offering great flexibility in utilizing meshes with (almost) arbitrary polygonal elements, and the possibility of constructing discrete spaces incorporating an arbitrary degree of inter-element continuity [33].

The discrete space used by the lowest order conforming virtual element method on each polygonal mesh element is simply the space of harmonic GBC. The functions in this space are implicitly defined through local Dirichlet problems. These problems must be solved before they can be used in the standard harmonic GBC method [215]. However, the virtual element method is formulated in such a way that these local problems *do not* have to be solved. Instead, projections of the functions onto a subspace of polynomials are computed, which do not rely on evaluating the local functions, and these projections are employed in the method.

The virtual element method also provides generalizations of the harmonic GBC of arbitrary order, in the sense that they contain a subspace of polynomials of specified degree. These spaces still define the functions implicitly through local Dirichlet problems, but by carefully choosing the set of functions taken as the basis of the space, the virtual element method can still be computed without ever having to solve (or even approximate) the local problems. Indeed, the method only requires the knowledge of a polynomial subspace of the local finite element space to provide stable and accurate computational methods. This is accomplished by separating the contributions of the polynomial subspace from that of the complementary non-polynomial virtual subspace through the introduction of suitable projection operators that can be computed using only the VEM degrees of freedom. The polynomial *consistency* terms, responsible for the convergence properties of the method, are computed accurately. The remaining terms are only required to ensure the *stability* of the method, and hence they can be roughly estimated from the degrees of freedom.

We present here a link between a *natural* basis of the local discrete space, associated with a parametrization of the data of the local Dirichlet problems, and the *virtual* basis that is used in practice. This construction also addresses the commonly asked question of how the virtual element spaces can use polynomial moments as degrees of freedom when their formal definition does not contain any explicit reference to the moments themselves.

In the rest of the chapter we discuss the construction of the local function spaces from the perspective of generalized barycentric coordinates (Section 15.2), give a short overview of the nodal VEM for elliptic problems (Section 15.3), and outline the connection with similar methods, like the nodal mimetic finite difference method and the BEM-based FEM of Chapter 14 (Section 15.4).

15.2 VIRTUAL ELEMENT SPACES AND GBC

15.2.1 Generalities

Let Ω be an open polygonal domain in $\mathbb{R}^2$, partitioned by a finite mesh Ω_h consisting of *simple polygonal elements.* By "simple," we mean that the boundary ∂P of each polygonal mesh element P is not self-intersecting and consists of a uniformly bounded number of straight line segments. We further assume that the intersection of any two distinct elements of Ω_h is either empty, a collection of mesh vertices, or a collection of mesh edges. In particular, neighboring elements are permitted to share multiple (possibly coplanar) edges. We denote the set of all edges of the element P by $\mathcal{E}_P$. We also introduce the notation $\mathbb{P}_k(P)$ to denote the space of physical-frame polynomials of total degree $\leq k$ on the element P.

15.2.2 Lowest order discrete space

The lowest order local virtual element function space on the mesh element P coincides with the space of harmonic generalized barycentric coordinates on P, and consists of functions satisfying the following properties:

- they are harmonic on P,
- they are linear on each edge,
- the space $\mathbb{P}_1(P)$ of linear polynomials on P is included as a subspace.

Specifically, this space is defined such that

$$V_P^1 := \Big\{ v_h \in H^1(P) : \Delta v_h = 0 \text{ in } P,\ v|_e \in \mathbb{P}_1(e) \text{ for each edge } e \text{ of } \partial P,\ v_{h|\partial P} \in C^0(\partial P) \Big\},$$

and we can define a natural Lagrangian basis for this space as the set of functions $\{\phi_i\}_{i=1}^n$, such that $\phi_i(\boldsymbol{v}_j) = \delta_{ij}$ for each vertex $\boldsymbol{v}_j$ of P. This definition ensures that these basis functions are linearly independent, and provides us with a natural way to identify V_P^1 with the space $\mathbb{R}^n$ by associating a vector of n numbers with the values of a function at the n vertices of P. These values will be referred to as the *degrees of freedom* of the space. Using this basis, any function $v \in V_P^1$ can be represented as:

$$v(\boldsymbol{x}) = \sum_{i=1}^{n} v(\boldsymbol{v}_i)\phi_i(\boldsymbol{x}) \quad \forall \boldsymbol{x} \in P. \tag{15.1}$$

In particular, using (15.1) with $v = 1, x, y$ we obtain the decompositions:

$$\sum_{i=1}^{n} \phi_i = 1, \quad \sum_{i=1}^{n} x_i\phi_i = x, \quad \sum_{i=1}^{n} y_i\phi_i = y, \tag{15.2}$$

where x_i and y_i are the x and y coordinates of the vertex $\boldsymbol{v}_i$ respectively. The first identity in (15.2) is the *partition of unity* property of barycentric coordinates, and together with the other identities expresses the *linear completeness* of V_P^1.

The global functional space is constructed from the local spaces as

$$V_h^1 = \left\{ v \in H^1(\Omega) \cap C^0(\overline{\Omega}) \, : \, v_{|P} \in V_P^1 \right\},$$

and as global degrees of freedom we can take the values of the functions at the vertices of the mesh, thus ensuring the continuity property of the ambient space.

While (in this case) the virtual element and harmonic GBC discrete spaces are identical, the crucial difference between the methods lies in how these functions are used. In particular, the virtual element method is designed in such a way that the local PDE problems, which define the space, *never have to be solved* because the method does not require evaluating the basis functions (cf. Section 15.3).

15.2.3 Generalization to arbitrary order discrete spaces

The virtual element technology provides a generalization of the local harmonic GBC spaces to include the space of polynomials of arbitrary degree k. Extensions to arbitrary global regularity that is more than just $C^0(\Omega)$ are also possible [33]. These extra polynomials can be used to improve the accuracy and convergence rate of the method, just like within the standard finite element method.

The general order local discrete space V_P^k is constructed on the element P for any integer $k \geq 1$ as solutions of local Poisson problems with a polynomial right-hand side and continuous piecewise polynomial Dirichlet boundary data, such that

$$\begin{aligned} V_P^k := \Big\{ v_h \in H^1(P) \, : & \Delta v_h \in \mathbb{P}_{k-2}(P), \\ & v|_e \in \mathbb{P}_k(e) \text{ for each edge } e \text{ of } P, \\ & v_{h|\partial P} \in C^0(\partial P) \Big\}, \end{aligned} \tag{15.3}$$

where we adopt the convention that $\mathbb{P}_{-1}(P) = \{0\}$. When $k = 1$, this definition simply produces the space described in the previous section. Moreover, this space will contain the space $\mathbb{P}_k(P)$ of physical-frame polynomials of degree k on P. As before, the global space is constructed from the local spaces as

$$V_h^k = \left\{ v \in H^1(\Omega) \cap C^0(\overline{\Omega}) \, : \, v_{|P} \in V_P^k \right\}. \tag{15.4}$$

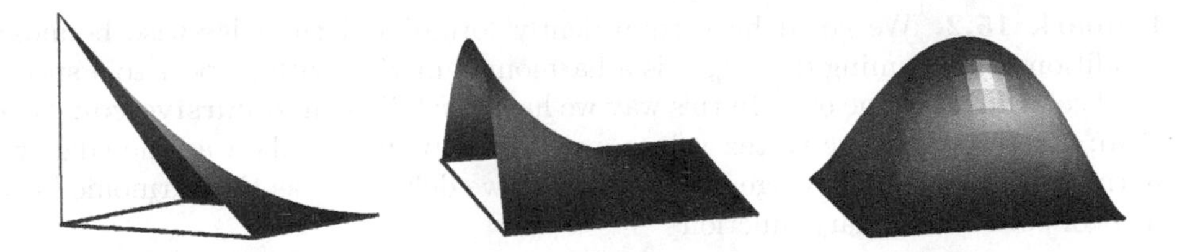

Figure 15.1 Vertex (left), edge (middle), and interior (right) polygonal basis functions for $k = 2$ on a square element. See color insert.

Remark 15.1. Using a Neumann-type boundary condition in the local definition (15.3) leads to a different class of virtual element spaces, known as the *non-conforming virtual elements* [21, 82]. In this setting, the formal definition in (15.4) must be changed, since non-conforming functions can be discontinuous over Ω and, consequently, are not in $H^1(\Omega) \cap C^0(\overline{\Omega})$. However, we will not pursue this method any further here.

15.2.4 The natural basis

An intuitive choice of basis for the general local space V_P^k is to use the degrees of freedom of the data that define the Dirichlet problem on each element P. This yields the set of basis functions

$$\boldsymbol{\psi} = \Big\{ \{\psi_{\boldsymbol{v}_i}\}_{i=1}^{n},\ \{\psi_{e,\alpha}\}_{\alpha=1}^{\mathcal{N}_e} \text{ for each } e \in \mathcal{E}_P,\ \{\psi_{P,\alpha}\}_{\alpha=1}^{\mathcal{N}_P} \Big\},$$

where each function is the formal solution of an independent problem associated with a vertex, edge, or the interior of P. Figure 15.1 shows an example of these basis functions for $k = 2$ on a square element. From the definition of the VE spaces, we may note that $k = 2$ is the lowest order case for which edge and interior basis functions are present. Indeed, for $k = 1$ we only have vertex basis functions, which are also the same for every $k > 1$. Here, n still denotes the number of vertices of P, and

$$\begin{aligned} \mathcal{N}_e &:= \dim\big(\mathbb{P}_{k-2}(e)\big) = k - 1; \\ \mathcal{N}_P &:= \dim\big(\mathbb{P}_{k-2}(P)\big) = k(k-1)/2. \end{aligned}$$

The basis functions are defined as follows:

Vertex basis functions $\psi_{\boldsymbol{v}_i}$

For every vertex $\boldsymbol{v}_i$, we define the function $\psi_{\boldsymbol{v}_i} \in V_P^k$ as the unique solution of the following Poisson problem:

$$\left\{ \begin{array}{lll} -\Delta\psi_{\boldsymbol{v}_i} & = 0 & \text{in } P, \\ \psi_{\boldsymbol{v}_i}(\boldsymbol{v}_j) & = \delta_{ij} & \text{for } i, j = 1, \ldots, n, \\ \psi_{\boldsymbol{v}_i}|_e \in \mathbb{P}_1(e) & & \forall e \in \mathcal{E}_P. \end{array} \right. \tag{15.5}$$

Remark 15.2. We could have equivalently formulated the edge-wise boundary conditions by assuming that $\psi_{\boldsymbol{v}_i}|_e$ is a harmonic function with respect to a specific local coordinate frame on e. In this way we have the following **recursive construction**: starting from the vertex values, we set the boundary data as the edge-wise harmonic lifting on each edge of ∂P and then we define $\psi_{\boldsymbol{v}_i}$ as the harmonic lifting in P of such a boundary function.

These vertex basis functions coincide with those considered in Section 15.2.2, although for $k > 1$ they are included alongside the extra edge and interior polygonal basis functions. In this sense, this natural basis can be seen as an augmentation of the standard basis of the harmonic GBC space.

Edge basis functions $\psi_{e,\alpha}$

If $k > 1$, for every edge $e \in \mathcal{E}_P$, let $\{q_{e,\alpha}\}_{\alpha=1}^{\mathcal{N}_e}$ be a basis for the polynomial space $\mathbb{P}_k(e)/\mathbb{P}_1(e)$ such that each $q_{e,\alpha}$ is zero at the end points of e. For every index $\alpha = 1, \ldots, \mathcal{N}_e$, the function $\psi_{e,\alpha} \in V_P^k$ is the solution of

$$\begin{cases} -\Delta\psi_{e,\alpha} = 0 & \text{in } P, \\ \psi_{e,\alpha|\partial P} = \begin{cases} q_{e,\alpha} & \text{on } e, \\ 0 & \text{otherwise.} \end{cases} \end{cases} \tag{15.6}$$

The crucial point is that the edge polynomials $q_{e,\alpha}$ together with the linear polynomials defining the vertex basis functions (15.5) form a basis for $\mathbb{P}_k(e)$, meaning that with these two sets of basis functions we have parametrized all of the possible choices of Dirichlet data for the local problems defining the space.

Interior basis functions $\psi_{P,\alpha}$

Let $\{q_{P,\alpha}\}_{\alpha=1}^{\mathcal{N}_P}$ be a basis for the polynomial space $\mathbb{P}_{k-2}(P)$. For every index $\alpha = 1, \ldots, \mathcal{N}_P$, we define $\psi_{P,\alpha} \in V_P^k$ to be the solution of

$$\begin{cases} -\Delta\psi_{P,\alpha} = q_{P,\alpha} & \text{in } P, \\ \psi_{P,\alpha|\partial P} = 0 & \text{in } \partial P. \end{cases} \tag{15.7}$$

Taken together, the functions $\psi_{P,\alpha}$, $\psi_{e,\alpha}$, and $\psi_{\boldsymbol{v}_i}$ are linearly independent and represent a basis for V_P^k. By counting them we conclude that the size of V_P^k is given by

$$\dim\left(V_P^k\right) = n + n(k-1) + \mathcal{N}_P.$$

As mentioned above, the vertex and edge basis functions fully determine the behavior of the functions of V_P^k on ∂P, because the interior polygonal basis functions $\psi_{P,\alpha}$ satisfy zero Dirichlet boundary conditions.

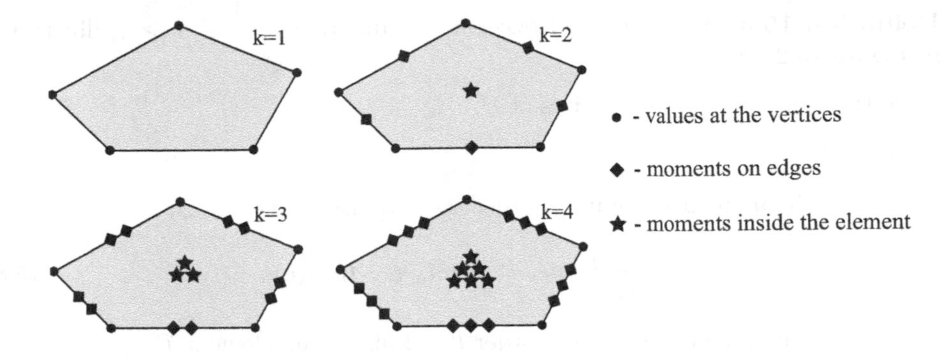

Figure 15.2 Illustration of the degrees of freedom of V_P^k for $k = 1, \dots, 4$.

15.2.5 A convenient basis

While the basis introduced in the previous section provides an intuitive splitting of the discrete space, using it in practice for solving PDEs requires (approximately) solving the associated Dirichlet problem for each basis function in order to evaluate the bilinear form of the global problem. Instead, we now describe an alternative basis for the virtual element spaces introduced in the previous section. With this new basis, we can gain access to partial information about the functions in V_P^k without needing to evaluate them directly, and hence without having to solve the local boundary-value problems. From the degrees of freedom, we will be able to compute local polynomial projections on each mesh element. In the context of the discretization of PDEs, these piecewise-polynomial projections can then be used to build a discrete bilinear form in such a way that the functions of V_P^k never have to be evaluated. Hence, this basis provides us with a method of using the generalized harmonic function spaces in such a way that we never need to even approximately solve the local boundary value problems.

For the special case of $k = 1$, we still take the standard vertex basis functions familiar from harmonic GBC. However, we still construct the discrete bilinear form in a virtual way that avoids ever having to evaluate the basis functions inside the element, and therefore avoids solving the local Dirichlet problems.

The basis for the general order space, on the other hand, is constructed differently from the *natural* basis. Instead of parametrizing the data of the local Dirichlet problems, we directly parametrize the functions in the space through their values at the vertices of the element, their moments on each edge, and their moments inside the element. Although not particularly intuitive, this choice of degrees of freedom is shown to be unisolvent for the space in [27] using formal arguments. In Section 15.2.6 we show a correspondence between this basis and the natural basis of the previous section, which can be seen as a constructive unisolvence proof.

Specifically, the basis is chosen to be the Lagrangian basis with respect to the degrees of freedom, and is illustrated in Figure 15.2:

Definition 15.3. The *degrees of freedom* of a function $v \in V_P^k$, $k \geq 1$, illustrated in Figure 15.2, are:

- the value of v at the vertices of P;
- for $k > 1$,
  - the moments of v up to order $k-2$ on each edge $e \in \mathcal{E}_P$:
$$\frac{1}{|e|}\int_e v m_\alpha \, dS \quad \forall m_\alpha \in \mathcal{M}_{k-2}(e); \tag{15.8}$$
  - the moments of v up to order $k-2$ inside the element P:
$$\frac{1}{|P|}\int_P v m_\alpha \, dS \quad \forall m_\alpha \in \mathcal{M}_{k-2}(P). \tag{15.9}$$

where $\mathcal{M}_k(P)$ and $\mathcal{M}_k(e)$ are the sets of *scaled monomials* of degree k, forming a basis of $\mathbb{P}_k(P)$ and $\mathbb{P}_k(e)$, respectively, and defined such that

$$\mathcal{M}_k(P) = \bigcup_{\ell \leq k} \mathcal{M}_\ell^*(P), \quad \text{with} \quad \mathcal{M}_\ell^*(P) := \left\{ \left(\frac{\boldsymbol{x} - \boldsymbol{x}_P}{h_P}\right)^{\mathrm{s}}, \quad |\mathrm{s}| = \ell \right\},$$

where $\boldsymbol{s} = (s_1, s_2)$ is a 2-index with the usual notation $|\boldsymbol{s}| = s_1 + s_2$ and $\boldsymbol{x}^{\boldsymbol{s}} = x_1^{s_1} x_2^{s_2}$. The scaled monomials $\mathcal{M}_k(e)$ are similarly defined by scaling back the previous definition to one spatial dimension.

The basis with respect to these degrees of freedom will be denoted by

$$\boldsymbol{\phi} = \left\{ \{\phi_{\boldsymbol{v}_i}\}_{i=1}^n, \ \{\phi_{e,\alpha}\}_{\alpha=1}^{k-1} \text{ for each } e \in \mathcal{E}_P, \ \{\phi_{P,\alpha}\}_{\alpha=1}^{\mathcal{N}_P} \right\},$$

associated with the vertex, edge, and internal degrees of freedom, respectively. While the vertex basis functions coincide with those of the natural basis in the lowest order case $k = 1$, this will not be true for $k > 1$ since these basis functions may no longer be linear on each edge.

This choice of degrees of freedom is particularly useful because it allows us to compute several different projections of functions in V_P^k onto the polynomial subspace $\mathbb{P}_k(P)$ *directly from the degrees of freedom* of the function, requiring knowledge only of the polynomial subspace. These projectors are crucial for the method presented in Section 15.3.

For instance, we can compute the $H^1(P)$-orthogonal projector $\Pi_k^\nabla : V_P^k \to \mathbb{P}_k(P)$, defined for $v \in V_P^k$ to be such that

$$\begin{cases} (\nabla(v - \Pi_k^\nabla v), \nabla p)_P = 0 & \forall p \in \mathbb{P}_k(P), \\ \int_{\partial P} v - \Pi_k^\nabla v \, dS = 0 \end{cases}$$

where $(\cdot,\cdot)_P$ denotes the $L^2(P)$ inner product. Integrating by parts, this can be rewritten as

$$(\nabla \Pi_k^\nabla v, \nabla p)_P = \int_{\partial P} v \boldsymbol{n} \cdot \nabla p \, dS - (v, \Delta p)_P,$$

where $\boldsymbol{n}$ denotes the unit vector normal to ∂P. The terms on the right-hand side of the above equation can be expressed through the degrees of freedom of v. The first term can be computed exactly because $\boldsymbol{n} \cdot \nabla p$ is a polynomial of degree $k-1$ on each edge of P, while Δp is a polynomial of degree $k-2$ inside P, so the first and the second terms are simply a sum of, respectively, the edge and internal degrees of freedom of v. Clearly, however, this projection would not be so straightforward to compute using the natural basis of Section 15.2.4. Expanding $\Pi_k^\nabla v$ in the basis of $\mathbb{P}_k(P)$ and varying p over the basis of $\mathbb{P}_k(P)$, it is clear that this provides a linear system of equations that can be solved for the coefficients of $\Pi_k^\nabla v$.

Similarly, we can compute the $L^2(P)$-orthogonal projection of the gradient onto polynomials, defined such that

$$(\nabla v - \Pi_{k-1}^0 \nabla v, \boldsymbol{p})_P = 0 \qquad \forall \boldsymbol{p} \in (\mathbb{P}_{k-1}(P))^2,$$

where $(\mathbb{P}_{k-1}(P))^2$ denotes the space of vector-valued polynomials of degree $k-1$. This projection is computed in a similar way, by integrating by parts to find that

$$(\Pi_{k-1}^0 \nabla v, \boldsymbol{p})_P = \int_{\partial P} v\boldsymbol{n} \cdot \boldsymbol{p}\, dS - (v, \nabla \cdot \boldsymbol{p})_P,$$

and observing that the terms on the right are exactly computable as before.

15.2.6 A link between the bases

We now establish a correspondence between the natural basis ψ of Section 15.2.4 and the virtual basis ϕ of Section 15.2.5. To this end, we define a linear transformation that maps the ψ basis to the ϕ basis:

$$\phi_{P,\alpha} = \sum_{1\le\gamma\le\mathcal{N}_P} C_{P,\gamma}^{P,\alpha} \psi_{P,\gamma} \qquad 1 \le \alpha \le \mathcal{N}_P, \tag{15.10}$$

$$\phi_{e,\alpha} = \sum_{1\le\gamma\le\mathcal{N}_P} C_{P,\gamma}^{e,\alpha} \psi_{P,\gamma} + \sum_{\substack{e'\in\mathcal{E}_P,\\ 1\le\gamma\le k-1}} C_{e',\gamma}^{e,\alpha} \psi_{e',\gamma} \qquad \forall e \in \mathcal{E}_P,\ 1 \le \alpha \le k-1, \tag{15.11}$$

$$\phi_{\boldsymbol{v}_i} = \sum_{1\le\gamma\le\mathcal{N}_P} C_{P,\gamma}^{\boldsymbol{v}_i} \psi_{P,\gamma} + \sum_{\substack{e'\in\mathcal{E}_P,\\ 1\le\gamma\le k-1}} C_{e',\gamma}^{\boldsymbol{v}_i} \psi_{e',\gamma} + \psi_{\boldsymbol{v}_i} \qquad 1 \le i \le n. \tag{15.12}$$

The system (15.10)–(15.12) may then be solved for the "C" coefficients in the three consecutive steps explained in the following paragraphs. Note the *reverse order* of this construction: we first solve for the interior basis functions $\phi_{P,\alpha}$, then for the edge basis functions $\phi_{e,\alpha}$, and finally for the vertex basis functions $\phi_{\boldsymbol{v}_i}$. This derivation can also be seen as an alternative constructive proof of the unisolvency of the degrees of freedom of Definition 15.3.

Interior basis functions $\phi_{P,\alpha}$

These are the basis functions associated with the internal degrees of freedom in (15.9). We determine the coefficients $C^{P,\alpha}_{P,\gamma}$ in (15.10) for $1 \leq \alpha, \gamma \leq \mathcal{N}_P$ by explicitly requiring that

$$\int_P \phi_{P,\alpha} q_{P,\beta} \, dV = \delta_{\alpha,\beta}, \qquad 1 \leq \alpha, \beta \leq \mathcal{N}_P.$$

Edge basis functions $\phi_{e,\alpha}$

These are the basis functions associated with the edge degrees of freedom in (15.8). In (15.11), we determine the coefficients $C^{e,\alpha}_{e',\gamma}$ by explicitly requiring that

$$\int_{e'} \phi_{e,\alpha} q_{e',\beta} \, dS = \begin{cases} \delta_{\alpha,\beta}, & \text{for } e = e', \\ 0, & \text{otherwise} \end{cases} \qquad \forall e, e' \in \mathcal{E}_P, \quad 1 \leq \alpha, \quad \beta \leq k - 1,$$

and the coefficients $C^{e,\alpha}_{P,\gamma}$ by explicitly requiring that

$$\int_P \phi_{e,\alpha} q_{P,\beta} \, dV = 0 \qquad \forall e \in \mathcal{E}_P, \quad 1 \leq \alpha \leq k-1, \quad 1 \leq \beta \leq \mathcal{N}_P.$$

Vertex basis functions $\phi_{\boldsymbol{v}_i}$

These are the basis functions associated with the cell vertices according to Definition 15.3. The *canonical condition* $\phi_{\boldsymbol{v}_i}(\boldsymbol{x}_j) = \delta_{ij}$ is automatically satisfied by the definition given in (15.12) since $\psi_{\boldsymbol{v}_i}(\boldsymbol{x}_j) = \delta_{ij}$, while $\psi_{e,\gamma}(\boldsymbol{x}_j) = 0$ for every $e \in \mathcal{E}_P$ and $1 \leq \gamma \leq k-1$, and $\psi_{P,\gamma}(\boldsymbol{x}_j) = 0$ for every $1 \leq \gamma \leq \mathcal{N}_P$. Hence, the coefficients "C" on the right-hand side of (15.12) must be determined so that all internal moments and edge moments of each $\phi_{\boldsymbol{v}_i}$ are zero. To this end, we require that the coefficients $C^{\boldsymbol{v}_i}_{e',\gamma}$ of (15.12) satisfy the condition

$$\int_e \phi_{\boldsymbol{v}_i} q_{e,\beta} \, dS = 0 \qquad \forall \boldsymbol{v}_i \in \partial P, \quad \forall e \in \mathcal{E}_P, \quad 1 \leq \beta \leq k-1,$$

and the coefficients $C^{\boldsymbol{v}_i}_{P,\gamma}$ satisfy the condition:

$$\int_P \phi_{\boldsymbol{v}_i} q_{P,\beta} \, dV = 0 \qquad \forall \boldsymbol{v}_i \in \partial P, 1 \leq \beta \leq \mathcal{N}_P.$$

It is easy to prove that the above set of conditions is unisolvent and hence the two bases represent the same space. However, we stress that the VEM uses the virtual basis *directly* through the corresponding degrees of freedom, and none of the local basis functions are ever computed or evaluated in practice.

15.2.7 Extension to three dimensions

We now briefly consider the situation when Ω is a subset of $\mathbb{R}^3$. In this case, Ω_h is assumed to be a mesh consisting of very general shaped polyhedra with planar faces and straight edges, and the virtual space (15.3) is defined by taking

$$V_P^k := \Big\{ v_h \in H^1(P) : \Delta v_h \in \mathbb{P}_{k-2}(P); \quad v|_e \in V_f^k \text{ for each face } f \text{ of } P; \quad v_{h|\partial P} \in C^0(\partial P) \Big\}, \tag{15.13}$$

where V_f^k is the two-dimensional discrete space on the face f, introduced in Section 15.2.3 (with the convention that $\mathbb{P}_{-1}(P) = \{0\}$).

As for the two-dimensional case, two equivalent bases for this space can be considered: the natural basis, defined through a set of Dirichlet problems from the space definition (15.13); a virtual basis, defined through the degrees of freedom associated with the vertex values and the moments on the edges, the faces, and the interior of element P. However, in the 3D case the orthogonal projections of Section 15.2.5 cannot be computed directly using these degrees of freedom, since this would require the knowledge of the moments of order up to $k-1$ and k on the element faces, which are not available from (15.13).

To get around this problem, the space must be modified in such a way that these extra moments are available. One approach is to simply expand the space by relaxing the conditions required in its definition such that the extra moments can be taken as degrees of freedom. However, this results in a significant increase in the dimension of the space and therefore the size of the final matrix. Instead, a process known as *VEM enhancement* has been developed (see [6, 82]) to produce a space with the same virtual degrees of freedom as V_P^k, but in which these moments may be computed. We point out that although the virtual degrees of freedom remain unchanged, the natural basis for V_P^k is no longer valid for the enhanced space.

15.3 VIRTUAL ELEMENT METHOD FOR ELLIPTIC PDES

15.3.1 Model problem

Consider the diffusion of the scalar variable u, which satisfies the following Poisson problem (we refer to [29, 82] for a treatment of more general elliptic problems):

$$-\operatorname{div}(K\nabla u) = f \quad \text{in } \Omega, \tag{15.14}$$

$$u = g \quad \text{on } \partial\Omega, \tag{15.15}$$

where K is the diffusion tensor, f is the forcing term, and g defines the (typically non-homogeneous) Dirichlet boundary condition. We assume that:

(H1) Ω is a bounded, open, polytopal subset of $\mathbb{R}^d$;

(H2) the diffusion tensor $K : \Omega \rightarrow \mathbb{R}^{d\times d}$ is a $d \times d$ bounded, measurable, and symmetric tensor. Moreover, we assume that K is *strongly elliptic*, i.e., there exist two positive constants C_* and C^* such that, for almost every $\boldsymbol{x} \in \Omega$,

$$C_*||\boldsymbol{\xi}||^2 \leq \boldsymbol{\xi} \cdot K(\boldsymbol{x})\boldsymbol{\xi} \leq C^*||\boldsymbol{\xi}||^2 \qquad \forall \boldsymbol{\xi} \in \mathbb{R}^d,$$

where $||\boldsymbol{\xi}||^2 = \boldsymbol{\xi} \cdot \boldsymbol{\xi}$ is the square of the Euclidean 2-norm of $\boldsymbol{\xi}$;

(H3) the two functions f and g belong to $L^2(\Omega)$ and $H^{\frac{1}{2}}(\partial\Omega)$, respectively.

The variational form of the problem (15.14)–(15.15) reads:

find $u \in H^1_g(\Omega)$ such that

$$\mathcal{A}(u, v) := \int_\Omega K\nabla u \cdot \nabla v \, dV = \int_\Omega fv \, dV \qquad \forall v \in H^1_0(\Omega), \tag{15.16}$$

where

$$H^1_g(\Omega) := \{u \in H^1(\Omega) \,|\, u_{|\partial\Omega} = g\}.$$

The existence of a unique weak solution to (15.16) follows from the continuity and coercivity of the bilinear form $\mathcal{A}(\cdot,\cdot)$ under Assumptions (H1)–(H3).

15.3.2 Overview of the conforming VEM

In the conforming virtual element method we approximate the symmetric, bilinear form $\mathcal{A}(\cdot,\cdot)$ with the symmetric bilinear form $\mathcal{A}_h(\cdot,\cdot)$ that can be *exactly* computed without having to evaluate the functions of the virtual space. Moreover, $\mathcal{A}_h(\cdot,\cdot)$ should exactly evaluate $\mathcal{A}(\cdot,\cdot)$ when at least one of its entries is a polynomial of degree $k \geq 1$. This property is called *polynomial consistency* and can be stated as:

$$\mathcal{A}_h(v, p) = \mathcal{A}(v, p),$$

which holds for any $v \in V_h$ and piecewise k-degree polynomial function p defined on Ω_h. The polynomial consistency property determines the accuracy of the method. Beyond this, we merely require $\mathcal{A}_h(\cdot,\cdot)$ to be *stable* with respect to $\mathcal{A}(\cdot,\cdot)$ when the trial and test functions are non-polynomial. The right-hand side of (15.16) is also discretized through a similar process. Finally, the Dirichlet boundary conditions of the original problem are imposed in the virtual element formulation by defining $V^k_{h,g}$ as the affine subspace of V^k_h whose functions interpolate the boundary condition g on $\partial\Omega$. In a practical implementation, the boundary conditions on the virtual functions of $V^k_{h,g}$ are imposed by restricting their degrees of freedom associated with the vertices, and edges lying on $\partial\Omega$, to coincide with vertex values and boundary moments of g.

The virtual element approximation to (15.16) reads: find $u_h \in V^k_{h,g}$ such that:

$$\mathcal{A}_h(u_h, v_h) = (f, v_h)_h \qquad \forall v_h \in V^k_{h,0}. \tag{15.17}$$

The well-posedness of this problem follows from the well-posedness of (15.16) through the stability property mentioned above. The discrete bilinear form in (15.17) is defined as the sum of local contributions

$$\mathcal{A}_h(u_h, v_h) = \sum_{P\in\Omega_h} \mathcal{A}_h^P(u_h, v_h),$$

each elemental term of which is the sum of a *consistency* term and a *stability* term,

$$\mathcal{A}_h^P(u_h, v_h) = \mathcal{A}_{\text{cons}}^P(u_h, v_h) + \mathcal{A}_{\text{stab}}^P(u_h, v_h),$$

where

$$\mathcal{A}_{\text{cons}}^P(u_h, v_h) = \int_P K\,\Pi_{k-1}^0 \nabla u_h \cdot \Pi_{k-1}^0 \nabla v_h \, dV,$$

$$\mathcal{A}_{\text{stab}}^P(u_h, v_h) = \overline{K}_P \mathcal{S}_P\big((I - \Pi_k^\nabla)u_h, (I - \Pi_k^\nabla)v_h\big),$$

with Π_{k-1}^0 and Π_k^∇ defined as in Section 15.2.5. Here, $\overline{K}_P$ is a constant approximation of K on P, as for example, its cell average or its value at the barycenter of the cell, and $\mathcal{S}_P : V_P^k \times V_P^k \to \mathbb{R}$ is the bilinear form

$$\mathcal{S}_P(u_h, v_h) := \sum_{r=1}^{n} \text{dof}_r(u_h)\text{dof}_r(v_h),$$

where, for $w_h \in V_P^k$, we use $\text{dof}_r(w_h)$ to denote the value of the r-th degree of freedom of w_h with respect to some arbitrary but fixed ordering of the local degrees of freedom. This bilinear form is just the Euclidean inner product on the space $\mathbb{R}^n$ of vectors of degrees of freedom. The right-hand side is discretized as

$$(f, v_h)_h := \sum_{P\in\Omega_h} \int_P f \Pi_{k-2}^0 v_h \, dV.$$

Remark 15.4. This approximation is sub-optimal in the L^2-norm for $k = 1, 2$. However, for these cases it is possible to construct alternative approximations of the right-hand side, which makes the method optimal.

Under fairly general mesh assumptions [82], it may be shown that there exists a constant C independent of h and P such that

$$C^{-1}\mathcal{A}^P(v_h, v_h) \leq \mathcal{A}_h^P(v_h, v_h) \leq C\mathcal{A}^P(v_h, v_h)$$

for all $v_h \in V_P^k$, which provides the *stability* of the method and ensures the existence of a unique discrete solution. Moreover, only the first term of $\mathcal{A}_h^P(\cdot,\cdot)$ will be non-zero when either u_h or v_h is a polynomial, and when both are polynomials the bilinear form will exactly evaluate $\mathcal{A}^P(\cdot,\cdot)$.

All terms can be evaluated using only the degrees of freedom and the ability to evaluate and integrate polynomials. Implementation details can be found in [28, 386]. In particular, no pointwise evaluation of the virtual functions other than the polynomial subspace is ever required. Given the *virtual* nature of the method, the numerical solution is known in the form of its degrees of freedom, i.e., nodal values and moments, from which a global continuous solution is obtained by interpolation.

15.4 CONNECTION WITH OTHER METHODS

In recent years, there has been a growing interest in numerical methods for PDEs on general polygonal and polynomial meshes. We refer, for example, to the mimetic finite difference (MFD) method [32, 253], composite finite elements (CFEs) in both their conforming [184] and discontinuous Galerkin (dG) [11] version related to the agglomeration dG method [25], *hp*-dG [80], HDG [105], weak Galerkin methods [287], Hybrid high-order (HHO) methods [123], polygonal finite element method [378], BEM-based FEM [421], extended finite element method [158], and gradient scheme [126], to mention just a few. In this section, we discuss methods in the above list that are related to the VEM. In particular, PFEM and BEM-based FEM are based on similar discretizations spaces. The VEM approach can also be viewed as a variational analogue of the MFD method.

15.4.1 Polygonal and polyhedral finite element method

Assuming that we know how to compute the basis functions at any point of P, a finite element formulation is readily available for polygonal and polyhedral elements. For the lowest-order case (i.e., for $k = 1$), such a method coincides with an instance of the PFEM method. For a more detailed discussion, see, for instance, the review in [268, 378]. For $k > 1$, the calculation of the integrals of the stiffness matrix and the right-hand side would be very expensive, which would render this approach to be non-competitive with other methods, and in particular, with the VEM. A different strategy based on approximating the Poisson problems that provide the basis functions can be devised by reformulating such problems as boundary element problems [421]. More details are discussed in Section 15.4.3.

15.4.2 Nodal MFD method

In this section, we compare the formulation of the conforming VEM to that of the mimetic finite difference method [32, 253] for the model problem presented in Section 15.3.1. For simplicity of exposition we assume that the diffusion tensor K is piecewise constant on Ω_h and denote its restriction to P by K_P. The major difference between the two formulations is that the MFD method follows a discrete approach like that of finite volumes, whereas the VEM relies on a continuous variational formulation like that of the finite element method. In fact, the conforming VEM is a variational reformulation of the nodal MFD method proposed in [31, 71] and the two families of schemes are equivalent up to the stability term.

To ease the notation at the price of a minor inconsistency, we will use the symbols $\mathcal{A}_h$ and $\mathcal{A}_h^P$ to denote the global and local discrete operators provided by the MFD method. We will also continue referring to them as "*bilinear forms*" although these operators are matrices that act directly on the degrees of freedom through a matrix-vector product. Moreover, in this section u_h, v_h represent both continuous functions and the associated "*grid functions*," whose values are associated with geometric objects such as vertices, edges, faces, and cells. We represent u_h and v_h as the

N_{dofs}-sized column vectors U_h and V_h, respectively, where N_{dofs} is the total number of degrees of freedom.

Just like in the VEM approach, the global bilinear form $\mathcal{A}_h$ is assembled as the sum of local contributions $\mathcal{A}_h^P(u_h, v_h)$. The elemental bilinear form $\mathcal{A}_h^P(\cdot,\cdot)$ is now built to explicitly satisfy the two following conditions:

- (S1) *Polynomial Consistency:* for any $p \in \mathbb{P}_k(P)$ and any grid function u_h, it holds that
$$\mathcal{A}_h^P(u_h, p) = \mathcal{A}(u_h, p). \tag{15.18}$$

- (S2) *Stability:* there exist two constants $\alpha^* > \alpha_* > 0$ independent of the mesh, such that for any u_h it holds that
$$\alpha_* \mathcal{A}(u_h, u_h) \leq \mathcal{A}_h^P(u_h, u_h) \leq \alpha^* \mathcal{A}(u_h, u_h). \tag{15.19}$$

The conditions (S1)–(S2) are also satisfied in the VEM. However, these conditions are an explicit part of the MFD construction process, whereas in the VEM they are the consequence of the formulation of the method.

The action of the mimetic bilinear form on the degrees of freedom of u_h and v_h is expressed by
$$\mathcal{A}_h^P(v_h, u_h) = V_h^T \mathbf{A}^P U_h,$$

where $\mathbf{A}^P$ is a positive semidefinite matrix. After an integration by parts, the polynomial consistency condition (15.18) can be restated as

$$\begin{aligned} U_h^T \mathbf{A}^P N_\alpha &= \int_P \nabla u_h \cdot K_P \nabla m_\alpha \, dV \\ &= -\int_P u_h \operatorname{div}(K_P \nabla m_\alpha)\, dV + \sum_{e \in \mathcal{E}_P} \int_e u_h \, \boldsymbol{n}_e \cdot K_P \nabla m_\alpha \, dS \\ &= U_h^T R_\alpha, \end{aligned} \tag{15.20}$$

where m_α is a scaled monomial of $\mathcal{M}_k(P)$, N_α the column vector of its degrees of freedom, and R_α the corresponding column vector of coefficients determined by (15.20). Assembling all column vectors N_α and R_α into the column-partitioned matrices $\mathbf{N} = (N_\alpha)$ and $\mathbf{R} = (R_\alpha)$ and noting that U_h is arbitrary, we can rewrite the consistency conditions (15.20) in matrix form as
$$\mathbf{A}^P \mathbf{N} = \mathbf{R}. \tag{15.21}$$

On a general d-dimensional polytope, (15.21) provides an underdetermined linear system that has a non-unique solution. A possible solution to (15.21) is:
$$\mathbf{A}_{\text{cons}}^P := \mathbf{R}(\mathbf{R}^T \mathbf{N})^\dagger \mathbf{R}^T, \tag{15.22}$$

where $(\mathbf{R}^T\mathbf{N})^\dagger$ denotes the pseudo-inverse of $(\mathbf{R}^T\mathbf{N})$ (note that matrix $\mathbf{R}^T\mathbf{N}$ is the stiffness matrix of the scaled monomials $\mathcal{M}_k(P)$ and thus is singular).

Matrix $\mathbf{A}^P_{\text{cons}}$ as defined by (15.22) does not satisfy the stability condition (15.19) due to an enlarged kernel. Hence a stability term has to be added, which is given by

$$\mathbf{A}^P_{\text{stab}} := \mathbf{N}_\perp \mathbf{S} \mathbf{N}_\perp^T,$$

where $\mathbf{S}$ is a symmetric, positive definite matrix, and the columns of $\mathbf{N}_\perp$ are orthogonal to the columns of $\mathbf{N}$ and such that the span of $\mathbf{N}$ and $\mathbf{N}_\perp$ is $\mathbb{R}^{N_{\text{dofs}}}$. From the orthogonality $\mathbf{N}_\perp^T\mathbf{N} = 0$ it follows that $\mathbf{A}^P_{\text{stab}}\mathbf{N} = 0$. Thus, for any choice of $\mathbf{S}$, the combined matrix

$$\mathbf{A}^P = \mathbf{A}^P_{\text{cons}} + \mathbf{A}^P_{\text{stab}}$$

still satisfies the consistency condition (15.21). The stability condition is satisfied by taking $\mathbf{S}$ to be spectrally equivalent to the non-zero part of $\mathbf{A}^P_{\text{cons}}$. A particularly simple but rather effective choice for $\mathbf{S}$ is the scalar matrix $\text{tr}(\mathbf{A}^P_{\text{cons}})|P|\mathbf{I}$, where $\text{tr}(\mathbf{A}^P_{\text{cons}})$ is the trace of matrix $\mathbf{A}^P_{\text{cons}}$.

The choice of $\mathbf{S}$ can be optimized for a particular set of criteria [183]. This optimization process is dubbed "*mimetic adaptation*" or "*m-adaptation*" and can lead to certain improvements of the scheme such as attaining a discrete form of maximum/minimum principles [254] or reducing the numerical dispersion for wave problems by several orders of magnitude on structured meshes [59, 182].

15.4.3 BEM-based FEM

A direct evaluation of the basis functions of V_P^k for $k > 1$ would require the approximation of the variational problems that define them on P. A few attempts have been made along this direction by meshing each element and solving the Poisson problems through a local finite element approximation. However, such a direct approach is prohibitive for its complexity and the related computational costs. An interesting alternative is found in the BEM-based FEM first introduced in [107] and reported in Chapter 14. We also refer to [421] for the two-dimensional case and [330] for the three-dimensional case.

In the two-dimensional case, the set of basis functions considered by the BEM-based FEM is precisely the one in Section 15.2.4 for all polynomial degrees, cf. Chapter 14. Therefore, the evaluation of the basis functions can be dealt with by rewriting the Poisson boundary-value problems (15.5), (15.6), and (15.7) as boundary element problems for the conormal derivative defined on the elemental boundary. This latter problem is then solved by a standard finite element method, whence the coining of the method as BEM-based FEM. Computational costs and complexity are reduced since the approximation is one-dimensional. Indeed, although the boundary element formulation leads to full matrices, these are rather small as no additional finer discretization of the boundary is needed.

Here we summarize how the local bases (associated with each vertex, edge, or face) of Section 15.2.4 can be approximated using the BEM approach and their use within the BEM-based FEM. The problem for an interior basis can always be treated by transforming it into a problem with homogeneous right-hand-side and non-homogeneous boundary data (cf. [421]).

Therefore, we only need to consider the case of a vertex or edge basis function, which solves a problem such as:

$$\begin{cases} -\Delta\phi = 0 & \text{in } P, \\ \phi_{|\partial P} = g, \end{cases} \tag{15.23}$$

where $g \in C(\partial P)$ is polynomial on each edge.

Let $\gamma_0^P \phi$ and $\gamma_1^P \phi = \boldsymbol{n}_P \cdot \gamma_0^P \nabla\phi$ denote, respectively, the trace and the Neumann trace of ϕ. It is well known that the solution operator for (15.23) can be seen as a map of the trace into the Neumann trace (Dirichlet-to-Neumann or Steklov–Poincaré operator). Formally,

$$\gamma_1^P \phi = \mathbf{S}_P \gamma_0^P \phi. \tag{15.24}$$

The Steklov–Poincaré operator is obtained by considering the representation formula for the solution of (15.23):

$$\phi(x) = \int_{\partial P} U^*(x,y)\gamma_1^P \phi(y) ds_y - \int_{\partial P} \gamma_{1,y}^P U^*(x,y)\gamma_0^P \phi(y) ds_y \tag{15.25}$$

for $x \in P$, where $U^*(x,y) := -\frac{1}{2\pi} \ln|x-y|$ is the fundamental solution of the negative Laplacian in two dimensions. Taking the trace of the above yields the following problem for $\gamma_1^P \phi$:

$$\mathbf{V}_P \gamma_1^P \phi = (\frac{1}{2}\mathbf{I} + \mathbf{K}_P)\gamma_0^P \phi, \tag{15.26}$$

where $\mathbf{V}_P$ and $\mathbf{K}_P$ are the single-layer and double-layer potential operators and $\mathbf{I}$ is the identity operator, from which (15.24) is derived; see Chapter 14 or [367] for details. The BEM now consists of: (1) discretizing problem (15.26) with the use of finite element methods to compute numerical approximations $\widetilde{\gamma_1^P \phi}$, and (2) evaluating the approximation $\widetilde{\phi}$ inside P by the representation formula (15.25). A guide on fast numerical approaches to boundary integral equations is given in [328].

Now, suppose we seek the numerical solution of problem (15.16). Here we assume that the diffusion coefficient is a piecewise constant scalar, say κ. Then the BEM-based FEM operates as follows. First, following the approach described above, compute on each $P \in \Omega_h$ a numerical approximation to each basis function in (15.5), (15.6), and (15.7). Second, solve the discrete problem (15.17) with

$$\mathcal{A}_h(u_h, v_h) = \sum_{P\in\Omega_h} \mathcal{A}_h^P(u_h, v_h) = \sum_{P\in\Omega_h} \int_{\partial P} \widetilde{\gamma_1^P u_h}\, \gamma_0^P v_h \, ds - \int_P \Delta u_h \, \widetilde{v_h} \, dV$$

$$(f, v_h)_h = \sum_{P\in\Omega_h} \int_P f\widetilde{v_h} \, dV.$$

The advantage of this approach is that no element-wise sub-meshing is required, although some form of volumetric quadrature is still needed in order to compute the volumetric terms in the previous equations.

Bibliography

[1] A. Abdollahi, D. Millán, C. Peco, M. Arroyo, and I. Arias. Revisiting pyramid compression to quantify flexoelectricity: A three-dimensional simulation study. *Physical Review B*, 91(10):Article 104103, 8 pages, 2015.

[2] A. Abdollahi, C. Peco, D. Millán, M. Arroyo, G. Catalan, and I. Arias. Fracture toughening and toughness asymmetry induced by flexoelectricity. *Physical Review B*, 92(9):Article 094101, 6 pages, 2015.

[3] R. A. Adams and J. Fournier. Cone conditions and properties of Sobolev spaces. *Journal of Mathematical Analysis and Applications*, 61(3):713–734, 1977.

[4] R. A. Adams and J. J. F. Fournier. *Sobolev Spaces*, volume 140 of *Pure and Applied Mathematics*. Academic Press, Amsterdam, 2nd edition, 2003.

[5] L. V. Ahlfors. *Complex Analysis*. International Series in Pure and Applied Mathematics. McGraw-Hill, New York, 3rd edition, 1979.

[6] B. Ahmad, A. Alsaedi, F. Brezzi, L. D. Marini, and A. Russo. Equivalent projectors for virtual element methods. *Computers & Mathematics with Applications*, 66(3):376–391, 2013.

[7] N. Al Shenk. Uniform error estimates for certain narrow Lagrange finite elements. *Mathematics of Computation*, 63(207):105–119, 1994.

[8] M. Alexa and M. Wardetzky. Discrete Laplacians on general polygonal meshes. *ACM Transactions on Graphics*, 30(4):Article 102, 10 pages, 2011.

[9] P. Alliez, D. Cohen-Steiner, M. Yvinec, and M. Desbrun. Variational tetrahedral meshing. *ACM Transactions on Graphics*, 24(3):617–625, 2005.

[10] D. Anisimov, C. Deng, and K. Hormann. Subdividing barycentric coordinates. *Computer Aided Geometric Design*, 43:172–185, 2016.

[11] P. F. Antonietti, S. Giani, and P. Houston. *hp*-Version composite discontinuous Galerkin methods for elliptic problems on complicated domains. *SIAM Journal on Scientific Computing*, 35(3):A1417–A1439, 2013.

[12] D. N. Arnold, R. S. Falk, and R. Winther. Finite element exterior calculus, homological techniques, and applications. *Acta Numerica*, 15:1–155, 2006.

[13] V. I. Arnold and B. A. Khesin. *Topological Methods in Hydrodynamics*, volume 125 of *Applied Mathematical Sciences*. Springer, New York, 1998.

[14] M. Arroyo, L. Heltai, D. Millán, and A. DeSimone. Reverse engineering the euglenoid movement. *Proceedings of the National Academy of Sciences*, 109(44):17874–17879, 2012.

[15] M. Arroyo and M. Ortiz. Local maximum-entropy approximation schemes: A seamless bridge between finite elements and meshfree methods. *International Journal for Numerical Methods in Engineering*, 65(13):2167–2202, 2006.

[16] M. Arroyo and M. Ortiz. Local maximum-entropy approximation schemes. In M. Griebel and M. A. Schweitzer, editors, *Meshfree Methods for Partial Differential Equations III*, volume 57 of *Lecture Notes in Computational Science and Engineering*, pages 1–16. Springer, Berlin, 2007.

[17] S. N. Atluri and E. Reissner. On the formulation of variational theorems involving volume constraints. *Computational Mechanics*, 5(5):337–344, 1989.

[18] F. Aurenhammer. A criterion for the affine equivalence of cell complexes in $\mathbb{R}^d$ and convex polyhedra in $\mathbb{R}^{d+1}$. *Discrete & Computational Geometry*, 2(1):49–64, 1987.

[19] F. Aurenhammer. Power diagrams: properties, algorithms, and applications. *SIAM Journal on Computing*, 16(1):78–96, 1987.

[20] F. Aurenhammer. Recognising polytopical cell complexes and constructing projection polyhedra. *Journal of Symbolic Computation*, 3(3):249–255, 1987.

[21] B. Ayuso de Dios, K. Lipnikov, and G. Manzini. The nonconforming virtual element method. *ESAIM: Mathematical Modelling and Numerical Analysis*, 50(3):879–904, 2016.

[22] I. Babuška and A. K. Aziz. On the angle condition in the finite element method. *SIAM Journal on Numerical Analysis*, 13(2):214–226, 1976.

[23] K. Bagi. When Heyman's Safe Theorem of rigid block systems fails: Non-Heymanian collapse modes of masonry structures. *International Journal of Solids and Structures*, 51(14):2696–2705, 2014.

[24] B. R. Baliga and S. V. Patankar. A control volume finite-element method for two-dimensional fluid flow and heat transfer. *Numerical Heat Transfer*, 6(3):245–261, 1983.

[25] F. Bassi, L. Botti, A. Colombo, D. Di Pietro, and P. Tesini. On the flexibility of agglomeration based physical space discontinuous Galerkin discretizations. *Journal of Computational Physics*, 231(1):45–65, 2012.

[26] C. Batty, S. Xenos, and B. Houston. Tetrahedral embedded boundary methods for accurate and flexible adaptive fluids. *Computer Graphics Forum*, 29:695–704, 2010.

[27] L. Beirão da Veiga, F. Brezzi, A. Cangiani, G. Manzini, L. D. Marini, and A. Russo. Basic principles of virtual element methods. *Mathematical Models and Methods in Applied Sciences*, 23(1):199–214, 2013.

[28] L. Beirão da Veiga, F. Brezzi, L. D. Marini, and A. Russo. The hitchhiker's guide to the virtual element method. *Mathematical Models and Methods in Applied Sciences*, 24(8):1541–1573, 2014.

[29] L. Beirão da Veiga, F. Brezzi, L. D. Marini, and A. Russo. Virtual element method for general second-order elliptic problems on polygonal meshes. *Mathematical Models and Methods in Applied Sciences*, 26(4):729–750, 2016.

[30] L. Beirão da Veiga and K. Lipnikov. A mimetic discretization of the Stokes problem with selected edge bubbles. *SIAM Journal on Scientific Computing*, 32(2):875–893, 2010.

[31] L. Beirão da Veiga, K. Lipnikov, and G. Manzini. Arbitrary-order nodal mimetic discretizations of elliptic problems on polygonal meshes. *SIAM Journal on Numerical Analysis*, 49(5):1737–1760, 2011.

[32] L. Beirão da Veiga, K. Lipnikov, and G. Manzini. *The Mimetic Finite Difference Method for Elliptic Problems*, volume 11 of *MS&A – Modeling, Simulation and Applications*. Springer, Cham, 2014.

[33] L. Beirão da Veiga and G. Manzini. A virtual element method with arbitrary regularity. *IMA Journal of Numerical Analysis*, 34(2):759–781, 2014.

[34] V. V. Belikov, V. D. Ivanov, V. K. Kontorovich, S. A. Korytnik, and A. Y. Semenov. The non-Sibsonian interpolation: A new method of interpolation of the values of a function on an arbitrary set of points. *Computational Mathematics and Mathematical Physics*, 37(1):9–15, 1997.

[35] M. Belkin, J. Sun, and Y. Wang. Constructing Laplace operator from point clouds in $\mathbb{R}^d$. In *Proceedings of the 20th Annual ACM-SIAM Symposium on Discrete Algorithms (SODA 2009)*, pages 1031–1040, 2009.

[36] S. R. Bell. *The Cauchy Transform, Potential Theory and Conformal Mapping*. Studies in advanced Mathematics. CRC Press, Boca Raton, 1992.

[37] A. Belyaev. On transfinite barycentric coordinates. In *Proceedings of the 4th Eurographics Symposium on Geometry Processing (SGP 2006)*, pages 89–99, 2006.

[38] A. Belyaev and P.-A. Fayolle. On transfinite Gordon-Wixom interpolation schemes and their extensions. *Computers & Graphics*, 51:74–80, 2015.

[39] A. Belyaev and P.-A. Fayolle. On variational and PDE-based distance function approximations. *Computer Graphics Forum*, 34(8):104–118, 2015.

[40] A. Belyaev, P.-A. Fayolle, and A. Pasko. Signed L_p-distance fields. *Computer-Aided Design*, 45(2):523–528, 2013.

[41] T. Belytschko, W. K. Liu, B. Moran, and K. I. Elkhodary. *Nonlinear Finite Elements for Continua and Structures*. Wiley, Chichester, 2nd edition, 2014.

[42] M. Ben-Chen, O. Weber, and C. Gotsman. Variational harmonic maps for space deformation. *ACM Transactions on Graphics*, 28(3):Article 34, 11 pages, 2009.

[43] M. Berger. *A Panoramic View of Riemannian Geometry*. Springer, Berlin, 2003.

[44] M. Biot. Surface instability of rubber in compression. *Applied Scientific Research. Section A*, 12(2):168–182, 1963.

[45] J. E. Bishop. Rapid stress analysis of geometrically complex domains using implicit meshing. *Computational Mechanics*, 30(5):460–478, 2003.

[46] J. E. Bishop. Simulating the pervasive fracture of materials and structures using randomly close packed Voronoi tessellations. *Computational Mechanics*, 44(4):455–471, 2009.

[47] J. E. Bishop. A displacement-based finite element formulation for general polyhedra using harmonic shape functions. *International Journal for Numerical Methods in Engineering*, 97(1):1–31, 2014.

[48] J. E. Bishop, J. M. Emery, R. V. Field, C. R. Weinberger, and D. J. Littlewood. Direct numerical simulations in solid mechanics for understanding the macroscale effects of microscale material variability. *Computer Methods in Applied Mechanics and Engineering*, 287:262–289, 2015.

[49] J. E. Bishop, M. J. Martinez, and P. Newell. Simulating fragmentation and fluid-induced fracture in disordered media using random finite element meshes. *International Journal for Multiscale Computational Engineering*, 14(4):349–366, 2016.

[50] J. E. Bishop and O. E. Strack. A statistical method for verifying mesh convergence in Monte Carlo simulations with application to fragmentation. *International Journal for Numerical Methods in Engineering*, 88(3):279–306, 2011.

[51] A. Biswas, V. Shapiro, and I. Tsukanov. Heterogeneous material modeling with distance fields. *Computer Aided Geometric Design*, 21(3):215–242, 2004.

[52] P. Block. *Thrust Network Analysis: Exploring Three-Dimensional Equilibrium*. PhD thesis, Massachusetts Institute of Technology, 2009.

[53] P. Block and L. Lachauer. Closest-fit, compression-only solutions for free form shells. In *Proceedings of the Jointly Organized 35th Annual Symposium of IABSE and 52nd Annual Symposium of IASS*, 2011.

[54] P. Block and J. Ochsendorf. Thrust network analysis: A new methodology for three-dimensional equilibrium. *Journal of the International Association for Shell and Spatial Structures*, 48(3):167–173, 2007.

[55] T. Bobach, M. Hering-Bertram, and G. Umlauf. Comparison of Voronoi based scattered data interpolation schemes. In *Proceedings of Visualization, Imaging, and Image Processing (VIIP 2006)*, pages 342–349, 2006.

[56] A. I. Bobenko, H. Pottmann, and J. Wallner. A curvature theory for discrete surfaces based on mesh parallelity. *Mathematische Annalen*, 348(1):1–24, 2010.

[57] A. I. Bobenko and B. A. Springborn. A discrete Laplace–Beltrami operator for simplicial surfaces. *Discrete & Computational Geometry*, 38(4):740–756, 2007.

[58] D. Boffi, F. Brezzi, and M. Fortin. *Mixed Finite Element Methods and Applications*, volume 44 of *Springer Series in Computational Mathematics*. Springer, Heidelberg, 2013.

[59] V. A. Bokil, N. L. Gibson, V. Gyrya, and D. A. McGregor. Dispersion reducing methods for edge discretizations of the electric vector wave equation. *Journal of Computational Physics*, 287:88–109, 2015.

[60] J. E. Bolander and S. Saito. Fracture analyses using spring networks with random geometry. *Engineering Fracture Mechanics*, 61(5–6):569–591, 1998.

[61] D. Bommes, B. Lévy, N. Pietroni, E. Puppo, C. Silva, M. Tarini, and D. Zorin. Quad-mesh generation and processing: A survey. *Comput. Graph. Forum*, 32(6):51–76, 2013.

[62] A. Bompadre, L. E. Perotti, C. J. Cyron, and M. Ortiz. Convergent meshfree approximation schemes of arbitrary order and smoothness. *Computer Methods in Applied Mechanics and Engineering*, 221–222:83–103, 2012.

[63] J. Bonet and R. D. Wood. *Nonlinear Continuum Mechanics for Finite Element Analysis*. Cambridge University Press, Cambridge, 2nd edition, 2008.

[64] M. J. Borden, T. J. R. Hughes, C. M. Landis, and C. V. Verhoosel. A higher-order phase-field model for brittle fracture: Formulation and analysis within the isogeometric analysis framework. *Computer Methods in Applied Mechanics and Engineering*, 273:100–118, 2014.

[65] A. Bossavit. *Computational Electromagnetism*. Electromagnetism. Academic Press, San Diego, 1998.

[66] M. Botsch, L. Kobbelt, M. Pauly, P. Alliez, and B. Lévy. *Polygon Mesh Processing*. A K Peters, Natick, 2010.

[67] S. Boyd and L. Vandenberghe. *Convex Optimization.* Cambridge University Press, Cambridge, 2004.

[68] S. K. Boyd and R. Muller. Smooth surface meshing for automated finite element model generation from 3D image data. *Journal of Biomechanics*, 39(7):1287–1295, 2006.

[69] J. H. Bramble and S. R. Hilbert. Estimation of linear functionals on Sobolev spaces with application to Fourier transforms and spline interpolation. *SIAM Journal on Numerical Analysis*, 7(1):112–124, 1970.

[70] S. C. Brenner and L. R. Scott. *The Mathematical Theory of Finite Element Methods*, volume 15 of *Texts in Applied Mathematics.* Springer, New York, 3rd edition, 2008.

[71] F. Brezzi, A. Buffa, and K. Lipnikov. Mimetic finite differences for elliptic problems. *ESAIM: Mathematical Modelling and Numerical Analysis*, 43(2):277–295, 2009.

[72] F. Brezzi, R. S. Falk, and L. D. Marini. Basic principles of mixed virtual element methods. *ESAIM: Mathematical Modelling and Numerical Analysis*, 48(4):1227–1240, 2014.

[73] F. Brezzi, K. Lipnikov, and M. Shashkov. Convergence of mimetic finite difference methods for diffusion problems on polyhedral meshes. *SIAM Journal on Numerical Analysis*, 43(5):1872–1896, 2005.

[74] F. Brezzi, K. Lipnikov, and V. Simoncini. A family of mimetic finite difference methods on polygonal and polyhedral meshes. *Mathematical Models and Methods in Applied Sciences*, 15(10):1533–1551, 2005.

[75] U. Brink and E. Stein. On some mixed finite element methods for incompressible and nearly incompressible finite elasticity. *Computational Mechanics*, 19(1):105–119, 1996.

[76] R. Brooks. Isospectral graphs and isospectral surfaces. *Séminaire de théorie spectrale et géométrie*, 15:105–113, 1996–1997.

[77] S. Bruvoll and M. S. Floater. Transfinite mean value interpolation in general dimension. *Journal of Computational and Applied Mathematics*, 233(7):1631–1639, 2010.

[78] M. Budninskiy, B. Liu, Y. Tong, and M. Desbrun. Power coordinates: A geometric construction of barycentric coordinates on convex polytopes. *ACM Transactions on Graphics*, 35(6):Article 241, 11 pages, 2016.

[79] P. Buser. A note on the isoperimetric constant. *Annales Scientifiques de l'École Normale Supérieure*, 15(2):213–230, 1982.

[80] A. Cangiani, E. H. Georgoulis, and P. Houston. *hp*-Version discontinuous Galerkin methods on polygonal and polyhedral meshes. *Mathematical Models and Methods in Applied Sciences*, 24(10):2009–2041, 2014.

[81] A. Cangiani, G. Manzini, A. Russo, and N. Sukumar. Hourglass stabilization and the virtual element method. *International Journal for Numerical Methods in Engineering*, 102(3–4):404–436, 2015.

[82] A. Cangiani, G. Manzini, and O. J. Sutton. Conforming and nonconforming virtual element methods for elliptic problems. *IMA Journal of Numerical Analysis*, 2016. DOI: https://doi.org/10.1093/imanum/drw036.

[83] Y. Canzani. Analysis on manifolds via the Laplacian. Course notes, Math 253, Harvard University, 2013. http://canzani.web.unc.edu/files/2016/08/Laplacian.pdf.

[84] W. Carl. A Laplace operator on semi-discrete surfaces. *Foundations of Computational Mathematics*, 16(5):1115–1150, 2015.

[85] M. J. Carley. Analytical formulae for potential integrals on triangles. *Journal of Applied Mechanics*, 80(4):Article 041008, 7 pages, 2013.

[86] CGAL. Computational Geometry Algorithms Library (release 4.9), 2016. http://www.cgal.org.

[87] T. Y. P. Chang, A. F. Saleeb, and G. Li. Large strain analysis of rubber-like materials based on a perturbed Lagrangian variational principle. *Computational Mechanics*, 8(4):221–233, 1991.

[88] D. Chapelle and K. J. Bathe. The inf-sup test. *Computers & Structures*, 47(4–5):537–545, 1993.

[89] I. Chavel. *Eigenvalues in Riemannian Geometry*, volume 115 of *Pure and Applied Mathematics*. Academic Press, Orlando, 1984.

[90] J. S. Chen, W. Han, C. T. Wu, and W. Duan. On the perturbed Lagrangian formulation for nearly incompressible and incompressible hyperelasticity. *Computer Methods in Applied Mechanics and Engineering*, 142(3–4):335–351, 1997.

[91] J.-S. Chen, M. Hillman, and M. Rüter. An arbitrary order variationally consistent integration for Galerkin meshfree methods. *International Journal for Numerical Methods in Engineering*, 95(5):387–418, 2013.

[92] L. Chen and J. Xu. Optimal Delaunay triangulation. *Journal of Computational Mathematics*, 22(2):299–308, 2004.

[93] R. Chen and C. Gotsman. Complex transfinite barycentric mappings with similarity kernels. *Computer Graphics Forum*, 35(5):41–53, 2016.

[94] R. Chen and C. Gotsman. On pseudo-harmonic barycentric coordinates. *Computer Aided Geometric Design*, 44:15–35, 2016.

[95] R. Chen and O. Weber. Bounded distortion harmonic mappings in the plane. *ACM Transactions on Graphics*, 34(4):Article 73, 12 pages, 2015.

[96] R. Chen, Y. Xu, C. Gotsman, and L. Liu. A spectral characterization of the Delaunay triangulation. *Computer Aided Geometric Design*, 27(4):295–300, 2010.

[97] L. P. Chew. Constrained Delaunay triangulations. *Algorithmica*, 4(1):97–108, 1989.

[98] H. Chi, O. Lopez-Pamies, and G. H. Paulino. A variational formulation with rigid-body constraints for finite elasticity: Theory, finite element implementation, and applications. *Computational Mechanics*, 57(2):325–338, 2016.

[99] H. Chi, C. Talischi, O. Lopez-Pamies, and G. H. Paulino. Polygonal finite elements for finite elasticity. *International Journal for Numerical Methods in Engineering*, 101(4):305–328, 2015.

[100] H. Chi, C. Talischi, O. Lopez-Pamies, and G. H. Paulino. A paradigm for higher order polygonal elements in finite elasticity. *Computer Methods in Applied Mechanics and Engineering*, 306:216–251, 2016.

[101] E. Chien, R. Chen, and O. Weber. Bounded distortion harmonic shape interpolation. *ACM Transactions on Graphics*, 35(4):Article 105, 15 pages, 2016.

[102] E. B. Chin, J. B. Lasserre, and N. Sukumar. Numerical integration of homogeneous functions on convex and nonconvex polygons and polyhedra. *Computational Mechanics*, 56(6):967–981, 2015.

[103] N. H. Christ, R. Friedberg, and T. D. Lee. Weights of links and plaquettes in a random lattice. *Nuclear Physics B*, 210(3):337–346, 1982.

[104] F. Chung. Four proofs for the Cheeger inequality and graph partition algorithms. In L. Ji, K. Liu, L. Yang, and S.-T. Yau, editors, *Fourth International Congress of Chinese Mathematicians*, volume 48 of *AMS/IP Studies in Advanced Mathematics*, pages 331–349. American Mathematical Society/International Press, Providence/-Somerville, 2010.

[105] B. Cockburn, W. Qiu, and M. Solano. A priori error analysis for HDG methods using extensions from subdomains to achieve boundary conformity. *Mathematics of Computation*, 83(286):665–699, 2014.

[106] D. Cohen-Or, A. Solomovic, and D. Levin. Three-dimensional distance field metamorphosis. *ACM Transactions on Graphics*, 17(2):116–141, 1998.

[107] D. Copeland, U. Langer, and D. Pusch. From the boundary element domain decomposition methods to local Trefftz finite element methods on polyhedral meshes. In M. Bercovier, M. Gander, R. Kornhuber, and O. Widlund, editors, *Domain Decomposition Methods in Science and Engineering XVIII*, volume 70 of *Lecture Notes in Computational Science and Engineering*, pages 315–322. Springer, Berlin, 2009.

[108] J. A. Cottrell, T. J. R. Hughes, and Y. Bazilevs. *Isogeometric Analysis: Toward Integration of CAD and FEA*. Wiley, Chichester, 2009.

[109] K. Crane, F. de Goes, M. Desbrun, and P. Schröder. *Digital Geometry Processing with Discrete Exterior Calculus*. Number 7 in SIGGRAPH 2013 Course Notes. ACM, New York, 2013.

[110] M. Crouzeix and P.-A. Raviart. Conforming and nonconforming finite element methods for solving the stationary Stokes equations I. *Revue française d'automatique, informatique, recherche opérationelle. Analyse Numérique*, 7(3):33–76, 1973.

[111] C. J. Cyron, M. Arroyo, and M. Ortiz. Smooth, second order, non-negative meshfree approximants selected by maximum entropy. *International Journal for Numerical Methods in Engineering*, 79(13):1605–1632, 2009.

[112] E. F. D'Azevedo and R. B. Simpson. On optimal interpolation triangle incidences. *SIAM Journal on Scientific and Statistical Computing*, 10(6):1063–1075, 1989.

[113] F. de Goes, P. Alliez, H. Owhadi, and M. Desbrun. On the equilibrium of simplicial masonry structures. *ACM Transactions on Graphics*, 32(4):Article 93, 10 pages, 2013.

[114] F. de Goes, M. Desbrun, M. Meyer, and T. DeRose. Subdivision exterior calculus for geometry processing. *ACM Transactions on Graphics*, 35(4):Article 133, 11 pages, 2016.

[115] F. de Goes. *Geometric Discretization Through Primal-Dual Meshes*. PhD thesis, California Institute of Technology, 2014.

[116] H. de Monantheuil. *Aristotelis Mechanica*. Jeremias Perier, Paris, 1599. In Greek, corrected, translated into Latin and illustrated with commentary.

[117] S. Dekel and D. Leviatan. The Bramble–Hilbert lemma for convex domains. *SIAM Journal on Mathematical Analysis*, 35(5):1203–1212, 2004.

[118] M. Desbrun, A. N. Hirani, M. Leok, and J. E. Marsden. Discrete Exterior Calculus. arXiv:math/0508341 [math.DG], 2005. https://arxiv.org/pdf/math/0508341.pdf.

[119] M. Desbrun, E. Kanso, and Y. Tong. Discrete differential forms for computational modeling. In A. I. Bobenko, J. M. Sullivan, P. Schröder, and G. M. Ziegler, editors, *Discrete Differential Geometry*, volume 38 of *Oberwolfach Seminars*, pages 287–324. Birkhäuser, Basel, 2008.

[120] M. Desbrun, M. Meyer, P. Schröder, and A. H. Barr. Implicit fairing of irregular meshes using diffusion and curvature flow. In *Proceedings of the 26th Annual Conference on Computer Graphics and Interactive Techniques (SIGGRAPH 1999)*, pages 317–324, 1999.

[121] M. Deuss, D. Panozzo, E. Whiting, Y. Liu, P. Block, O. Sorkine-Hornung, and M. Pauly. Assembling self-supporting structures. *ACM Transactions on Graphics*, 33(6):Article 214, 10 pages, 2014.

[122] T. K. Dey, P. Ranjan, and Y. Wang. Convergence, stability, and discrete approximation of Laplace spectra. In *Proceedings of the 21st Annual ACM-SIAM Symposium on Discrete Algorithms (SODA 2010)*, pages 650–663, 2010.

[123] D. A. Di Pietro and A. Ern. Hybrid high-order methods for variable-diffusion problems on general meshes. *Comptes Rendus Mathematiques*, 353(1):31–34, 2015.

[124] J. Dodziuk and V. K. Patodi. Riemannian structures and triangulations of manifolds. *The Journal of the Indian Mathematical Society*, 40:1–52, 1976.

[125] W. Dörfler. A convergent adaptive algorithm for Poisson's equation. *SIAM Journal on Numerical Analysis*, 33(3):1106–1124, 1996.

[126] J. Droniou, R. Eymard, and R. Herbin. Gradient schemes: generic tools for the numerical analysis of diffusion equations, 2015. https://hal.archives-ouvertes.fr/hal-01150517v2/document.

[127] Q. Du, V. Faber, and M. Gunzburger. Centroidal Voronoi tessellations: applications and algorithms. *SIAM Review*, 41(4):637–676, 1999.

[128] Q. Du, C. Liu, and X. Wang. A phase field approach in the numerical study of the elastic bending energy for vesicle membranes. *Journal of Computational Physics*, 198(2):450–468, 2004.

[129] Q. Du and L. Zhu. Analysis of a mixed finite element method for a phase field bending elasticity model of vesicle membrane deformation. *Journal of Computational Mathematics*, 24(3):265–280, 2006.

[130] R. J. Duffin. Distributed and lumped networks. *Journal of Mathematics and Mechanics*, 8(5):793–826, 1959.

[131] D. A. Dunavant. High degree efficient symmetrical Gaussian quadrature rules for the triangle. *International Journal for Numerical Methods in Engineering*, 21(6):1129–1148, 1985.

[132] A. Düster, J. Parvizian, Z. Yang, and E. Rank. The finite cell method for three-dimensional problems of solid mechanics. *Computer Methods in Applied Mechanics and Engineering*, 197(45–48):3768–3782, 2008.

[133] C. Dyken and M. S. Floater. Transfinite mean value interpolation. *Computer Aided Geometric Design*, 26(1):117–134, 2009.

[134] G. Dziuk. Finite elements for the Beltrami operator on arbitrary surfaces. In S. Hildebrandt and R. Leis, editors, *Partial Differential Equations and Calculus of Variations*, volume 1357 of *Lecture Notes in Mathematics*, pages 142–155. Springer, Berlin, 1988.

[135] M. S. Ebeida, S. A. Mitchell, A. Patney, A. A. Davidson, and J. D. Owens. A simple algorithm for maximal Poisson-disk sampling in high dimensions. *Computer Graphics Forum*, 31(2):785–794, 2012.

[136] M. Eck, T. DeRose, T. Duchamp, H. Hoppe, M. Lounsbery, and W. Stuetzle. Multiresolution analysis of arbitrary meshes. In *Proceedings of the 22nd Annual Conference on Computer Graphics and Interactive Techniques (SIGGRAPH 1995)*, pages 173–182, 1995.

[137] H. Edelsbrunner. *Algorithms in Combinatorial Geometry*, volume 10 of *EATCS Monographs on Theoretical Computer Science*. Springer, Berlin, 1987.

[138] Y. Efendiev, J. Galvis, R. Lazarov, and S. Weißer. Mixed FEM for second order elliptic problems on polygonal meshes with BEM-based spaces. In I. Lirkov, S. Margenov, and J. Waśniewski, editors, *Large-Scale Scientific Computing*, volume 8353 of *Lecture Notes in Computer Science*, pages 331–338. Springer, Heidelberg, 2014.

[139] A. L. Eterovic and K.-J. Bathe. A hyperelastic-based large strain elasto-plastic constitutive formulation with combined isotropic-kinematic hardening using the logarithmic stress and strain measures. *International Journal for Numerical Methods in Engineering*, 30(6):1099–1114, 1990.

[140] L. C. Evans. *Partial Differential Equations*, volume 19 of *Graduate Studies in Mathematics*. American Mathematical Society, Providence, 1998.

[141] G. Ewald. *Combinatorial Convexity and Algebraic Geometry*, volume 168 of *Graduate Texts in Mathematics*. Springer, New York, 1996.

[142] J. L. Finney. Random packings and the structure of simple liquids. I. The geometry of random close packing. *Proceedings of the Royal Society of London. Series A, Mathematical and Physical Sciences*, 319(1539):479–493, 1970.

[143] M. S. Floater. Parametrization and smooth approximation of surface triangulations. *Computer Aided Geometric Design*, 14(3):231–250, 1997.

[144] M. S. Floater. Mean value coordinates. *Computer Aided Geometric Design*, 20(1):19–27, 2003.

[145] M. S. Floater. Generalized barycentric coordinates and applications. *Acta Numerica*, 24:161–214, 2015.

[146] M. S. Floater. On the monotonicity of generalized barycentric coordinates on convex polygons. *Computer Aided Geometric Design*, 42:34–39, 2016.

[147] M. S. Floater, A. Gillette, and N. Sukumar. Gradient bounds for Wachspress coordinates on polytopes. *SIAM Journal on Numerical Analysis*, 52(1):515–532, 2014.

[148] M. S. Floater and K. Hormann. Surface parameterization: A tutorial and survey. In N. A. Dodgson, M. S. Floater, and M. A. Sabin, editors, *Advances in Multiresolution for Geometric Modeling*, Mathematics and Visualization, pages 157–186. Springer, Berlin, 2005.

[149] M. S. Floater, K. Hormann, and G. Kós. A general construction of barycentric coordinates over convex polygons. *Advances in Computational Mathematics*, 24(1–4):311–331, 2006.

[150] M. S. Floater, G. Kós, and M. Reimers. Mean value coordinates in 3D. *Computer Aided Geometric Design*, 22(7):623–631, 2005.

[151] M. S. Floater and J. Kosinka. Barycentric interpolation and mappings on smooth convex domains. In *Proceedings of the 14th ACM Symposium on Solid and Physical Modeling (SPM 2010)*, pages 111–116, 2010.

[152] M. S. Floater and J. Kosinka. On the injectivity of Wachspress and mean value mappings between convex polygons. *Advances in Computational Mathematics*, 32(2):163–174, 2010.

[153] M. S. Floater and C. Schulz. Pointwise radial minimization: Hermite interpolation on arbitrary domains. *Computer Graphics Forum*, 27(5):1505–1512, 2008.

[154] H. Florez, R. Manzanilla-Morillo, J. Florez, and M. F. Wheeler. Spline-based reservoir's geometry reconstruction and mesh generation for coupled flow and mechanics simulation. *Computational Geosciences*, 18(6):949–967, 2014.

[155] T. Frankel. *The Geometry of Physics: An Introduction.* Cambridge University Press, Cambridge, 2nd edition, 2004.

[156] F. Fraternali. A thrust network approach to the equilibrium problem of unreinforced masonry vaults via polyhedral stress functions. *Mechanics Research Communications*, 37(2):198–204, 2010.

[157] M. Freytag, V. Shapiro, and I. Tsukanov. Finite element analysis in situ. *Finite Elements in Analysis and Design*, 47(9):957–972, 2011.

[158] T.-P. Fries and T. Belytschko. The extended/generalized finite element method: An overview of the method and its applications. *International Journal for Numerical Methods in Engineering*, 84(3):253–304, 2010.

[159] T.-P. Fries and H.-G. Matthies. Classification and overview of meshfree methods. Technical Report 2003-3, Institute of Scientific Computing, Technische Universität Braunschweig, July 2004.

[160] A. L. Gain, G. H. Paulino, L. S. Duarte, and I. F. M. Menezes. Topology optimization using polytopes. *Computer Methods in Applied Mechanics and Engineering*, 293:411–430, 2015.

[161] A. L. Gain, C. Talischi, and G. H. Paulino. On the Virtual Element Method for three-dimensional linear elasticity problems on arbitrary polyhedral meshes. *Computer Methods in Applied Mechanics and Engineering*, 282:132–160, 2014.

[162] F. Gao, D. M. Ingram, D. M. Causon, and C. G. Mingham. The development of a Cartesian cut cell method for incompressible viscous flows. *International Journal for Numerical Methods in Fluids*, 54(9):1033–1053, 2007.

[163] A. Gillette and A. Rand. Interpolation error estimates for harmonic coordinates on polytopes. *ESAIM: Mathematical Modelling and Numerical Analysis*, 50(3):651–676, 2016.

[164] A. Gillette, A. Rand, and C. Bajaj. Error estimates for generalized barycentric coordinates. *Advances in Computational Mathematics*, 37(3):417–439, 2012.

[165] A. Gillette, A. Rand, and C. Bajaj. Construction of scalar and vector finite element families on polygonal and polyhedral meshes. *Computational Methods in Applied Mathematics*, 16(4):667–683, 2016.

[166] D. Glickenstein. Geometric triangulations and discrete Laplacians on manifolds. arXiv:math/0508188 [math.MG], 2005. https://arxiv.org/pdf/math/0508188.pdf.

[167] R. Goldman. *Pyramid Algorithms: A Dynamic Programming Approach to Curves and Surfaces for Geometric Modeling.* The Morgan Kaufmann Series in Computer Graphics and Geometric Modeling. Morgan Kaufmann Publishers, Amsterdam, 2003.

[168] E. Goldstein and C. Gotsman. Polygon morphing using a multiresolution representation. In *Proceedings of Graphics Interface (GI 1995)*, pages 247–254, 1995.

[169] C. Gordon, D. L. Webb, and S. Wolpert. One cannot hear the shape of a drum. *Bulletion of the American Mathematical Society*, 27(1):134–138, 1992.

[170] W. J. Gordon and C. A. Hall. Transfinite element methods: Blending-function interpolation over arbitrary curved element domains. *Numerische Mathematik*, 21(2):109–129, 1973.

[171] W. J. Gordon and J. A. Wixom. Pseudo-harmonic interpolation on convex domains. *SIAM Journal on Numerical Analysis*, 11(5):909–933, 1974.

[172] S. J. Gortler, C. Gotsman, and D. Thurtson. Discrete one-forms on meshes and applications to 3D mesh parameterization. *Computer Aided Geometric Design*, 23(2):83–112, 2006.

[173] C. Gotsman and V. Surazhsky. Guaranteed intersection-free polygon morphing. *Computers & Graphics*, 25(1):67–75, 2001.

[174] T. Goudarzi. *Iterative and Variational Homogenization Methods for Filled Elastomers.* PhD thesis, University of Illinois at Urbana-Champaign, 2014.

[175] T. Goudarzi, D. W. Spring, G. H. Paulino, and O. Lopez-Pamies. Filled elastomers: A theory of filler reinforcement based on hydrodynamic and interphasial effects. *Journal of the Mechanics and Physics of Solids*, 80:37–67, 2015.

[176] L. J. Grady and J. R. Polimeni. *Discrete Calculus: Applied Analysis on Graphs for Computational Science.* Springer, London, 2010.

[177] F. Greco and N. Sukumar. Derivatives of maximum-entropy basis functions on the boundary: Theory and computations. *International Journal for Numerical Methods in Engineering*, 94(12):1123–1149, 2013.

[178] A. E. Green and W. Zerna. *Theoretical Elasticity.* Clarendon Press, Oxford, 2nd edition, 1968.

[179] B. Grünbaum. *Convex Polytopes*, volume 221 of *Graduate Texts in Mathematics.* Springer, New York, 2nd edition, 2003.

[180] L. Guibas, J. Hershberger, and S. Suri. Morphing simple polygons. *Discrete & Computational Geometry*, 24(1):1–34, 2000.

[181] V. Guillemin and A. Pollack. *Differential Topology.* Prentice-Hall, Englewood Cliffs, 1974.

[182] V. Gyrya and K. Lipnikov. M-adaptation method for acoustic wave equation on square meshes. *Journal of Computational Acoustics*, 20(4):Article 1250022, 23 pages, 2012.

[183] V. Gyrya, K. Lipnikov, G. Manzini, and D. Svyatskiy. M-adaptation in the mimetic finite difference method. *Mathematical Models and Methods in Applied Sciences*, 24(8):1621–1663, 2014.

[184] W. Hackbusch and S. A. Sauter. Composite finite elements for the approximation of PDEs on domains with complicated micro-structures. *Numerische Mathematik*, 75(4):447–472, 1997.

[185] A. Hannukainen, S. Korotov, and M. Křížek. The maximum angle condition is not necessary for convergence of the finite element method. *Numerische Mathematik*, 120(1):79–88, 2012.

[186] B. Hashemian, D. Millán, and M. Arroyo. Charting molecular free-energy landscapes with an atlas of collective variables. *The Journal of Chemical Physics*, 145(17):Article 174109, 13 pages, 2016.

[187] P. Herholz, J. E. Kyprianidis, and M. Alexa. Perfect Laplacians for polygon meshes. *Computer Graphics Forum*, 34(5):211–218, 2015.

[188] L. R. Herrmann. Elasticity equations for incompressible and nearly incompressible materials by a variational theorem. *AIAA Journal*, 3(10):1896–1900, 1965.

[189] J. Heyman. The stone skeleton. *International Journal of Solids and Structures*, 2(2):249–279, 1966.

[190] J. Heyman. *The Stone Skeleton: Structural Engineering of Masonry Architecture.* Cambridge University Press, Cambridge, 1995.

[191] K. Hildebrandt, K. Polthier, and M. Wardetzky. On the convergence of metric and geometric properties of polyhedral surfaces. *Geometriae Dedicata*, 123(1):89–112, 2006.

[192] A. N. Hirani. *Discrete Exterior Calculus.* PhD thesis, California Institute of Technology, 2003.

[193] H. Hiyoshi and K. Sugihara. Two generalizations of an interpolant based on Voronoi diagrams. *International Journal of Shape Modelling*, 5(2):219–231, 1999.

[194] H. Hiyoshi and K. Sugihara. Voronoi-based interpolation with higher continuity. In *Proceedings of the 16th Annual Symposium on Computational Geometry (SCG 2000)*, pages 242–250, 2000.

[195] C. Hofreither. L_2 error estimates for a nonstandard finite element method on polyhedral meshes. *Journal of Numerical Mathematics*, 19(1):27–39, 2011.

[196] C. Hofreither, U. Langer, and C. Pechstein. Analysis of a non-standard finite element method based on boundary integral operators. *Electronic Transactions on Numerical Analysis*, 37:413–436, 2010.

[197] C. Hofreither, U. Langer, and C. Pechstein. A non-standard finite element method for convection-diffusion-reaction problems on polyhedral meshes. In M. D. Todorov and C. I. Christov, editors, *Application of Mathematics in Technical and Natural Sciences*, number 1404 in AIP Conference Proceedings, pages 397–404. American Insitute of Physics, Melville, 2011.

[198] C. Hofreither, U. Langer, and C. Pechstein. FETI solvers for non-standard finite element equations based on boundary integral operators. In J. Erhel, M. J. Gander, L. Halpern, G. Pichot, T. Sassi, and O. Widlund, editors, *Domain Decomposition Methods in Science and Engineering XXI*, volume 98 of *Lecture Notes in Computational Science and Engineering*, pages 729–737. Springer, Cham, 2014.

[199] C. Hofreither, U. Langer, and S. Weißer. Convection-adapted BEM-based FEM. *ZAMM*, 96(12):1467–1481, 2016.

[200] K. Hormann and M. S. Floater. Mean value coordinates for arbitrary planar polygons. *ACM Transactions on Graphics*, 25(4):1424–1441, 2006.

[201] K. Hormann, B. Lévy, and A. Sheffer. *Mesh Parameterization: Theory and Practice.* Number 2 in SIGGRAPH 2007 Course Notes. ACM, New York, 2007.

[202] K. Hormann and N. Sukumar. Maximum entropy coordinates for arbitrary polytopes. *Computer Graphics Forum*, 27(5):1513–1520, 2008.

[203] S. Huerta. Mechanics of masonry vaults: The equilibrium approach. In *Proceedings of the 3rd International Conference on Structural Analysis of Historical Constructions (SAHC 2001)*, pages 47–69, 2001.

[204] T. J. R. Hughes. *The Finite Element Method: Linear Static and Dynamic Finite Element Analysis.* Dover Publications, Mineola, 2000.

[205] T. J. R. Hughes, J. A. Cottrell, and Y. Bazilevs. Isogeometric analysis: CAD, finite elements, NURBS, exact geometry and mesh refinement. *Computer Methods in Applied Mechanics and Engineering*, 194(39–41):4135–4195, 2005.

[206] H. N. Iben, J. F. O'Brien, and E. D. Demaine. Refolding planar polygons. *Discrete & Computational Geometry*, 41(3):444–460, 2009.

[207] D. M. Ingram, D. M. Causon, and C. G. Mingham. Developments in Cartesian cut cell methods. *Mathematics and Computers in Simulation*, 61(3–6):561–572, 2003.

[208] A. Jacobson. Bijective mappings with generalized barycentric coordinates: A counterexample. *Journal of Graphics Tools*, 17(1–2):1–4, 2013.

[209] P. Jamet. Estimations d'erreur pour des éléments finis droits presque dégénérés. *Revue française d'automatique, informatique, recherche opérationelle. Analyse Numérique*, 10(1):43–60, 1976.

[210] L. Jardine. Monuments and microscopes: Scientific thinking on a grand scale in the early Royal Society. *Notes and Records of the Royal Society of London*, 55(2):289–308, 2001.

[211] E. T. Jaynes. Information theory and statistical mechanics. *Physical Review*, 106(4):620–630, 1957.

[212] E. T. Jaynes. Information theory and statistical mechanics. In K. W. Ford, editor, *Statistical Physics*, volume 3 of *Brandeis University Summer Institute Lectures in Theoretical Physics*, pages 181–218. W. A. Benjamin, New York, 1963.

[213] M. Jirásek. Comparative study on finite elements with embedded discontinuities. *Computer Methods in Applied Mechanics and Engineering*, 188(1–3):307–330, 2000.

[214] M. W. Jones, J. A. Bærentzen, and M. Sramek. 3D distance fields: A survey of techniques and applications. *IEEE Transactions on Visualization and Computer Graphics*, 12(4):581–599, 2006.

[215] P. Joshi, M. Meyer, T. DeRose, B. Green, and T. Sanocki. Harmonic coordinates for character articulation. *ACM Transactions on Graphics*, 26(3):Article 71, 9 pages, 2007.

[216] T. Ju, P. Liepa, and J. Warren. A general geometric construction of coordinates in a convex simplicial polytope. *Computer Aided Geometric Design*, 24(3):161–178, 2007.

[217] T. Ju, S. Schaefer, and J. Warren. Mean value coordinates for closed triangular meshes. *ACM Transactions on Graphics*, 24(3):561–566, 2005.

[218] T. Ju, S. Schaefer, J. Warren, and M. Desbrun. A geometric construction of coordinates for convex polyhedra using polar duals. In *Proceedings of the 3rd Eurographics Symposium on Geometry Processing (SGP 2005)*, pages 181–186, Vienna, Austria, 2005.

[219] M. Juntunen and M. F. Wheeler. Two-phase flow in complicated geometries: Modeling the Frio data using improved computational meshes. *Computational Geosciences*, 17(2):239–247, 2013.

[220] M. Kac. Can one hear the shape of a drum? *The American Mathematical Monthly*, 73(4, Part 2):1–23, 1966.

[221] J. A. Kalman. Continuity and convexity of projections and barycentric coordinates in convex polyhedra. *Pacific Journal of Mathematics*, 11(3):1017–1022, 1961.

[222] R. Kannan, S. Vempala, and A. Vetta. On clusterings: Good, bad and spectral. *Journal of the ACM*, 51(3):497–515, 2004.

[223] A. Y. Khinchin. *Mathematical Foundations of Information Theory.* Dover Publications, New York, 1957.

[224] H.-G. Kim and D. Sohn. A new finite element approach for solving three-dimensional problems using trimmed hexahedral elements. *International Journal for Numerical Methods in Engineering*, 102(9):1527–1553, 2015.

[225] K. Kobayashi and T. Tsuchiya. A Babuška–Aziz type proof of the circumradius condition. *Japan Journal of Industrial and Applied Mathematics*, 31(1):193–210, 2014.

[226] A. Kooharian. Limit analysis of Voussoir (segmental) and concrete arches. *Proc. Am. Concr. Inst.*, 49(12):317–328, 1952.

[227] J. Kosinka and M. Bartoň. Convergence of barycentric coordinates to barycentric kernels. *Computer Aided Geometric Design*, 43:200–210, 2016.

[228] T. Kotnik and M. Weinstock. Material, form and force. *Architectural Design*, 82(2):104–111, 2012.

[229] Y. Krongauz and T. Belytschko. Consistent pseudo-derivatives in meshless methods. *Computer Methods in Applied Mechanics and Engineering*, 146(3–4):371–386, 1997.

[230] S. Kullback and R. A. Leibler. On information and sufficiency. *The Annals of Mathematical Statistics*, 22(1):79–86, 1951.

[231] M. Křížek. On the maximum angle condition for linear tetrahedral elements. *SIAM Journal on Numerical Analysis*, 29(2):513–520, 1992.

[232] T. Langer, A. Belyaev, and H.-P. Seidel. Spherical barycentric coordinates. In *Proceedings of the 4th Eurographics Symposium on Geometry Processing (SGP 2006)*, pages 81–88, 2006.

[233] C. L. Lawson. Software for C^1 surface interpolation. In J. R. Rice, editor, *Mathematical Software III*, volume 39 of *Publication of the Mathematics Research Center, University of Wisconsin-Madison*, pages 161–194. Academic Press, New York, 1977.

[234] C. L. Lawson and R. J. Hanson. *Solving Least Squares Problems*, volume 15 of *Classics in Applied Mathematics.* SIAM, Philadelphia, 1995.

[235] P. Le Tallec. Existence and approximation results for nonlinear mixed problems: Application to incompressible finite elasticity. *Numerische Mathematik*, 38(3):365–382, 1982.

[236] J. L. Leblanc. *Filled Polymers: Science and Industrial Applications.* CRC Press, Boca Raton, 2010.

[237] C. W. Lee. Regular triangulations of convex polytopes. In *Applied Geometry and Discrete Mathematics: The Victor Klee Festschrift*, volume 4 of *DIMACS Series in Discrete Mathematics and Theoretical Computer Science*, pages 443–456. American Mathematical Society/Association for Computing Machinery, Providence/Baltimore, 1991.

[238] S.-B. Lee, G. S. Rohrer, and A. D. Rollett. Three-dimensional digital approximations of grain boundary networks in polycrystals. *Modelling and Simulation in Materials Science and Engineering*, 22(2):Article 025017, 21 pages, 2014.

[239] S. E. Leon, D. W. Spring, and G. H. Paulino. Reduction in mesh bias for dynamic fracture using adaptive splitting of polygonal finite elements. *International Journal for Numerical Methods in Engineering*, 100(8):555–576, 2014.

[240] G. Leoni. *A First Course in Sobolev Spaces*, volume 105 of *Graduate Studies in Mathematics.* American Mathematical Society, Providence, 2009.

[241] Z. Levi and O. Weber. On the convexity and feasibility of the bounded distortion harmonic mapping problem. *ACM Transactions on Graphics*, 35(4):Article 106, 15 pages, 2016.

[242] B. Lévy and H. Zhang. *Spectral Mesh Processing.* Number 8 in SIGGRAPH 2010 Course Notes. ACM, New York, 2010.

[243] B. Li, F. Habbal, and M. Ortiz. Optimal transportation meshfree approximation schemes for fluid and plastic flows. *International Journal for Numerical Methods in Engineering*, 83(12):1541–1579, 2010.

[244] B. Li, C. Peco, D. Millán, I. Arias, and M. Arroyo. Phase-field modeling and simulation of fracture in brittle materials with strongly anisotropic surface energy. *International Journal for Numerical Methods in Engineering*, 102(3–4):711–727, 2015.

[245] S. Li and W. K. Liu. Meshfree and particle methods and their applications. *Applied Mechanics Reviews*, 55(1):1–34, 2002.

[246] X. S. Li. An overview of SuperLU: Algorithms, implementation, and user interface. *ACM Transactions on Mathematical Software*, 31(3):302–325, 2005.

[247] X.-Y. Li and S.-M. Hu. Poisson coordinates. *IEEE Transactions on Visualization and Computer Graphics*, 19(2):344–352, 2013.

[248] X.-Y. Li, T. Ju, and S.-M. Hu. Cubic mean value coordinates. *ACM Transactions on Graphics*, 32(4):Article 126, 10 pages, 2013.

[249] H. Lim, F. Abdeljawad, S. J. Owen, B. W. Hanks, J. W. Foulk, and C. C. Battaile. Incorporating physically-based microstructures in materials modeling: Bridging phase field and crystal plasticity frameworks. *Modelling and Simulation in Materials Science and Engineering*, 24(4):Article 045016, 19 pages, 2016.

[250] E. Lindelöf. Sur l'application de la méthode des approximations successives aux équations différentielles ordinaires du premier ordre. *Comptes Rendus des Séances de l'Académie des Sciences*, 118(9):454–457, 1894.

[251] Y. Lipman, J. Kopf, D. Cohen-Or, and D. Levin. GPU-assisted positive mean value coordinates for mesh deformations. In *Proceedings of the 5th Eurographics Symposium on Geometry Processing (SGP 2007)*, pages 117–123, 2007.

[252] Y. Lipman, D. Levin, and D. Cohen-Or. Green coordinates. *ACM Transactions on Graphics*, 27(3):Article 78, 10 pages, 2008.

[253] K. Lipnikov, G. Manzini, and M. Shashkov. Mimetic finite difference method. *Journal of Computational Physics*, 257, Part B:1163–1227, 2014.

[254] K. Lipnikov, G. Manzini, and D. Svyatskiy. Analysis of the monotonicity conditions in the mimetic finite difference method for elliptic problems. *Journal of Computational Physics*, 230(7):2620–2642, 2011.

[255] Y. Liu, H. Pan, J. Snyder, W. Wang, and B. Guo. Computing self-supporting surfaces by regular triangulation. *ACM Transactions on Graphics*, 32(4):Article 92, 10 pages, 2013.

[256] Y. Liu, A. A. Saputra, J. Wang, F. Tin-Loi, and C. Song. Automatic polyhedral mesh generation and scaled boundary finite element analysis of STL models. *Computer Methods in Applied Mechanics and Engineering*, 313:106–132, 2017.

[257] C. T. Loop and T. D. DeRose. A multisided generalization of Bézier surfaces. *ACM Transactions on Graphics*, 8(3):204–234, 1989.

[258] O. Lopez-Pamies. An exact result for the macroscopic response of particle-reinforced neo-Hookean solids. *Journal of Applied Mechanics*, 77(2):Article 021016, 5 pages, 2010.

[259] O. Lopez-Pamies. A new I_1-based hyperelastic model for rubber elastic materials. *Comptes Rendus Mécanique*, 338(1):3–11, 2010.

[260] O. Lopez-Pamies, T. Goudarzi, and K. Danas. The nonlinear elastic response of suspensions of rigid inclusions in rubber: II—a simple explicit approximation for finite-concentration suspensions. *Journal of the Mechanics and Physics of Solids*, 61(1):19–37, 2013.

[261] W. Lorensen and H. Cline. Marching cubes: A high resolution 3D surface construction algorithm. *ACM SIGGRAPH Computer Graphics*, 21(4):163–169, 1987.

[262] R. MacNeal. *The Solution of Partial Differential Equations by Means of Electrical Networks*. PhD thesis, California Institute of Technology, 1949.

[263] M. Majeed and F. Cirak. Isogeometric analysis using manifold-based smooth basis functions. *Computer Methods in Applied Mechanics and Engineering*, 316(1):547–567, 2017.

[264] E. A. Malsch and G. Dasgupta. Interpolations for temperature distributions: A method for all non-concave polygons. *International Journal of Solids and Structures*, 41(8):2165–2188, 2004.

[265] E. A. Malsch, J. J. Lin, and G. Dasgupta. Smooth two-dimensional interpolants: A recipe for all polygons. *Journal of Graphics Tools*, 10(2):27–39, 2005.

[266] J. Manson, K. Li, and S. Schaefer. Positive Gordon–Wixom coordinates. *Computer-Aided Design*, 43(11):1422–1426, 2011.

[267] J. Manson and S. Schaefer. Moving least squares coordinates. *Computer Graphics Forum*, 29(5):1517–1524, 2010.

[268] G. Manzini, A. Russo, and N. Sukumar. New perspectives on polygonal and polyhedral finite element methods. *Mathematical Models and Methods in Applied Sciences*, 24(8):1665–1699, 2014.

[269] S. Martin, P. Kaufmann, M. Botsch, M. Wicke, and M. Gross. Polyhedral finite elements using harmonic basis functions. *Computer Graphics Forum*, 27(5):1521–1529, 2008.

[270] J. C. Maxwell. On reciprocal figures, frames, and diagrams of forces. *Transactions of the Royal Society of Edinburgh*, 26:1–40, 1870.

[271] S. F. McCormick. *Multilevel Adaptive Methods for Partial Differential Equations*, volume 6 of *Frontiers in Applied Mathematics*. SIAM, Philadelphia, 1989.

[272] G. H. Meisters and C. Olech. Locally one-to-one mappings and a classical theorem on the schlicht functions. *Duke Mathematical Journal*, 30(1):63–80, 1963.

[273] T. V. Mele and P. Block. A novel form finding method for fabric formwork for concrete shells. *Journal of the International Association for Shell and Spatial Structures*, 52(4):217–224, 2011.

[274] P. Memari, P. Mullen, and M. Desbrun. Parametrization of generalized primal-dual triangulations. In W. R. Quadros, editor, *Proceedings of the 20th International Meshing Roundtable*, pages 237–253. Springer, Berlin, 2012.

[275] M. Meyer, A. Barr, H. Lee, and M. Desbrun. Generalized barycentric coordinates on irregular polygons. *Journal of Graphics Tools*, 7(1):13–22, 2002.

[276] M. Meyer, M. Desbrun, P. Schröder, and A. H. Barr. Discrete differential-geometry operators for triangulated 2-manifolds. In H.-C. Hege and K. Polthier, editors, *Visualization and Mathematics III*, Mathematics and Visualization, pages 35–57. Springer, Berlin, 2003.

[277] C. Miehe, M. Hofacker, L.-M. Schänzel, and F. Aldakheel. Phase field modeling of fracture in multi-physics problems. Part II. Coupled brittle-to-ductile failure criteria and crack propagation in thermo-elastic-plastic solids. *Computer Methods in Applied Mechanics and Engineering*, 294(1):486–522, 2015.

[278] C. Miehe, L.-M. Schänzel, and H. Ulmer. Phase field modeling of fracture in multi-physics problems. Part I. Balance of crack surface and failure criteria for brittle crack propagation in thermo-elastic solids. *Computer Methods in Applied Mechanics and Engineering*, 294(1):449–485, 2015.

[279] D. Millán, A. Rosolen, and M. Arroyo. Thin shell analysis from scattered points with *maximum-entropy* approximants. *International Journal for Numerical Methods in Engineering*, 85(6):723–751, 2011.

[280] D. Millán, A. Rosolen, and M. Arroyo. Nonlinear manifold learning for meshfree finite deformation thin-shell analysis. *International Journal for Numerical Methods in Engineering*, 93(7):685–713, 2013.

[281] D. Millán, N. Sukumar, and M. Arroyo. Cell-based maximum-entropy approximants. *Computer Methods in Applied Mechanics and Engineering*, 284:712–731, 2015.

[282] A. F. Möbius. *Der barycentrische Calcul*. J. A. Barth, Leipzig, 1827.

[283] C. Montani, R. Scateni, and R. Scopigno. A modified look-up table for implicit disambiguation of Marching Cubes. *The Visual Computer*, 10(6):353–355, 1994.

[284] J. Moraleda, J. Segurado, and J. Llorca. Finite deformation of incompressible fiber-reinforced elastomers: A computational micromechanics approach. *Journal of the Mechanics and Physics of Solids*, 57(9):1596–1613, 2009.

[285] S. E. Mousavi and N. Sukumar. Numerical integration of polynomials and discontinuous functions on irregular convex polygons and polyhedrons. *Computational Mechanics*, 47(5):535–554, 2011.

[286] S. E. Mousavi, H. Xiao, and N. Sukumar. Generalized Gaussian quadrature rules on arbitrary polygons. *International Journal for Numerical Methods in Engineering*, 82(1):99–113, 2010.

[287] L. Mu, J. Wang, and X. Ye. Weak Galerkin finite element methods on polytopal meshes. *International Journal of Numerical Analysis and Modeling*, 12(1):31–53, 2015.

[288] L. Mu, X. Wang, and Y. Wang. Shape regularity conditions for polygonal/polyhedral meshes, exemplified in a discontinuous Galerkin discretization. *Numer. Methods Partial Differential Equations*, 31(1):308–325, 2015.

[289] P. Mullen, P. Memari, F. de Goes, and M. Desbrun. HOT: Hodge-optimized triangulations. *ACM Transactions on Graphics*, 30(4):Article 103, 12 pages, 2011.

[290] J. B. C. Neto, P. A. Wawrzynek, M. T. M. Carvalho, L. F. Martha, and A. R. Ingraffea. An algorithm for three-dimensional mesh generation for arbitrary regions with cracks. *Engineering with Computers*, 17(1):75–91, 2001.

[291] T. S. Newman and Y. Hong. A survey of the marching cubes algorithm. *Computers & Graphics*, 30(5):854–879, 2006.

[292] K. Nissen, C. J. Cyron, V. Gravemeier, and W. A. Wall. Information-flux method: a meshfree maximum-entropy Petrov–Galerkin method including stabilised finite element methods. *Computer Methods in Applied Mechanics and Engineering*, 241–244:225–237, 2012.

[293] J. T. Oden. Recent developments in the theory of finite element approximations of boundary value problems in nonlinear elasticity. *Computer Methods in Applied Mechanics and Engineering*, 17–18(1):183–202, 1979.

[294] R. W. Ogden. *Non-linear Elastic Deformations*. Dover, Mineola, 1997.

[295] A. Okabe, B. Boots, K. Sugihara, and S. N. Chiu. *Spatial Tessellations: Concepts and Applications of Voronoi Diagrams*. Wiley Series in Probability and Statistics. Wiley, Chichester, 2nd edition, 2000.

[296] A. Ortiz, M. A. Puso, and N. Sukumar. Maximum-entropy meshfree method for compressible and near-incompressible elasticity. *Computer Methods in Applied Mechanics and Engineering*, 199(25–28):1859–1871, 2010.

[297] S. Osher and R. Fedkiw. *Level Set Methods and Dynamic Implicit Surfaces*, volume 153 of *Applied Mathematical Sciences*. Springer, Berlin, 2003.

[298] A. Paluszny, S. K. Matthäi, and M. Hohmeyer. Hybrid finite-element finite volume discretization of complex geologic structures and a new simulation workflow demonstrated on fractured rocks. *Geofluids*, 7(2):186–208, 2007.

[299] A. Pandolfi, P. Krysl, and M. Ortiz. Finite element simulation of ring expansion and fragmentation: The capturing of length and time scales through cohesive models of fracture. *International Journal of Fracture*, 95(1):279–297, 1999.

[300] A. Pandolfi and M. Ortiz. An efficient adaptive procedure for three-dimensional fragmentation simulations. *Engineering with Computers*, 18(2):148–159, 2002.

[301] D. Panozzo, P. Block, and O. Sorkine-Hornung. Designing unreinforced masonry models. *ACM Transactions on Graphics*, 32(4):Article 91, 12 pages, 2013.

[302] K. D. Papoulia, S. A. Vavasis, and P. Ganguly. Spatial convergence of crack nucleation using a cohesive finite-element model on a pinwheel-based mesh. *International Journal for Numerical Methods in Engineering*, 67(1):1–16, 2006.

[303] K. Park, G. H. Paulino, and J. R. Roesler. A unified potential-based cohesive model of mixed-mode fracture. *Journal of the Mechanics and Physics of Solids*, 57(6):891–908, 2009.

[304] G. H. Paulino, K. Park, W. Celes, and R. Espinha. Adaptive dynamic cohesive fracture simulation using nodal perturbation and edge-swap operators. *International Journal for Numerical Methods in Engineering*, 84(11):1303–1343, 2010.

[305] C. Peco, D. Millán, A. Rosolen, and M. Arroyo. Efficient implementation of Galerkin meshfree methods for large-scale problems with an emphasis on maximum entropy approximants. *Computers & Structures*, 150:52–62, 2015.

[306] C. Peco, A. Rosolen, and M. Arroyo. An adaptive meshfree method for phase-field models of biomembranes. Part II: A Lagrangian approach for membranes in viscous fluids. *Journal of Computational Physics*, 249:320–336, 2013.

[307] D. Peng, S. Osher, B. Merriman, and H.-K. Zhao. The geometry of Wulff crystal shapes and its relations with Riemann problems. In G.-Q. Chen and E. DiBenedetto, editors, *Nonlinear Partial Differential Equations*, volume 238 of *Contemporary Mathematics*, pages 251–303. American Mathematical Society, Providence, 1999.

[308] É. Picard. Sur l'application des méthodes d'approximations successives á l'étude de certaines équations différentielles ordinaires. *Journal de Mathématiques Pures et Appliquées*, 9:217–271, 1893.

[309] U. Pinkall and K. Polthier. Computing discrete minimal surfaces and their conjugates. *Experimental Mathematics*, 2(1):15–36, 1993.

[310] H. Pottmann, P. Grohs, and N. J. Mitra. Laguerre minimal surfaces, isotropic geometry and linear elasticity. *Advances in Computational Mathematics*, 31(4):391–419, 2009.

[311] H. Pottmann and Y. Liu. Discrete surfaces in isotropic geometry. In R. Martin, M. Sabin, and J. Winkler, editors, *Mathematics of Surfaces XII*, volume 4647 of *Lecture Notes in Computer Science*, pages 341–363. Springer, Berlin, 2007.

[312] P. L. Powar. Minimal roughness property of the Delaunay triangulation: a shorter approach. *Computer Aided Geometric Design*, 9(6):491–494, 1992.

[313] F. P. Preparata and M. I. Shamos. *Computational Geometry: An Introduction.* Texts and Monographs in Computer Science. Springer, New York, 1985.

[314] V. Rajan. Optimality of the Delaunay triangulation in $\mathbb{R}^d$. *Discrete & Computational Geometry*, 12(1):189–202, 1994.

[315] J. Ramier. *Comportement mécanique d'élastomèrs chargés, influence de l'adhésion charge – polymére, influence de la morphologie.* PhD thesis, L'Institut National des Sciences Appliquées de Lyon, 2004.

[316] L. Ramshaw. Blossoming: A connect-the-dots approach to splines. Technical Report 19, Digital Equipment Corporation, Systems Research Centre, Palo Alto, June 1987.

[317] A. Rand. *Delaunay Refinement Algorithms for Numerical Methods.* PhD thesis, Carnegie Mellon University, 2009.

[318] A. Rand. Average interpolation under the maximum angle condition. *SIAM Journal on Numerical Analysis*, 50(5):2538–2559, 2012.

[319] A. Rand, A. Gillette, and C. Bajaj. Interpolation error estimates for mean value coordinates over convex polygons. *Advances in Computational Mathematics*, 39(2):327–347, 2013.

[320] A. Rand, A. Gillette, and C. Bajaj. Quadratic serendipity finite elements on polygons using generalized barycentric coordinates. *Mathematics of Computation*, 83(290):2691–2716, 2014.

[321] M. M. Rashid and A. Sadri. The partitioned element method in computational solid mechanics. *Computer Methods in Applied Mechanics and Engineering*, 237–240:152–165, 2012.

[322] M. M. Rashid and M. Selimotic. A three-dimensional finite element method with arbitrary polyhedral elements. *International Journal for Numerical Methods in Engineering*, 67(2):226–252, 2006.

[323] N. Ray, B. Vallet, L. Alonso, and B. Lévy. Geometry-aware direction field processing. *ACM Transactions on Graphics*, 29(1):Article 1, 11 pages, 2009.

[324] N. Ray, B. Vallet, W. C. Li, and B. Lévy. N-symmetry direction field design. *ACM Transactions on Graphics*, 27(2):Article 10, 13 pages, 2008.

[325] J. J. Rimoli and J. J. Rojas. Meshing strategies for the alleviation of mesh-induced effects in cohesive element models. *International Journal of Fracture*, 193(1):29–42, 2015.

[326] S. Rippa. Minimal roughness property of the Delaunay triangulation. *Computer Aided Geometric Design*, 7(6):489–497, 1990.

[327] S. Rippa. Long and thin triangles can be good for linear interpolation. *SIAM Journal on Numerical Analysis*, 29(1):257–270, 1992.

[328] S. Rjasanow and O. Steinbach. *The Fast Solution of Boundary Integral Equations.* Mathematical and Analytical Techniques with Applications to Engineering. Springer, New York, 2007.

[329] S. Rjasanow and S. Weißer. Higher order BEM-based FEM on polygonal meshes. *SIAM Journal on Numerical Analysis*, 50(5):2357–2378, 2012.

[330] S. Rjasanow and S. Weißer. FEM with Trefftz trial functions on polyhedral elements. *Journal of Computational and Applied Mathematics*, 263:202–217, 2014.

[331] A. Rodríguez-Ferran, T. Bennett, H. Askes, and E. Tamayo-Mas. A general framework for softening regularisation based on gradient elasticity. *International Journal of Solids and Structures*, 48(9):1382–1394, 2011.

[332] B. Roget and J. Sitaraman. Wall distance search algorithm using voxelized marching spheres. *Journal of Computational Physics*, 241:76–94, 2013.

[333] S. Rosenberg. *The Laplacian on a Riemannian Manifold*, volume 31 of *London Mathematical Society Student Texts.* Cambridge University Press, Cambridge, 1997.

[334] A. Rosolen and M. Arroyo. Blending isogeometric analysis and local *maximum entropy* meshfree approximants. *Computer Methods in Applied Mechanics and Engineering*, 264:95–107, 2013.

[335] A. Rosolen, D. Millán, and M. Arroyo. On the optimum support size in meshfree methods: A variational adaptivity approach with maximum-entropy approximants. *International Journal for Numerical Methods in Engineering*, 82(7):868–895, 2010.

[336] A. Rosolen, D. Millán, and M. Arroyo. Second-order convex *maximum entropy* approximants with applications to high-order PDE. *International Journal for Numerical Methods in Engineering*, 94(2):150–182, 2013.

[337] A. Rosolen, C. Peco, and M. Arroyo. An adaptive meshfree method for phase-field models of biomembranes. Part I: Approximation with maximum-entropy approximants. *Journal of Computational Physics*, 249:303–319, 2013.

[338] Y. Rouchdy and L. D. Cohen. Geodesic voting methods: overview, extensions and application to blood vessel segmentation. *Computer Methods in Biomechanics and Biomedical Engineering: Imaging & Visualization*, 1(2):79–88, 2013.

[339] G. Ruiz, M. Ortiz, and A. Pandolfi. Three-dimensional finite-element simulation of the dynamic Brazilian tests on concrete cylinders. *International Journal for Numerical Methods in Engineering*, 48(7):963–994, 2000.

[340] G. Ruiz, A. Pandolfi, and M. Ortiz. Three-dimensional cohesive modeling of dynamic mixed-mode fracture. *International Journal for Numerical Methods in Engineering*, 52(1–2):97–120, 2001.

[341] M. Rumpf and M. Wardetzky. Geometry processing from an elastic perspective. *GAMM Mitteilungen*, 37(2):184–216, 2014.

[342] R. M. Rustamov. Boundary element formulation of harmonic coordinates. Technical report, Department of Mathematics, Purdue University, Nov. 2008.

[343] M. Saba, T. Schneider, K. Hormann, and R. Scateni. Curvature-based blending of closed planar curves. *Graphical Models*, 76(5):263–272, 2014.

[344] M. Sabin. Transfinite surface interpolation. In *Proceedings of the 6th IMA Conference on the Mathematics of Surfaces*, pages 517–534, 1996.

[345] L. A. Santaló. *Integral Geometry and Geometric Probability.* Cambridge Mathematical Library. Cambridge University Press, Cambridge, 2nd edition, 2004.

[346] S. Schaefer, T. Ju, and J. Warren. A unified, integral construction for coordinates over closed curves. *Computer Aided Geometric Design*, 24(8–9):481–493, 2007.

[347] A. Schiftner. Planar quad meshes from relative principal curvature lines. Master's thesis, Technische Universität Wien, 2007.

[348] D. Schillinger, L. Dedè, M. A. Scott, J. A. Evans, M. J. Borden, E. Rank, and T. J. R. Hughes. An isogeometric design-through-analysis methodology based on adaptive hierarchical refinement of NURBS, immersed boundary methods, and T-spline CAD surfaces. *Computer Methods in Applied Mechanics and Engineering*, 249-252:116–150, 2012.

[349] D. Schillinger and M. Ruess. The finite cell method: A review in the context of higher-order structural analysis of CAD and image-based geometric models. *Archives of Computational Methods in Engineering*, 22(3):391–455, 2015.

[350] T. Schneider, K. Hormann, and M. S. Floater. Bijective composite mean value mappings. *Computer Graphics Forum*, 32(5):137–146, 2013.

[351] G. Schwarz. *Hodge Decomposition: A Method for Solving Boundary Value Problems*, volume 1607 of *Lecture Notes in Mathematics*. Springer, Berlin, 1995.

[352] T. W. Sederberg, P. Gao, G. Wang, and H. Mu. 2-D shape blending: an intrinsic solution to the vertex path problem. In *Proceedings of the 20th Annual Conference on Computer Graphics and Interactive Techniques (SIGGRAPH 1993)*, pages 15–18, 1993.

[353] T. W. Sederberg and S. R. Parry. Free-form deformation of solid geometric models. *ACM SIGGRAPH Computer Graphics*, 20(4):151–160, 1986.

[354] D. Shepard. A two-dimensional interpolation function for irregularly-spaced data. In *Proceedings of the 23rd ACM national conference*, pages 517–524, 1968.

[355] J. R. Shewchuk. Triangle: Engineering a 2D quality mesh generator and Delaunay triangulator. In M. C. Lin and D. Manocha, editors, *Applied Computational Geometry. Towards Geometric Engineering*, volume 1148 of *Lecture Notes in Computer Science*, pages 203–222. Springer, Berlin, 1996.

[356] H. V. Shin, C. F. Porst, E. Vouga, J. Ochsendorf, and F. Durand. Reconciling elastic and equilibrium methods for static analysis. *ACM Transactions on Graphics*, 35(2):Article 13, 16 pages, 2016.

[357] J. E. Shore and R. W. Johnson. Axiomatic derivation of the principle of maximum entropy and the principle of minimum cross-entropy. *IEEE Transactions on Information Theory*, 26(1):26–37, 1980.

[358] R. Shu, C. Zhou, and M. S. Kankanhalli. Adaptive marching cubes. *The Visual Computer*, 11(4):202–217, 1995.

[359] R. Sibson. Locally equiangular triangulations. *The Computer Journal*, 21(3):243–245, 1978.

[360] R. Sibson. A vector identity for the Dirichlet tessellation. *Mathematical Proceedings of the Cambridge Philosophical Society*, 87(1):151–155, 1980.

[361] R. Sibson. A brief description of natural neighbor interpolation. In V. Barnet, editor, *Interpreting Multivariate Data*, Wiley Series in Probability and Mathematical Statistics, pages 21–36. Wiley, Chichester, 1981.

[362] J. C. Simo, R. L. Taylor, and K. S. Pister. Variational and projection methods for the volume constraint in finite deformation elasto-plasticity. *Computer Methods in Applied Mechanics and Engineering*, 51(1–3):177–208, 1985.

[363] I. Simonovski and L. Cizelj. Automatic parallel generation of finite element meshes for complex spatial structures. *Computational Materials Science*, 50(5):1606–1618, 2011.

[364] J. Smith and S. Schaefer. Selective degree elevation for multi-sided Bézier patches. *Computer Graphics Forum*, 34(2):609–615, 2015.

[365] D. Sohn, Y.-S. Cho, and S. Im. A novel scheme to generate meshes with hexahedral elements and poly-pyramid elements: The carving technique. *Computer Methods in Applied Mechanics and Engineering*, 201–204:208–227, 2012.

[366] D. Sohn, J. Han, Y.-S. Cho, and S. Im. A finite element scheme with the aid of a new carving technique combined with smoothed integration. *Computer Methods in Applied Mechanics and Engineering*, 254:42–60, 2013.

[367] O. Steinbach. *Numerical Approximation Methods for Elliptic Boundary Value Problems: Finite and Boundary Elements.* Springer, New York, 2008.

[368] E. Steinitz. Polyeder und Raumeinteilungen. In W. F. Meyer and H. Mohrmann, editors, *Encyklopädie der mathematischen Wissenschaften mit Einschluss ihrer Anwendungen*, volume III, part 1, 2nd half, chapter 12, pages 1–139. Teubner, Leipzig, 1922.

[369] G. Strang and G. J. Fix. *An Analysis of the Finite Element Method.* Prentice-Hall Series in Automatic Computation. Prentice-Hall, Englewood Cliffs, 1973.

[370] K. Sugihara. Surface interpolation based on new local coordinates. *Computer-Aided Design*, 31(1):51–58, 1999.

[371] N. Sukumar. Voronoi cell finite difference method for the diffusion operator on arbitrary unstructured grids. *International Journal for Numerical Methods in Engineering*, 57(1):1–34, 2003.

[372] N. Sukumar. Construction of polygonal interpolants: a maximum entropy approach. *International Journal for Numerical Methods in Engineering*, 61(12):2159–2181, 2004.

[373] N. Sukumar. Quadratic maximum-entropy serendipity shape functions for arbitrary planar polygons. *Computer Methods in Applied Mechanics and Engineering*, 263:27–41, 2013.

[374] N. Sukumar, J. E. Dolbow, and N. Moës. Extended finite element method in computational fracture mechanics: a retrospective examination. *International Journal of Fracture*, 196(1):189–206, 2015.

[375] N. Sukumar and E. A. Malsch. Recent advances in the construction of polygonal finite element interpolants. *Archives of Computational Methods in Engineering*, 13(1):129–163, 2006.

[376] N. Sukumar, B. Moran, and T. Belytschko. The natural element method in solid mechanics. *International Journal for Numerical Methods in Engineering*, 43(5):839–887, 1998.

[377] N. Sukumar, B. Moran, A. Y. Semenov, and V. V. Belikov. Natural neighbor Galerkin methods. *International Journal for Numerical Methods in Engineering*, 50(1):1–27, 2001.

[378] N. Sukumar and A. Tabarraei. Conforming polygonal finite elements. *International Journal for Numerical Methods in Engineering*, 61(12):2045–2066, 2004.

[379] N. Sukumar and R. J.-B. Wets. Deriving the continuity of maximum-entropy basis functions via variational analysis. *SIAM Journal on Optimization*, 18(3):914–925, 2007.

[380] N. Sukumar and R. W. Wright. Overview and construction of meshfree basis functions: from moving least squares to entropy approximants. *International Journal for Numerical Methods in Engineering*, 70(2):181–205, 2007.

[381] R. W. Sumner and J. Popović. Deformation transfer for triangle meshes. *ACM Transactions on Graphics*, 23(3):399–405, 2004.

[382] R. W. Sumner, M. Zwicker, C. Gotsman, and J. Popović. Mesh-based inverse kinematics. *ACM Transactions on Graphics*, 24(3):488–495, 2005.

[383] T. Sunada. Riemannian coverings and isospectral manifolds. *Annals of Mathematics*, 121(1):169–186, 1985.

[384] T. Sunada. Discrete geometric analysis. In P. Exner, J. P. Keating, P. Kuchment, T. Sunada, and A. Teplyaev, editors, *Analysis on Graphs and Its Applications*, volume 77 of *Proceedings of Symposia in Pure Mathematics*, pages 51–83. American Mathematical Society, Providence, 2008.

[385] T. Sussman and K.-J. Bathe. A finite element formulation for nonlinear incompressible elastic and inelastic analysis. *Computers & Structures*, 26(1–2):357–409, 1987.

[386] O. J. Sutton. The Virtual Element Method in 50 lines of MATLAB. arXiv: 1604.06021 [math.NA], 2016. https://arxiv.org/pdf/1604.06021.pdf.

[387] H. Talebi, A. Saputra, and C. Song. Stress analysis of 3D complex geometries using the scaled boundary polyhedral finite elements. *Computational Mechanics*, 58(4):697–715, 2016.

[388] C. Talischi and G. H. Paulino. Addressing integration error for polygonal finite elements through polynomial projections: A patch test connection. *Mathematical Models and Methods in Applied Sciences*, 24(8):1701–1727, 2014.

[389] C. Talischi, G. H. Paulino, and C. H. Le. Honeycomb Wachspress finite elements for structural topology optimization. *Structural and Multidisciplinary Optimization*, 37(6):569–583, 2009.

[390] C. Talischi, G. H. Paulino, A. Pereira, and I. F. M. Menezes. Polygonal finite elements for topology optimization: A unifying paradigm. *International Journal for Numerical Methods in Engineering*, 82(6):671–698, 2010.

[391] C. Talischi, G. H. Paulino, A. Pereira, and I. F. M. Menezes. PolyMesher: a general-purpose mesh generator for polygonal elements written in Matlab. *Structural and Multidisciplinary Optimization*, 45(3):309–328, 2012.

[392] C. Talischi, G. H. Paulino, A. Pereira, and I. F. M. Menezes. PolyTop: a Matlab implementation of a general topology optimization framework using unstructured polygonal finite element meshes. *Structural and Multidisciplinary Optimization*, 45(3):329–357, 2012.

[393] C. Talischi, A. Pereira, I. F. M. Menezes, and G. H. Paulino. Gradient correction for polygonal and polyhedral finite elements. *International Journal for Numerical Methods in Engineering*, 102(3–4):728–747, 2015.

[394] C. Talischi, A. Pereira, G. H. Paulino, I. F. M. Menezes, and M. S. Carvalho. Polygonal finite elements for incompressible fluid flow. *International Journal for Numerical Methods in Fluids*, 74(2):134–151, 2014.

[395] C. Taylor and P. Hood. A numerical solution of the Navier–Stokes equations using the finite element technique. *Computers & Fluids*, 1(1):73–100, 1973.

[396] S. Torabi and J. Lowengrub. Simulating interfacial anisotropy in thin-film growth using an extended Cahn–Hilliard model. *Physical Review E*, 85(4):Article 041603, 16 pages, 2012.

[397] J. Tournois, C. Wormser, P. Alliez, and M. Desbrun. Interleaving Delaunay refinement and optimization for practical isotropic tetrahedron mesh generation. *ACM Transactions on Graphics*, 28(3):Article 75, 9 pages, 2009.

[398] N. G. Trillos, D. Slepčev, J. von Brecht, T. Laurent, and X. Bresson. Consistency of Cheeger and ratio graph cuts. *Journal of Machine Learning Research*, 17:Article 181, 46 pages, 2016.

[399] P. G. Tucker. Hybrid Hamilton–Jacobi–Poisson wall distance function model. *Computers & Fluids*, 44(1):130–142, 2011.

[400] W. T. Tutte. How to draw a graph. *Proceedings of the London Mathematical Society*, 13(3):743–767, 1963.

[401] A. van Gelder and J. Wilhelms. Topological considerations in isosurface generation. *ACM Transactions on Graphics*, 13(4):337–375, 1994.

[402] R. Verfürth. A note on polynomial approximation in Sobolev spaces. *ESAIM: Mathematical Modelling and Numerical Analysis*, 33(4):715–719, 1999.

[403] R. Verfürth. *A Posteriori Error Estimation Techniques for Finite Element Methods*. Numerical Mathematics and Scientific Computation. Oxford University Press, Oxford, 2013.

[404] G. Voronoi. Nouvelles applications des paramètres continus à la théorie de formes quadratiques. *Journal für die reine und angewandte Mathematik*, 134:198–287, 1908.

[405] E. Vouga, M. Höbinger, J. Wallner, and H. Pottmann. Design of self-supporting surfaces. *ACM Transactions on Graphics*, 31(4):Article 87, 11 pages, 2012.

[406] E. Wachspress. *Rational Bases and Generalized Barycentrics*. Springer, Cham, 2016.

[407] E. L. Wachspress. *A Rational Finite Element Basis*, volume 114 of *Mathematics in Science and Engineering*. Academic Press, New York, 1975.

[408] S. Waldron. Affine generalised barycentric coordinates. *Jaen Journal on Approximation*, 3(2):209–226, 2011.

[409] M. Wardetzky. *Discrete Differential Operators on Polyhedral Surfaces: Convergence and Approximation*. PhD thesis, Freie Universität Berlin, 2006.

[410] M. Wardetzky, S. Mathur, F. Kälberer, and E. Grinspun. Discrete Laplace operators: No free lunch. In *Proceedings of the 5th Eurographics Symposium on Geometry Processing (SGP 2007)*, pages 33–37, 2007.

[411] J. Warren. Barycentric coordinates for convex polytopes. *Advances in Computational Mathematics*, 6(1):97–108, 1996.

[412] J. Warren. On the uniqueness of barycentric coordinates. In R. Goldman and R. Krasauskas, editors, *Topics in Algebraic Geometry and Geometric Modeling*, volume 334 of *Contemporary Mathematics*, pages 93–100. American Mathematical Society, Providence, 2003.

[413] J. Warren, S. Schaefer, A. N. Hirani, and M. Desbrun. Barycentric coordinates for convex sets. *Advances in Computational Mathematics*, 27(3):319–338, 2007.

[414] O. Weber. *Hybrid Methods for Interactive Shape Manipulation*. PhD thesis, Technion – Israel Institute of Technology, 2010.

[415] O. Weber, M. Ben-Chen, and C. Gotsman. Complex barycentric coordinates with applications to planar shape deformation. *Computer Graphics Forum*, 28(2):587–597, 2009.

[416] O. Weber, M. Ben-Chen, C. Gotsman, and K. Hormann. A complex view of barycentric mappings. *Computer Graphics Forum*, 30(5):1533–1542, 2011.

[417] O. Weber and C. Gotsman. Controllable conformal maps for shape deformation and interpolation. *ACM Transactions on Graphics*, 29(4):Article 78, 11 pages, 2010.

[418] O. Weber, R. Poranne, and C. Gotsman. Biharmonic coordinates. *Computer Graphics Forum*, 31(8):2409–2422, 2012.

[419] O. Weber and D. Zorin. Locally injective parametrization with arbitrary fixed boundaries. *ACM Transactions on Graphics*, 33(4):Article 75, 12 pages, 2014.

[420] S. Weißer. Residual error estimate for BEM-based FEM on polygonal meshes. *Numerische Mathematik*, 118(4):765–788, 2011.

[421] S. Weißer. Arbitrary order Trefftz-like basis functions on polygonal meshes and realization in BEM-based FEM. *Computers & Mathematics with Applications*, 67(7):1390–1406, 2014.

[422] S. Weißer. BEM-based finite element method with prospects to time dependent problems. In *Proceedings of the Jointly Organized 11th World Congress on Computational Mechanics (WCCM XI), 5th European Conference on Computational Mechanics (ECCM V), and 6th European Conference on Computational Fluid Dynamics (ECFD VI)*, pages 4420–4427, 2014.

[423] S. Weißer. Residual based error estimate and quasi-interpolation on polygonal meshes for high order BEM-based FEM. *Computers & Mathematics with Applications*, 73(2):187–202, 2017.

[424] R. Westermann, L. Kobbelt, and T. Ertl. Real-time exploration of regular volume data by adaptive reconstruction of isosurfaces. *The Visual Computer*, 15(2):100–111, 1999.

[425] E. Whiting, J. Ochsendorf, and F. Durand. Procedural modeling of structurally-sound masonry buildings. *ACM Transactions on Graphics*, 28(5):Article 112, 9 pages, 2009.

[426] H. Whitney. *Geometric Integration Theory*. Number 21 in Princeton Mathematical Series. Princeton University Press, Princeton, 1957.

[427] S. O. Wilson. Cochain algebra on manifolds and convergence under refinement. *Topology and its Applications*, 154(9):1898–1920, 2007.

[428] T. Winkler, J. Drieseberg, M. Alexa, and K. Hormann. Multi-scale geometry interpolation. *Computer Graphics Forum*, 29(2):309–318, 2010.

[429] P. Wriggers. *Nonlinear Finite Element Methods*. Springer, Berlin, 2008.

[430] G. Yang, D. M. Causon, and D. M. Ingram. Calculation of compressible flows about complex moving geometries using a three-dimensional Cartesian cut cell method. *International Journal for Numerical Methods in Fluids*, 33(8):1121–1151, 2000.

[431] R. Zallen. *The Physics of Amorphous Solids*, chapter 2. Wiley, New York, 1983.

[432] W. Zeng, R. Guo, F. Luo, and X. Gu. Discrete heat kernel determines discrete Riemannian metric. *Graphical Models*, 74(4):121–129, 2012.

[433] J. Zhang, B. Deng, Z. Liu, G. Patanè, S. Bouaziz, K. Hormann, and L. Liu. Local barycentric coordinates. *ACM Transactions on Graphics*, 33(6):Article 188, 12 pages, 2014.

[434] Y. Zhang, C. Bajaj, and B.-S. Sohn. 3D finite element meshing from imaging data. *Computer Methods in Applied Mechanics and Engineering*, 194(48–49):5083–5106, 2005.

[435] S. W. Zucker. Distance images and the enclosure field: applications in intermediate-level computer and biological vision. In M. Breuß, A. Bruckstein, and P. Maragos, editors, *Innovations for Shape Analysis: Models and Algorithms*, Mathematics and Visualization, pages 301–323. Springer, New York, 2013.

Index